T0212095

Birkhäuser

Computer Science Foundations and Applied Logic

Computer Science Foundations and Applied Logic is a growing series that focuses on the foundations of computing and their interaction with applied logic, including how science overall is driven by this. Thus, applications of computer science to mathematical logic, as well as applications of mathematical logic to computer science, will yield many topics of interest. Among other areas, it will cover combinations of logical reasoning and machine learning as applied to AI, mathematics, physics, as well as other areas of science and engineering. The series (previously known as *Progress in Computer Science and Applied Logic*) welcomes proposals for research monographs, textbooks and polished lectures, and professional text/references. The scientific content will be of strong interest to both computer scientists and logicians.

William M. Farmer

Simple Type Theory

A Practical Logic for Expressing
and Reasoning About Mathematical Ideas

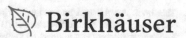 Birkhäuser

William M. Farmer
Department of Computing and Software
McMaster University
Hamilton, ON, Canada

ISSN 2731-5754   ISSN 2731-5762   (electronic)
Computer Science Foundations and Applied Logic
ISBN 978-3-031-21114-0   ISBN 978-3-031-21112-6   (eBook)
https://doi.org/10.1007/978-3-031-21112-6

This book is published under the imprint Birkhäuser, www.birkhauser-science.com by the registered
company Springer Nature Switzerland AG
The registered company address is: Gewerbestrasse 11, 6330 Cham, Switzerland

To the memory of my father, William D. Farmer (1931–2019)

Contents

Preface

Overview

This textbook is an introduction to *simple type theory* [47], a classical higher-order version of predicate logic that extends first-order logic. It presents a practice-oriented logic called *Alonzo* that is based on Alonzo Church's formulation of simple type theory known as *Church's type theory* [13]. Unlike traditional predicate logics, Alonzo admits undefined expressions. The book illustrates using Alonzo how simple type theory is exceptionally well suited for expressing and reasoning about mathematical ideas.

Organization

The book begins with an introduction, answers to questions that readers are likely to have, and a survey of mathematical concepts and notation needed for the study of simple type theory (Chapters 1–3). Then the syntax and semantics of Alonzo are presented (Chapters 4–7). After that, a proof system for Alonzo named 𝔄 is introduced (Chapter 8 and Appendices A–C). Next the tools — theories and theory morphisms — required for building libraries of mathematical knowledge using Alonzo in accordance with the *little theories method* [52] are presented (Chapters 9–13). The book ends with three variants of Alonzo (Chapter 14) and a survey of the various kinds of support that software could provide for Alonzo (Chapter 15).

Audience

The book is aimed at students of mathematics and computing at the graduate or upper undergraduate level. It is also intended for mathematicians, computing professionals, engineers, and scientists who need a practical logic for expressing and reasoning about mathematical ideas.

Suggested Uses

The book is designed to serve as the text for a stand-alone course on simple type theory. However, it can alternatively be used as the text for the second course in a two-course series on predicate logic where the first course is on first-order logic or the first course in a two-course series on type theory where the second course is on dependent type theory. Parts of the book may be useful to people who are interested in higher-order model theory (Chapters 5, 8, and 9), higher-order proof theory (Chapter 8 and Appendices A–C), libraries of mathematical knowledge (Chapters 9–13), subtype systems (Section 14.2), and quotation and evaluation operators (Section 14.3).

Acknowledgments

Peter Andrews presents $\mathcal{Q}_0$, an elegant version of Church's type theory, in his superb textbook *An Introduction to Mathematical Logic and Type Theory: To Truth Through Proof* [5]. He also presents in [5] an equally elegant proof system for $\mathcal{Q}_0$. The design and presentation of Alonzo is greatly influenced by the design and presentation of $\mathcal{Q}_0$. The key definitions associated with Alonzo are modifications of Andrews' definitions. And $\mathfrak{A}$, the proof system for Alonzo, is derived from the proof system for $\mathcal{Q}_0$.

Many of the ideas exhibited in this book originated in the IMPS proof assistant [53] developed by Joshua Guttman, Javier Thayer, and the author at The MITRE Corporation in the early 1990s. Alonzo is closely related to LUTINS [38, 39, 40], the logic of IMPS; it handles undefined expressions in essentially the same way as LUTINS. The implementation of the little theories method in Alonzo is modeled on the implementation in IMPS, the first such implementation in a proof assistant. The IMPS system demonstrates that a logic like Alonzo with undefinedness as well as the little theories method can be effectively implemented in a proof assistant.

I would like to thank all of the many people who have contributed to this book in one way or another. However, I give special thanks to my colleagues Jacques Carette, Joshua Guttman, Michael Kohlhase, Leonard Monk, Mark Nadel, Russell O'Connor, David Parnas, Florian Rabe, John Ramsdell, Javier Thayer, Martin von Mohrenschildt, and Ronald Watro from whom I have learned a great deal of logic, mathematics, and computing. Finally, I would like to thank the students at McMaster University for their questions and comments that helped iron out the ideas in this book.

Hamilton, Ontario, Canada   William M. Farmer
October 2022

Chapter 1

Introduction

The overarching objective of this book is to transform how logic is taught in university to students who will actually need to use logic in a practical way. The book seeks to achieve this objective by pursuing the following six goals:

1. Offer a practical, general-purpose predicate logic L for expressing and reasoning about mathematical ideas that can be readily employed in mathematics, computing, engineering, and science.

2. Introduce the key concepts of predicate logic and type theory using L.

3. Show that mathematical ideas can be expressed in L in a manner that is very close to how they are usually expressed in mathematical practice.

4. Show that L is ideally suited for reasoning about mathematical structures and constructing libraries of mathematical knowledge.

5. Focus on the model theory of L instead of on the proof theory of L.

6. Provide a foundation for the study of higher-order logic and type theory and the use of programming languages, proof assistants, and other mathematical software systems based on higher-order logic and type theory.

So what logic is L? We will answer this question after giving some background.

Bertrand Russell proposed in 1908 [110] a logic called the *theory of types* which is now known as the *ramified theory of types*. Designed to be a foundation for mathematics that is free of the set-theoretic and semantic paradoxes, the ramified theory of types is the logical basis for Alfred North Whitehead and Russell's monumental, three-volume *Principia Mathematica* [127] published in 1910–13, the first attempt to formalize a significant portion of mathematics starting from first principles. In the 1920s, Leon Chwistek [30] and Frank Ramsey [106] proposed a simplified version of this logic called the *simple theory of types* or, more briefly, *simple type theory*. In the 1930s, Rudolf Carnap [26, 27], Kurt Gödel [56], Willard Van Orman Quine [102], and Alfred Tarski [119, 120, 121] published detailed formulations of simple type theory. (See [47] for an introduction to simple type theory and [31] or [55] for the history of simple type theory.)

Simple type theory is a classical higher-order version of predicate logic. Statements in predicate logic are expressed using predicates, functions, Boolean operators, and quantifiers. First-order logic is predicate logic restricted to *first-order statements* in which the predicates, functions, and quantifiers are over so-called *individuals*. Simple type theory extends first-order logic by allowing *higher-order statements* in which predicates, functions, and quantifiers can be over predicates and functions as well as individuals. With access to high-order statements, simple type theory is more useful and natural than first-order logic for expressing and reasoning about mathematical ideas and, in particular, mathematical structures.

Alonzo Church introduced a version of simple type theory with lambda notation in a paper entitled "A Formulation of the Simple Theory of Types" published in 1940 in the *Journal of Symbolic Logic* [29]. Church's system, which is known as *Church's type theory* [13], is a very practical form of simple type theory that is tailored for reasoning with functions.[1] The influence of Church's type theory on mathematics has been, up until now, almost negligible, but its influence on computing has been profound, particularly in the areas of functional programming, formal methods, theorem proving, and constructive logic.

We have developed a version of simple type theory named *Alonzo* in honor of Church. Alonzo is based on Church's type theory and includes improvements made to Church's original system by his students Leon Henkin and Peter Andrews. The semantics of Alonzo is derived from Henkin's general models semantics. Henkin famously proved in his 1950 Ph.D. thesis that the proof system for Church's type theory is sound and complete with respect to this semantics [61].[2] Henkin later showed in 1963 that Church's

---

[1]In this book we view Church's type theory as a version of predicate logic. It can also be viewed as a version of simply typed lambda calculus (e.g., see [64]).

[2]Andrews showed in [3] that the proof system for Church's type theory is actually not sound with respect to Henkin's original general models semantics. However, it is sound

type system can be axiomatized using just equality, function application, and function abstraction [62]. Also in 1963, working with Henkin, Andrews produced a very simple and elegant proof system for Henkin's system [2]. Alonzo also admits, unlike traditional predicate logics, undefined expressions (resulting from partial functions and definite descriptions) in accordance with the approach employed in mathematical practice that we call the *traditional approach to undefinedness* [45].

This book presents Alonzo as the vehicle for achieving the six goals listed above. The book is intended to be an introduction to the use of logic as a practical tool for working with mathematical ideas.

As much as possible, we have tried to make this book self-contained, but we do assume that the reader is familiar with university-level mathematics, set theory (including cardinality), first-order logic, recursive definitions, mathematical proof (including proof by induction), and decidability. The theorems presented in this book are divided into four kinds: *propositions* that express immediately or easily proved facts; *lemmas*, usually of a technical nature, that are used to prove other theorems; *theorems* that express results of fundamental importance; and *corollaries*, usually of fundamental importance, that follow immediately from other theorems. Following standard convention, the statements of theorems and formal remarks are written in italics. The end of a proof or example is designated by a □ symbol.

Summary of the Contents

We will now briefly summarize the fifteen chapters and three appendices of the book.

Chapter 1 presents the *overarching objective* of the book and *six goals* for achieving the objective. It also introduces *Alonzo*, a version of simple type theory that is the vehicle for achieving the six goals. Alonzo is based on Church's type theory and admits undefined expressions.

Chapter 2 answers the questions readers are likely to have about the rationale behind the six goals of the book and the choice of Alonzo.

Chapter 3 presents mathematical concepts and notation that are needed for the study of simple type theory. There is a short discussion of *what is mathematics*. The common basic mathematical values — *sets*, *sequences*, *relations*, *functions*, *Boolean values*, and *predicates* — are surveyed. The notion of a *binder* — a special operator applied to a variable, a set, and an

---

if Henkin's definition of a general model is modified to guarantee that a general model contains, for every type α, the identity relation over the domain of α.

expression — is defined and illustrated with several examples. *Undefined expressions* are discussed, and the *traditional approach to undefinedness* is presented and its benefits are described. Finally, the important notion of a *mathematical structure* is precisely defined and several examples of mathematical structures are given.

Chapter 4 presents the *syntax* of Alonzo. The *types* and *expressions* of Alonzo are defined. There are two notations for types and expressions: a *formal notation* for machines and a *compact notation* for humans that is introduced by a set of notational definitions and conventions. Various basic syntactic concepts are defined such as *bound and free variables*, *substitution*, and the *free-for-in* concept. And a *language* of Alonzo, consisting of set of base types and a set of constants, and related concepts are defined.

Chapter 5 presents the *semantics* of Alonzo. A *general model* of a language is defined. Concepts related to general models are defined such as a *standard model* and an *expansion* of a general model as well as basic logic concepts such as *satisfiability*, *validity*, and *semantic consequence*. Two semantics are given for Alonzo: a logic-oriented *general semantics* based on general models and a mathematics-oriented *standard semantics* based on standard models. Several examples of standard models are given corresponding to the examples of mathematical structures given in Chapter 3.

Chapter 6 extends the *compact notation* for Alonzo types and expressions by introducing notation for Boolean operators, binary operators, quantifiers, expressions involving definedness, sets, tuples, functions, quasitypes (subtypes), and dependent quasitypes.

Chapter 7 introduces and proves the *beta-reduction theorem* for Alonzo. The *universal instantiation* and *alpha-conversion theorems* are also proved.

Chapter 8 discusses *proof systems* for Alonzo and presents $\mathfrak{A}$, a proof system for Alonzo based on Peter Andrews' proof system for Q_0 [5], his version of Church's type theory. $\mathfrak{A}$ is shown to be *sound* and *complete* with respect to the semantics defined in Chapter 5 using results proved in the appendices. The *compactness theorem* for Alonzo is proved.

Chapter 9 presents the central notion of a *theory* of Alonzo. An *extension* of a theory is defined and several kinds of *conservative extensions* are defined. Two kinds of maximal theories are discussed: *categorical theories* and *complete theories*. *Mathematical knowledge modules* are introduced for presenting theories and theory extensions, and several examples of theories and theory extensions are given using the modules. The *fundamental form of a mathematical problem* is presented. The chapter ends with a discussion of the *model theory* of simple type theory with the general models semantics.

Chapter 10 introduces *compact notation for infinite and finite sequences* in theories that include the natural numbers.

Chapter 11 presents a *development* of a theory which represents a narrative presentation of a theory in which definitions are made and theorems are proved. Mathematical knowledge modules are introduced for presenting theory developments and theory development *extensions*. As an example, a development of the theory *Peano Arithmetic* defined in Chapter 9 is given using the modules. The development includes definitions and theorems for a commutative semiring, a weak total order, and a divides lattice.

Chapter 12 defines a *theory of complete ordered fields* in Alonzo and then gives a theory development of this theory using the modules defined in Chapter 11. The development includes definitions and theorems for natural numbers, integers, rational numbers, iterated sum and product operators, calculus, and Euclidean space. *Skolem's paradox* for this theory is discussed.

Chapter 13 presents the *little theories method* [52] for organizing mathematical knowledge and defines a *theory morphism* from one theory to another and a *development morphism* from one development to another. *Extensions* of theory and development morphisms are defined. *Transportations* of definitions and theorems are also defined. Several examples of morphisms and transportations are presented using additional mathematical knowledge modules. Three kinds of graphical structures — *theory graphs*, *development graphs*, and *realm graphs* — are presented as suitable architectures for libraries of mathematical knowledge. Finally, *combinators* for constructing theory graphs are discussed.

Chapter 14 presents three variants of Alonzo. *AlonzoID* is Alonzo plus an indefinite description operator. *AlonzoS* is Alonzo with its type system replaced by a sort system that includes subtypes as well as types. And *AlonzoQE* is Alonzo with the addition of quotation and evaluation operators similar to quote and eval in the Lisp programming language.

Chapter 15 surveys the various kinds of support that *software* could provide for Alonzo.

Appendix A proves the basic *metatheorems* of 𝔄.

Appendix B proves that 𝔄 is *sound*.

Appendix C proves *Henkin's theorem* for 𝔄 that says every theory of Alonzo that is consistent in 𝔄 has a frugal general model. This theorem is used in Chapter 8 to prove that 𝔄 is *complete*.

Chapter 2
Answers to Readers' Questions

Before presenting the main content of the book, we will answer some questions many readers are likely to have. Some readers may want to skip this chapter and come back it later after they are more familiar with Alonzo.

2.1 Why Logic?

Why do mathematicians, computing professionals, engineers, and scientists need to know about logic? Logic, the study of sound reasoning, is an ancient discipline that goes back to at least the work of Aristotle (384–322 BCE) and Chrysippus (c. 280–207 BCE). Logic is a crucial component of critical thinking and an essential part of the foundation underlying mathematics, computing, engineering, and science. Thus every mathematician, computing professional, engineer, and scientist needs to have a basic understanding of logic.

2.2 Why a Practical Logic?

Why do mathematicians, computing professionals, engineers, and scientists need to be introduced to a practical logic? A *logic* can be defined as a family of formal languages that have a precise common syntax plus a precise common semantics with a notion of *logical consequence*. Predicate logics like first-order logic and simple type theory provide a rigorous framework for expressing and reasoning about mathematical ideas. The ideas are expressed as unambiguous statements in the formal language, and results about the ideas are obtained using the notion of logical consequence. However, most predicate logics are not suitable for practical applications because expressing mathematical ideas in them requires significant effort and produces statements that are often very unwieldy. A *practical logic* is one in which mathematical ideas can be directly expressed and reasoned about with effort comparable to not using a logic at all. Having access to a practical logic provides a mathematics practitioner with the option of doing mathematics within a logic whenever a high level of rigor is needed or desired.

2.3 Why Simple Type Theory?

Why have we chosen simple type theory as a practical logic for mathematicians, computing professionals, engineers, and scientists? In the paper "The Seven Virtues of Simple Type Theory" [47], we argue that simple type theory is endowed with the following virtues:

1. Simple type theory has a simple and highly uniform syntax.

2. The semantics of simple type theory is based on a small collection of well-established ideas.

3. Simple type theory is a highly expressive logic.

4. Simple type theory admits categorical theories of infinite structures.

5. There is a proof system for simple type theory that is simple, elegant, and powerful.

6. Henkin's general models semantics enables the techniques of first-order model theory to be applied to simple type theory and illuminates the distinction between standard and nonstandard models.

7. There are practical extensions of simple type theory that can be effectively implemented.

These virtues make simple type theory an attractive logic for practical-minded mathematicians, computing professionals, engineers, and scientists. We will discuss the first three virtues in detail here.

The core of simple type theory has a very simple and uniform syntax. There are just two kinds of syntactic entities: *types* that denote nonempty sets of mathematical values and *expressions* that denote mathematical values of particular types. The syntax is an extension of the syntax of first-order logic in which expressions include both first-order terms and formulas. Expressions can be built from variables and constants using only three kinds of construction: equality, function application, and function abstraction. Notational definitions can represent a very wide range of mathematical concepts including Boolean operators and quantifiers.

The semantics of simple type theory is based on exactly the same ideas as the semantics of many-sorted first-order logic: domains of individuals, truth values, functions of various types, interpretations for languages, variable assignments, and valuation functions recursively defined on the syntactic structure of expressions. The only significant difference between the two semantics is that the semantics of simple type theory includes a much richer set of functions.

The expressivity of a logic can be measured in two ways. The *theoretical expressivity* of a logic is the measure of what ideas can be expressed in the logic without regard to how the ideas are expressed. The *practical expressivity* of a logic is the measure of how readily ideas can be expressed in the logic. For example, first-order logic with predicate symbols but no function symbols has the same theoretical expressivity as standard first-order logic (with both predicate and function symbols) but has much lower practical expressivity than standard first-order logic.

Simple type theory has both high theoretical and high practical expressivity. It supports reasoning that is very close to mathematical practice, and almost any kind of mathematics can be expressed in it. In particular, simple type theory is ideally suited for reasoning about mathematical structures; it has just the machinery that is needed to describe and explore the mathematical structures that are typically found in mathematical practice.

Several proof assistants have been developed that implement versions of Church's type theory including HOL [58], HOL Light [60], IMPS [53], Isabelle/HOL [98], ProofPower [101], PVS [95], and TPS [6]. These systems show that simple type theory, in the form of Church's type theory, is indeed an effective logic for practical use. Hundreds of theories have been formulated and tens of thousands of theorems have been proved using these systems.

2.4   Why not First-Order Logic?

Why have we not chosen first-order logic? First-order logic and simple type theory have comparable theoretical expressivity, but the practical expressivity of first-order logic is very low compared to that of simple type theory.

First-order logic has little built-in support for higher-order objects like sets and functions. In particular, quantification over sets and functions is not possible, so statements requiring such quantification — e.g., the induction principle for the natural numbers and the completeness axiom for the real numbers — cannot be *directly* expressed in first-order logic. There is also no built-in support for describing domains of higher-order objects, so mathematical structures involving such domains — e.g., topological spaces and power set algebras — cannot be *directly* described in first-order logic. And mathematical statements that are succinct in mathematical practice often become verbose and unwieldy when expressed in first-order logic because first-order logic lacks an abstraction mechanism for defining functions and a definite description mechanism for describing values.

Although first-order logic lacks the practical expressivity of simple type theory, it has practical applications in computer-supported mathematics. For example, *SMT solvers* determine whether a given mathematical formula is satisfiable in a certain first-order theory using decision procedures, and *first-order automated theorem provers* determine whether a given mathematical formula is valid in a certain first-order theory using proof search.

In summary, first-order logic is an important and useful logic for both theoretical and practical purposes, but it is much less suitable than simple type theory for expressing and reasoning about many common mathematical ideas.

2.5   Why not Set Theory?

Why have we not chosen set theory? The most celebrated set theories — Zermelo-Fraenkel (ZF), von-Neumann-Bernays-Gödel (NBG), Morse-Kelly (MK), New Foundations (NF), and Tarski-Grothendieck (TG) — have much greater theoretical expressivity than simple type theory, but they lack the practical mechanisms — types and support for reasoning with functions — offered by simple type theory. To make a set theory suitable for practical use, these mechanisms from simple type theory need to be added to it. For example, the leading proof assistants based on set theory, Metamath/ZFC [85] and Mizar [88], utilize set theories with added support for types and functions.

In summary, set theory has much more theoretical power and much less practical power than what the typical mathematics practitioner needs. It does not provide the balance between theoretical and practical expressivity that simple type theory does.

2.6   Why not Dependent Type Theory?

Why have we not chosen dependent type theory? Dependent type theories such as *Martin-Löf type theory* [84] or the *calculus of constructions* [33] have high theoretic and practical expressivity. They have rich type systems, exploit the Curry-Howard isomorphism [70], and adhere to constructive reasoning principles. They are ideally suited to explore the frontier where programming and proving meet. Interest in dependent type theory is growing, and several proof assistants and programming languages are based on versions of dependent type theory including Agda [15, 93], Automath [90], Coq [124], Epigram [36], F* [37], Idris [72], Lean [34], and Nuprl [32]. However, the special features of dependent type theory are, at best, a major distraction and, at worst, a "bridge too far" for the vast majority of mathematics practitioners.

A *dependent type* is a type whose construction depends on an individual value. For example, the type $\mathbb{R}^n$ of tuples of real numbers of length n (n-tuples) depends on the natural number n. The hallmark of dependent type theories is that they admit dependent types unlike simple type theory. Many dependent types are subtypes of a nondependent type. For instance, each type $\mathbb{R}^n$ of n-tuples is a subtype of $\mathbb{R}^*$, the type of tuples of real numbers of unrestricted length. These kinds of dependent types, *dependent subtypes*, are commonplace in mathematics. Fortunately, they can be represented in simple type theory as certain expressions we call *dependent quasitypes*. Hence dependent subtypes can be utilized in simple type theory without embracing the tenets of dependent type theory [73, 108]. See Section 6.9 for details.

In summary, dependent type theory offers interesting and useful capabilities that are not available in simple type theory, but these capabilities are not needed by most users of mathematics and the learning curve for a logic with these capabilities is much steeper than the learning curve for simple type theory. Moreover, it is possible to represent commonplace dependent subtypes in simple type theory.

This book, as an introduction to simple type theory, can serve as a stepping stone for readers who want to study dependent type theory. There is a rich literature on dependent type theory; see [33, 59, 68, 84, 89, 126] for some sample articles and books.

2.7 Why Undefinedness?

Why does Alonzo admit undefined expressions? Partial functions like division and definite descriptions as employed in the definition of a limit are commonplace in mathematics. They produce undefined expressions like $1/0$ and $\lim_{x \to 0} \sin(1/x)$. Mathematicians have developed a practical approach for dealing with undefinedness — what we call the *traditional approach to undefinedness (TATU)* [45] — that enables statements about undefined expressions produced by partial functions and definite descriptions to be expressed very concisely. (TATU is presented and its benefits are described in Section 3.4.) We have shown that TATU can be formalized in Church's type theory by slightly modifying the logic [38, 46]. The result is a version of simple type theory that is much closer to mathematical practice than other standard versions. Moreover, Church's type theory with undefinedness can be effectively implemented as demonstrated by the development of the IMPS proof assistant [53], whose logic LUTINS [38, 39, 40] is a close relative of Alonzo.

In summary, providing Alonzo with support for undefinedness enables partial functions and definite descriptions to be handled in essentially the same way they are handled in mathematical practice, but it does not make Alonzo significantly more complex or more difficult to implement than traditional versions of Church's type theory.

2.8 Why Model Theory instead of Proof Theory?

Why do we focus on the model theory of Alonzo instead of on the proof theory of Alonzo? Like all predicate logics, Alonzo provides a framework for understanding what *logical consequence* is and for exploring how logical consequence is woven into the fabric of mathematics. There are three ways we can prove that a sentence B is a logical consequence of a set $\{A_1, \ldots, A_n\}$ of sentences in Alonzo: (1) We can give a model-theoretic argument that B is a *semantic consequence* of $\{A_1, \ldots, A_n\}$, i.e., that B is valid in every model of $\{A_1, \ldots, A_n\}$. (2) We can give a proof-theoretic argument that B is a *syntactic consequence* of $\{A_1, \ldots, A_n\}$, i.e., that there is a derivation of B from $\{A_1, \ldots, A_n\}$ in a sound proof system for Alonzo. (3) We can construct an actual derivation of B from $\{A_1, \ldots, A_n\}$ in a sound proof system for Alonzo.

A proof of form (1) is usually much more perspicuous than a proof of form (2) or (3); (1) is thus usually the easier form to read and write. Writing proofs of form (2) is challenging since it requires an understanding of both the semantics of the sentences and the machinery of the proof system. Except for very simple proofs, proofs of form (3) are impossible to write without software support. However, their correctness can be mechanically checked unlike proofs of form (1) and (2).

In this book we focus primarily on proofs of form (1) that embody model-theoretic arguments because we believe that reading and writing this kind of proof is more effective in helping the reader to develop intuition about the semantics of Alonzo. Nevertheless, we present a proof system $\mathfrak{A}$ for Alonzo in Section 8.2, prove a series of metatheorems of $\mathfrak{A}$ in Appendix A, prove that $\mathfrak{A}$ is sound in Appendix B, and finally prove Henkin's theorem for $\mathfrak{A}$ in Appendix C from which we show that $\mathfrak{A}$ is complete in Section 8.5.

Our approach is the opposite of the approach Andrews employs in his book *An Introduction to Mathematical Logic and Type Theory: To Truth Through Proof* [5]. He develops there a proof system for $\mathcal{Q}_0$, his version of Church's type theory, before even giving the semantics of $\mathcal{Q}_0$, and his proofs of logical consequences in $\mathcal{Q}_0$ are almost entirely proof-theoretic. His approach makes sense because it develops the reader's intuition for how formulas are proved in the proof system for $\mathcal{Q}_0$ and thus prepares the reader for using his TPS automatic theorem prover for $\mathcal{Q}_0$ [6]. For this book, there is much less justification for developing the reader's intuition for how $\mathfrak{A}$ works since there is currently no proof assistant based on $\mathfrak{A}$ and the existing proof assistants based on Church's type theory implement proof systems significantly different from $\mathfrak{A}$.

Chapter 3

Preliminary Concepts

We have argued that simple type theory is a kind of logic that is exceptionally well suited for reasoning about mathematics and, in particular, mathematical structures. So before we begin our study of Alonzo, our version of simple type theory, we need to look at several preliminary mathematical concepts. For many readers this chapter will largely be a review. Nevertheless, we urge all readers to read this chapter since it establishes the mathematical ideas, definitions, and notation that will be used throughout the rest of the book.

3.1 What is Mathematics?

Many people would say that mathematics is a huge body of objects, concepts, and facts. Objects like the natural numbers, the irrational numbers, triangles, and the trigonometric functions. Concepts like addition, square root, the measure of an angle, and the limit of a function. Facts like "$2 + 2 = 4$", "$\sqrt{2}$ is irrational", "the sum of the angles of a triangle is 180 degrees", and "$\lim_{x \to 0} \sin(1/x)$ does not exist".

But mathematics is actually something very different: it is a *process* for understanding the mathematical aspects of the world involving such things as counting, time, space, measure, pattern, and logical consequence. Here is how it works. A *mathematical model* consisting of objects, concepts, and facts is *created* to describe a particular mathematical phenomenon exhibited in the world. The model is *explored* in various ways to discover new objects, concepts, and facts related to the model. The resulting enriched model then provides a deeper understanding of the mathematical phenomenon being modeled.

As an example of a mathematical model, consider the phenomenon of *counting* which is used to determine the "size" of a set of things. The standard model for this phenomenon is the mathematical structure of natural number arithmetic with the concepts of successor, addition, and multiplication and the facts that define these concepts. Basic counting is modeled as the repeated application of a successor function, and more sophisticated counting is modeled using the addition and multiplication functions. Exploring this model leads to new objects like the negative numbers, new concepts like subtraction and division, and new facts like the associativity and commutativity of addition and multiplication. Then the model enriched with these new objects, concepts, and facts both gives us a deeper understanding of counting and leads us to the discovery of new mathematical ideas, some of which are well removed from counting. *Arithmetic* is the study of the model of counting.

As another example, consider the notion of a *linear transformation*. A linear transformation is any function that maps objects in one set to objects in another set in such a way that the graph of the function is a straight line. All kinds of linear transformations are found in the world. Linear transformations are modeled as homomorphisms between two vector spaces and are represented as matrices. So the model contains a variety of mathematical structures including fields, vector spaces, homomorphisms, and matrices. *Linear algebra* is the study of the model of linear transformations.

Mathematical models are explored by *inference* (e.g., proving a conjecture), *computation* (e.g., simplifying a mathematical expression), *concretization* (e.g., collecting examples), *narration* (e.g., presenting a linear development of a model in natural language), and *organization* (e.g., identifying common structure in a collection of models). These are the five aspects of the *tetrapod model* of mathematical knowledge [22]. As shown in Figure 3.1, it is convenient to arrange the aspects in a three-dimensional tetrapodal representation with the first four aspects at the ends of a tetrapod (vertices of a tetrahedron) and the organization aspect at the center of the tetrapod (tetrahedron).

As we can see from the two examples above, the building blocks for mathematical models are *mathematical structures* (which we discuss below in Section 3.5). If we consider mathematics to be the creation and exploration of mathematical models, then (1) the study of mathematical structures is a key component of the study of mathematics and (2) simple type theory, as a logic for reasoning about mathematical structures, is a key tool.

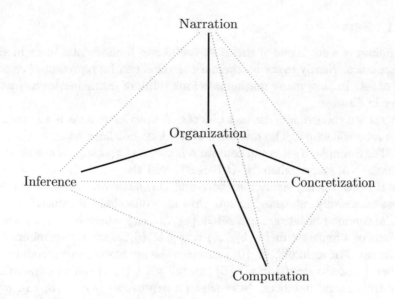

Figure 3.1: The Tetrapod Model of Mathematical Knowledge[1]

3.2 Mathematical Values

A *mathematical value* is any object — concrete or abstract — that is used
in mathematics. In particular, variables, constants, and other mathematical
expressions denote mathematical values; sets and their members are mathematical values; functions and their inputs and outputs are mathematical
values; and truth values are mathematical values. A mathematical value is
atomic (or *first-order*) if it is not composed of other mathematical values
and is *compound* (or *higher-order*) otherwise. For example, the number 2 is
atomic and the singleton set $\{2\}$ containing 2 is compound.

Whether a mathematical value is atomic or compound depends on the
context. Numbers are usually considered atomic, but in some contexts they
may be considered compound. For example, the rational number $1/2$ may be
considered to be the set of pairs (m, n) of integers such that $n = 2m$, which
is a compound value. For the sake of brevity, we will refer to mathematical
values from now on as simply *values*.

Most of this chapter is a survey of different kinds of values. At the end
of the chapter, we will see the importance of values as the building blocks
of mathematical structures.

---

[1]This figure is taken from [22].

3.2.1   Sets

The notion of a set is one of the most useful and fundamental ideas in all of
mathematics. Nearly every mathematical value can be represented as some
kind of set. In fact, many mathematicians think of mathematics as just set
theory in disguise.

A *set* is a collection of distinct objects. A *member* of a set is any member
of the set's collection. The members of a set do not have to be of the same
kind. For example, a set could contain a horse, a sawhorse, a horse chestnut,
the basketball game called "H-O-R-S-E", and the word "horse". A set is
finite if it has finitely many members and is *infinite* otherwise. We will be
almost exclusively interested in sets that are collections of values.[2]

The common notation for a set is $\{a_1, \ldots, a_n\}$ where $a_1, \ldots, a_n$ are the
members of a finite set or $\{b_1, b_2, \ldots\}$ where $b_1, b_2, \ldots$ are the members of an
infinite set. For example, $\{0, \{0\}\}$ denotes the set having two members, the
number 0 and the singleton set $\{0\}$, and $\mathbb{N} = \{0, 1, \ldots\}$ denotes the infinite
set of the natural numbers. Note that an expression $\{a_1, \ldots, a_n\}$ in which
the values $a_1, \ldots, a_n$ are not distinct denotes the set of the distinct values
in $a_1, \ldots, a_n$. For instance, $\{0, 0\}$ denotes the set $\{0\}$. Also note that not
all sets can be presented using this notation. For instance, $\mathbb{R}$, the set of
real numbers, cannot be written as an enumeration since it is uncountable.
$\mathbb{Q}$, the set of rational numbers, can be written as enumeration since it is
countable, but an enumeration of $\mathbb{Q}$ is more complex than other ways of
presenting $\mathbb{Q}$. Stretching the notation a bit, we can define $\mathbb{Z}$, the set of
integers, as $\{\ldots, -1, 0, 1, \ldots\}$.

Let S and T be sets. The statement $a \in S$ asserts that the value a is a
member of S. S is a *subset* of T, written $S \subseteq T$, if, for all $a \in S$, $a \in T$.
S is a *proper subset* of a set T, written $S \subset T$, if $S \subseteq T$ and $S \neq T$. A set is
often defined as the subset of some set S whose values satisfy a particular
property P. The set is then written as $\{x \in S \mid P(x)\}$ and read as "the set
of x in S such that P holds for x". For example,

$$\{n \in \mathbb{N} \mid n \text{ is prime}\}$$

denotes the set of prime numbers and

$$\{n \in \mathbb{N} \mid n \neq n\}$$

denotes the *empty set*, written $\emptyset$ or $\{\,\}$, the set containing no members at all.

---

[2]Russell's paradox shows that some collections of sets are "too big" to be a set. A
proper class is a collection of sets that is not a set.

Now let S, T, and U be sets such that $S, T \subseteq U$. The *complement of S with respect to U*, written $\overline{S}$ (when U is understood), is the set

$$\{x \in U \mid x \notin S\}.$$

The *union* of S and T, written $S \cup T$, is the set

$$\{x \in U \mid x \in S \text{ or } x \in T\}.$$

The *intersection* of S and T, written $S \cap T$, is the set

$$\{x \in U \mid x \in S \text{ and } x \in T\}.$$

And the *difference* of S and T, written $S \setminus T$, is the set

$$\{x \in S \mid x \notin T\}.$$

We will use the notation $\{x \mid P(x)\}$ to define the set of values that satisfy the property P when it is inconvenient to specify a domain from which the values are chosen. The *power set* $\mathcal{P}(S)$ of S, the set of all subsets of S, can thus be defined as the set

$$\{x \mid x \subseteq S\}.$$

So, for example, $\mathcal{P}(\{0, 1\}) = \{\{\,\}, \{0\}, \{1\}, \{0, 1\}\}$. We will also use the notation $\{p \mid P(x_1, \ldots, x_n)\}$ to define the set of values denoted by the pattern p when the components $x_1, \ldots, x_n$ of p satisfy the property P. Thus we can define $\mathbb{Q}$ as

$$\left\{\frac{m}{n} \mid m, n \in \mathbb{Z} \text{ are relatively prime and } n > 0\right\}.$$

Finally, the *cardinality* of a set S, written $|S|$, is the cardinal number κ such that there is a bijection $f : \kappa \to S$ (where a *bijection* is defined in Subsection 3.2.4).

3.2.2 Sequences

A *sequence* is either a finite enumeration $a_1, a_2, \ldots, a_n$ (called a *finite sequence*) or an infinite enumeration $b_1, b_2, \ldots$ (called an *infinite sequence*) of values in which a value may occur more than once. For example, $0, 1$; $1, 0$; and $0, 1, 0$ are distinct finite sequences. The *empty sequence* is the enumeration having no members. The *length* of a finite sequence is the number of members in the sequence counting duplicates.

A *tuple* is a finite sequence $a_1, a_2, \ldots, a_n$ that is usually written as $(a_1, a_2, \ldots, a_n)$. An *n-tuple* is a finite sequence of length n. An *ordered pair* is a 2-tuple. The *empty tuple*, written ( ), is the empty sequence.

Let $n \geq 1$. The *n-ary Cartesian product over sets* $A_1, \ldots, A_n$, written

$$A_1 \times \cdots \times A_n,$$

is the set

$$\{(a_1, \ldots, a_n) \mid a_1 \in A_1, \ldots, a_n \in A_n\}$$

of n-tuples. A unary Cartesian product over A is written as A^1 to distinguish it from A itself. The *n-ary Cartesian power over the set* A, written A^n, is the n-ary Cartesian product $A \times \cdots \times A$. Define $A^0 = \{()\}$. The *Kleene star* of A, written A^*, is the set

$$A^0 \cup A^1 \cup A^2 \cup \cdots.$$

A *list over a set* A is a member of A^*, i.e., a sequence of values taken from the set A. A list $(a_1, \ldots, a_n)$ is often be written as $[a_1, \ldots, a_n]$ to emphasize that the a_i are taken from the same set. The *empty list* (i.e., the empty tuple) is written []. The *list construction*

$$a_1 :: [a_2, \ldots, a_n]$$

denotes the list $[a_1, \ldots, a_n]$, and the *list concatenation*

$$[a_1, \ldots, a_m] \mathbin{+\!\!+} [a_{m+1}, \ldots, a_n]$$

denotes the list $[a_1, \ldots, a_m, a_{m+1}, \ldots, a_n]$.

3.2.3 Relations

Relationships among values are represented in mathematics by compound values called "relations".

Let $n \geq 1$. An *n-ary relation over sets* $A_1, \ldots, A_n$ is a set R such that

$$R \subseteq A_1 \times \cdots \times A_n.$$

R is *over the set* A if

$$A = A_1 = \cdots = A_n.$$

If $A_1 \subseteq B_1$, ..., $A_n \subseteq B_n$, then R is also a relation over $B_1, \ldots, B_n$. For $(a_1, \ldots, a_n) \in A_1 \times \cdots \times A_n$, the statement $R(a_1, \ldots, a_n)$ asserts $(a_1, \ldots, a_n) \in R$. When R is binary, $a\,R\,b$ is often written instead of $R(a, b)$. Unary relations and sets can be used interchangeably in most contexts.

Let R be a binary relation over A. Then the following are important properties that R may have:

- R is *reflexive (on A)* if $a\,R\,a$ for all $a \in A$.

- R is *irreflexive* if not $a\,R\,a$ for all $a \in A$.

- R is *symmetric* if $a\,R\,b$ implies $b\,R\,a$ for all $a, b \in A$.

- R is *asymmetric* if $a\,R\,b$ implies not $b\,R\,a$ for all $a, b \in A$.

- R is *antisymmetric* if $a\,R\,b$ and $b\,R\,a$ implies $a = b$ for all $a, b \in A$.

- R is *transitive* if $a\,R\,b$ and $b\,R\,c$ implies $a\,R\,c$ for all $a, b, c \in A$.

- R is *total (on A)* if $a\,R\,b$ or $b\,R\,a$ for all $a, b \in A$.

- R is *trichotomous (on A)* if exactly one of $a\,R\,b$ or $b\,R\,a$ or $a = b$ holds for all $a, b \in A$.

3.2.4 Functions

The notion of a function is one of the most important and powerful ideas in mathematics. It is the fundamental concept in Alonzo and all other versions of Church's type theory.

A *function* from a set I of *inputs* to a set O of *outputs* is a rule f that assigns, to each input in I, at most one output in O. The notation $f : I \to O$ designates that f is a function from I to O, and $f(i)$ denotes the $o \in O$ assigned to $i \in I$ by f. A function is not required to assign an output to each possible input, and so $f(i)$ may not denote any member of O. We say, in this case, that $f(i)$ is *undefined*. The *function space from I to O*, written $I \to O$, is the set of functions $f : I \to O$.

Let $f : I \to O$ be a function. The *domain* of f, written $\mathsf{dom}(f)$, is the set

$$\{i \in I \mid f(i) \text{ is defined}\}$$

and the *range* of f, written $\mathsf{ran}(f)$, is the set

$$\{o \in O \mid o = f(i) \text{ for some } i \in I\}.$$

f is *empty* if $\mathsf{dom}(f) = \emptyset$. f is *total* if $\mathsf{dom}(f) = I$ and *partial* if f is not total. f is *surjective* if $\mathsf{ran}(f) = O$. f is *injective* if, for all $i, j \in I$, $f(i) = f(j)$ implies $i = j$. (We assume that $f(i) = f(j)$ is false if $f(i)$ or $f(j)$ is undefined.) And f is *bijective* if f is total, surjective, and injective. f is a *surjection*, an *injection*, or a *bijection* if it is surjective, injective, or bijective, respectively.

Let $A \subseteq I$. The *image of A under f*, written $f[A]$, is the set

$$\{f(a) \mid a \in A\}.$$

The *restriction* of f to A, written $g{\restriction}_A$, is the function $f' : I \to O$ such that, for all $x \in A$, $f'(x) = f(x)$ if $f(x)$ is defined and $f'(x)$ is undefined if $f(x)$ is undefined and, for all $x \in I \setminus A$, $f'(x)$ is undefined.

Let $f_i : I_i \to O_i$ be a function for $i \in \{1, 2\}$ with $I_1 \subseteq I_2$ and $O_1 \subseteq O_2$. f_2 is an *extension* of f_1 (or f_1 is a *subfunction* of f_2), written $f_1 \sqsubseteq f_2$, if $f_1(x) = f_2(x)$ for all $x \in \mathsf{dom}(f_1)$. Notice that $f_1 \sqsubseteq f_2$ implies $\mathsf{dom}(f_1) \subseteq \mathsf{dom}(f_2)$. If $I_1 \cap I_2 = \emptyset$, the *union* of f_1 and f_2, written $f_1 \cup f_2$, is the function $g : I_1 \cup I_2 \to O_1 \cup O_2$ such that

$$g(x) = \begin{cases} f_1(x) & \text{if } x \in I_1; \\ f_2(x) & \text{if } x \in I_2. \end{cases}$$

For $n \geq 0$, an *n-ary function* $f : I_1, \ldots, I_n \to O$ is a rule that assigns, to each sequence of n inputs $i_1 \in I_1$, $\ldots$, $i_n \in I_n$, at most one output $f(i_1, \ldots, i_n) \in O$. A function $f : I \to O$ is a *1-ary* or *unary* function. A *0-ary* or *nullary* function $f : \; \to O$ is effectively a constant that denotes a single, fixed value. The application of a *2-ary* or *binary* function $f : I_1, I_2 \to O$ is often written using infix notation as $i_1 \, f \, i_2$ instead of as $f(i_1, i_2)$. For example, if $+$ is the usual addition function, then $+$ applied to 2 and 3 is usually written as $2 + 3$ instead of $+(2, 3)$.

For $n \geq 1$, an *n-ary function* $f : I_1, \ldots, I_n \to O$ can be represented as the relation $R_f \subseteq I_1 \times \cdots I_n \times O$ such that

$$R_f(i_1, \ldots, i_n, o) \quad \text{iff} \quad f(i_1, \ldots, i_n) = o.$$

"iff" is an abbreviation for "if and only if". A iff B asserts that A holds whenever B holds and B holds whenever A holds.

When $n \geq 2$, an *n-ary function* $f : I_1, \ldots, I_n \to O$ can be represented as a unary function in two different ways. First, f can be represented as the function $g : (I_1 \times \cdots \times I_n) \to O$ on n-tuples such that, for all $i_1 \in I_1$, $\ldots$, $i_n \in I_n$,

$$g((i_1, \ldots, i_n)) = f(i_1, \ldots, i_n).$$

A	B	$\neg A$	$A \Rightarrow B$	$A \Leftrightarrow B$	$A \vee B$	$A \wedge B$
F	F	T	T	T	F	F
F	T	T	T	F	T	F
T	F	F	F	F	T	F
T	T	F	T	T	T	T

Table 3.1: Truth Table for Boolean Operators

This representation is so natural that there is often no distinction made between n-ary functions and unary functions on n-tuples. We will write the application $g((i_1, \ldots, i_n))$ as simply $g(i_1, \ldots, i_n)$ when g is known to be a function on n-tuples.

Second, f can be represented as the higher-order function

$$h : I_1 \rightarrow (I_2 \rightarrow (\cdots \rightarrow (I_n \rightarrow O) \cdots))$$

such that, for all $i_1 \in I_1$, ..., $i_n \in I_n$,

$$h(i_1) \cdots (i_n) = f(i_1, \ldots, i_n)$$

where function application associates to the left, i.e., $k(a)(b)(c)$ means $((k(a))(b))(c)$. h is called the *Curryed form* of f (in honor of Haskell Curry).

3.2.5 Boolean Values and Predicates

The *Boolean values* are *false* and *true*, which we will write as F and T. These are the standard truth values. Let $\mathbb{B} = \{F, T\}$. For $n \geq 1$, an n-ary *predicate* is any function of the form $p : I_1, \ldots, I_n \rightarrow \mathbb{B}$. If $i_1 \in I_1$, ..., $i_n \in I_n$, then the expression $p(i_1, \ldots, i_n)$ can be used both to denote its value and to assert that $p(i_1, \ldots, i_n)$ is a true statement.

The Boolean operators "not" ($\neg$), "implies" ($\Rightarrow$), "iff" ($\Leftrightarrow$) "or" ($\vee$), and "and" ($\wedge$) are predicates that map Boolean values to Boolean values. They are defined by the truth table given in Table 3.1.

Every set A is represented by a unary predicate $p_A : A \rightarrow \mathbb{B}$ such that

$$a \in A \text{ iff } p_A(a) = T$$

(or, equivalently, "$a \in A$ iff $p_A(a)$") for all $a \in A$. Similarly, for $n \geq 1$, every relation n-ary $R \subseteq A_1 \times \cdots \times A_n$ is represented by an n-ary predicate $p_R : A_1, \ldots, A_n \rightarrow \mathbb{B}$ such that

$$R(a_1, \ldots, a_n) \text{ iff } p_R(a_1, \ldots, a_n) = T$$

(or, equivalently, "$R(a_1, \ldots, a_n)$ iff $p_R(a_1, \ldots, a_n)$") for all $(a_1, \ldots, a_n) \in A_1 \times \cdots \times A_n$. So sets and unary predicates as well as n-ary relations and n-ary predicates can be used interchangeably in most contexts.

Predicate logics, such as first-order logic and simple type theory, are designed for reasoning about properties of values and relations over values. These logics are called predicate logics because the properties and relations are expressed by predicates.

3.3   Binders

A *binder* B is a special operator that is applied to a variable x, a set S of values, and an expression E in which x may occur. Let $\mathsf{val}(E, x, a)$ denote the value of E when the value of x is a. For example,

$$\mathsf{val}(x + 1, x, 2) = 3.$$

The application of B to x, S, and E is written

$$B\, x \in S \,.\, E.$$

The value of $B\, x \in S \,.\, E$ is derived from the set

$$\{(a, \mathsf{val}(E, x, a)) \mid a \in S\}$$

of ordered pairs according to the definition of B.

Binders are common in mathematical practice, but they are often not made explicit. We will present several examples commonly found in mathematics and logic.

Example 3.1 (Set Abstraction) The *set abstraction operator* Ⓢ is a binder that is applied to x, S, and a Boolean expression E. The value of a *set abstraction*

$$Ⓢ\, x \in S \,.\, E$$

is the set of all $a \in S$ such that $\mathsf{val}(E, x, a) = \text{T}$. That is, the value of Ⓢ $x \in S \,.\, E$ is the set $\{x \in S \mid E\}$. For example,

$$(Ⓢ\, x \in \mathbb{R} \,.\, x + x = x * x) = \{x \in \mathbb{R} \mid x + x = x * x\} = \{0, 2\}.$$

Thus the $\{x \in S \mid P(x)\}$ notation is a disguised application of the set abstraction binder.   □

Example 3.2 (Function Abstraction) The *function abstraction opera-tor* λ is a binder that is applied to x, S, and an expression E such that, for some set T, $\mathsf{val}(E, x, a) \in T$ for all $a \in S$. The value of a *function abstraction*

$$\lambda x \in S . E$$

(for the set T) is the function $f : S \to T$ such that, for all $a \in S$, $f(a) = \mathsf{val}(E, x, a)$. For example,

$$\lambda x \in \mathbb{R} . e^x$$

denotes the exponential function, and

$$\lambda x \in \mathbb{R} . \lambda y \in \mathbb{R} . x + y$$

denotes the Curried form of the addition function $+ : \mathbb{R}, \mathbb{R} \to \mathbb{R}$. A function abstraction $\lambda x \in S . E$ is also called a *lambda-expression*. □

Example 3.3 (Universal Quantification) The *universal quantifier* $\forall$ is a binder that is applied to x, S, and a Boolean expression E. The value of a *universal statement*

$$\forall x \in S . E$$

is T if, for all $a \in S$, $\mathsf{val}(E, x, a) = $ T and is F otherwise. For example, the value of

$$\forall x \in \mathbb{R} . e^x \geq 0,$$

which asserts that the exponential function is nonnegative, is T. □

Example 3.4 (Existential Quantification) The *existential quantifier* $\exists$ is a binder that is applied to x, S, and a Boolean expression E. The value of an *existential statement*

$$\exists x \in S . E$$

is T if, for some $a \in S$, $\mathsf{val}(E, x, a) = $ T and is F otherwise. For example, the value of

$$\exists x \in \mathbb{R} . x^2 = -2,$$

which asserts that -2 has a real number square root, is F. □

Example 3.5 (Definite Description) The *definite description operator* I is a binder that is applied to x, S, and a Boolean expression E. The value of a *definite description*

$$\mathrm{I}\,x \in S\,.\,E$$

is *the* (unique) $a \in S$ such that $\mathsf{val}(E, x, a) = \mathrm{T}$ and is undefined if there is no such value or more than one such value. For example,

$$\mathrm{I}\,x \in \mathbb{R}\,.\,x \geq 0 \wedge x^2 = 2$$

denotes $\sqrt{2}$, the positive square root of 2, and

$$\mathrm{I}\,x \in \mathbb{R}\,.\,x^2 = -2$$

and

$$\mathrm{I}\,x \in \mathbb{R}\,.\,x^2 = 2$$

are undefined. The use of definite description is common in mathematical writing, but definite descriptions are rarely written explicitly as the application of a binder. For example, $\sqrt{2}$ is commonly defined as "the nonnegative real number x such that $x^2 = 2$". The definite description operator is often called the *iota operator* and is represented by a lower case iota (ι). Bertrand Russell represents it by an inverted lowercase iota ($\imath$). We represent it by an inverted uppercase iota (I).   □

Example 3.6 (Indefinite Description) The *indefinite description operator* ϵ is a binder that is applied to x, S, and a Boolean expression E. The value of an *indefinite description*

$$\epsilon\,x \in S\,.\,E$$

is *some* $a \in S$ such that $\mathsf{val}(E, x, a) = \mathrm{T}$ and is undefined if there is no such value. For example,

$$\epsilon\,x \in \mathbb{R}\,.\,x^2 = 2$$

denotes either $-\sqrt{2}$ or $\sqrt{2}$, and

$$\epsilon\,x \in \mathbb{R}\,.\,x^2 = -2$$

is undefined. The use of indefinite description is much more common in computing than in mathematics. The indefinite description operator is very often called the *Hilbert epsilon operator* because it is the chief operator in Hilbert's *epsilon calculus* [1], where it is used to define universal and existential quantification.   □

3.4 Undefinedness

A mathematical expression is *undefined* if it has no prescribed meaning or if it denotes a value that does not exist. Undefined expressions are commonplace in mathematics. They arise wherever there are partial functions. More precisely, if $f : A \to B$ is a partial function, then $\mathsf{dom}(f) \subset A$ and $f(a)$ is undefined for all $a \in A \setminus \mathsf{dom}(f)$. For example, consider the square root function $\sqrt{\cdot} : \mathbb{R} \to \mathbb{R}$, where $\mathbb{R}$ denotes the set of real numbers. $\sqrt{\cdot}$ is a partial function; $\mathsf{dom}(\sqrt{\cdot}) = \{x \in \mathbb{R} \mid x \geq 0\}$. So $\sqrt{2}$ is defined and $\sqrt{-2}$ is undefined. We have seen in the previous section that definite and indefinite description are two other sources of undefined expressions.

There is a *traditional approach to undefinedness (TATU)* [45] that is widely practiced in mathematics. This approach treats undefined expressions as legitimate, nondenoting expressions that can be components of meaningful statements. TATU is based on three principles:

1. Atomic expressions (i.e., variables and constants) are always defined — they always denote something.

2. Compound expressions may be undefined. A function application $f(a)$ is undefined if f is undefined, a is undefined, or $a \notin \mathsf{dom}(f)$. A definite description $I\,x \in S\,.\,E$ is undefined if there is no $a \in S$ or more than one $a \in S$ such that $\mathsf{val}(E, x, a) = \mathrm{T}$. An indefinite description $\epsilon\,x \in S\,.\,E$ is undefined if there is no $a \in S$ such that $\mathsf{val}(E, x, a) = \mathrm{T}$.

3. Formulas are always true or false and hence always defined. So, by convention, an "undefined" predicate application is given the value F. That is, a predicate application $p(a) = \mathrm{F}$ if p is undefined, a is undefined, or $a \notin \mathsf{dom}(p)$.

The second and third principles are extended to n-ary functions and predicates, for $n \geq 2$, in the obvious way. In particular, $a = b$ is false if a or b is undefined since $=$ is a binary predicate.

The third principle requires some discussion. Mathematicians strongly prefer to avoid nonstandard truth values to represent the value of an undefined statement. In mathematical practice, the value of a statement is thus either T or F. So an intuitively undefined predicate application must be either T or F, with F being the better choice.

With this approach to undefinedness, there are two natural notions of equality. a and b are *equal*, written $a = b$, if a and b are both defined and denote the same value. a and b are *quasi-equal*, written $a \simeq b$, if either $a = b$

or a and b are both undefined. Hence $1/0 = 1$, $1/0 = 1/0$, and $1/0 \simeq 1$ are false, but $1/0 \simeq 1/0$ is true.

TATU is commonly employed in mathematics because it offers the following benefits:

1. Meaningful statements can involve undefined expressions. For example, the statement

   $$\forall x \in \mathbb{R} . \, 0 \leq x \Rightarrow (\sqrt{x})^2 = x$$

   is meaningful even though

   $$0 \leq -2 \Rightarrow (\sqrt{-2})^2 = -2$$

   is a substitution instance of it containing the undefined expression $\sqrt{-2}$.

2. Function domains can be implicit. For example, the following statements define functions using quasi-equality without stating their domains:

   $$k(x) \simeq \frac{1}{x} + \frac{1}{x-1}.$$
   $$\left(\frac{f}{g}\right)(x) \simeq \frac{f(x)}{g(x)}.$$

3. Definedness assumptions can be implicit, and as a result, expressions involving undefinedness can be very concise. For example, there is no need to explicitly say $y \neq 0$ in the antecedent of the implication in

   $$\forall x, y, z \in \mathbb{R} . \, \frac{x}{y} = z \Rightarrow x = y * z$$

   since this is implied by the condition $\frac{x}{y} = z$.

4. Values can be defined implicitly using definite or indefinite description. For example, the square root function over the real numbers, which is partial, can be defined by the following statement using quasi-equality and definite description:

   $$\sqrt{x} \simeq I \, y . \, 0 \leq y \wedge y^2 = x.$$

In addition to facilitating the formalization of partial functions, definite descriptions, and indefinite descriptions, TATU also facilitates the formalization of the following mathematical values:

1. *Function subtypes.* If $A \subseteq A'$ and $B \subseteq B'$, then the subtype $A \to B$ of the type $A' \to B'$ can be formalized as the set of functions $f : A' \to B'$ such that $f(x)$ is defined implies $x \in A$ and $f(x) \in B$. In particular, a dependent function type $\Pi x \in A . B_x$, where $B_x \subseteq B'$ for all $x \in A$, can be formalized as a subtype of the function type $A' \to B'$. See Section 6.9 for details.

2. *Finite sequences.* A finite sequence of values in A can be formalized as a function $s : \mathbb{N} \to A$ such that, for some $n \in \mathbb{N}$, $s(m)$ is defined iff $m \leq n$. See Chapter 10 for details.

3. *Recursively defined functions.* A recursively defined function $f : A \to B$ can be formalized as the least (partial or total) function that is a fixed point of a monotone functional $F : (A \to B) \to (A \to B)$. See [42] for details.

Let a *logic with undefinedness (LwU)* be a logic that admits undefined expressions and statements about definedness. LwUs are also called *free logics* [92] since they are free from existence assumptions. Let a *logic with traditional undefinedness (LwTU)* be an LwU that satisfies the three principles of TATU. LwTUs are also called *negative free logics* since undefined predicate applications are considered to be false assertions.

TATU can be formalized in a standard logic by slightly modifying the logic. For example, [46] shows in detail how the LwTU $\mathcal{Q}_0^u$ is obtained from $\mathcal{Q}_0$ [5] — Peter Andrews' version of Church's type theory — by making just a few changes to the syntax and semantics of $\mathcal{Q}_0$. Alonzo is another example of an LwTU which is obtained from a variant of $\mathcal{Q}_0$. We believe the first published example of an LwTU is the version of first-order logic presented by Rolf Schock in [112]. The literature includes several other LwTUs derived from first-order logic [11, 12, 16, 17, 51, 86, 97, 122], Church's type theory [38, 39, 40, 46], and set theory [44, 51].

3.5 Mathematical Structures

Loosely speaking, a mathematical structure is a set of values on which a structure is placed using a set of distinguished elements, functions, and relations (or predicates). More precisely, a *mathematical structure* (*structure* for short) is a pair $S = (\mathcal{D}, \mathcal{A})$ where:

1. $\mathcal{D}$ is a nonempty finite set of *base domains* that are nonempty sets of values.

2. $\mathcal{A}$ is a set of *distinguished values* that are members of the domains in $\{\mathbb{B}\} \cup \mathcal{D}$ or domains constructed from these domains by the function space, power set, Cartesian product, and Kleene star operations.

The base domains are the *explicit domains* of the structure S, while $\mathbb{B}$ and the domains constructed by the function space, power set, Cartesian product, and Kleene star operations are the *implicit domains* of S. The members of $\mathbb{B}$ and the base domains are usually considered to be atomic, but the members of the implicit domains other than $\mathbb{B}$ are considered to be compound. The distinguished values are usually (1) elements of the base domains, (2) functions whose inputs and outputs reside in the base domains, and (3) relations (or predicates) over the base domains.

If the values in the explicit and implicit domains are available, why are distinguished values needed? Without distinguished values, we have only the means to refer to a very limited set of values in the explicit and implicit domains. For example, if we choose $\mathbb{N}$, the set of natural numbers, as a base domain, we can refer to the identity function on $\mathbb{N}$ as $\lambda x \in \mathbb{N} . x$ using function abstraction, but we have no way of referring to 0 or any of the other members of $\mathbb{N}$. However, if 0 and S, the successor function on $\mathbb{N}$, are distinguished values, then we can refer to the first natural number as 0, the second as $S(0)$, the third as $S(S(0))$, etc. We can also refer to the addition and multiplication functions on $\mathbb{N}$ by defining them using 0 and S. The distinguished values thus give a structure to the unstructured values of the base domains by providing a language to access selected members of the explicit and implicit domains.

When the set of distinguished values is finite, we may write a structure

$$S = (\{D_1, D_2, \ldots, D_m\}, \{a_1, a_2, \ldots, a_n\})$$

as the tuple

$$(D_1, D_2, \ldots, D_m, a_1, a_2, \ldots, a_n).$$

The *size* of S, written $|S|$, is the cardinality of $D_1 \cup \cdots \cup D_m$. S is *finite* if its size is finite and is *infinite* otherwise.

Let $S_i = (\mathcal{D}_i, \mathcal{A}_i)$ be a structure for $i \in \{1, 2\}$. S_2 is an *expansion* of S_1 (or S_1 is a *reduct* of S_2), written $S_1 \leq S_2$, if $\mathcal{D}_1 \subseteq \mathcal{D}_2$ and $\mathcal{A}_1 \subseteq \mathcal{A}_2$. The expansion is *creative* if $\mathcal{D}_1 \subset \mathcal{D}_2$. When S_2 is a noncreative expansion of S_1, we can say that S_2 inherits the structure of S_1 since S_2 has the same base domains as S_1 and includes the distinguished values of S_1. In mathematical practice, two structures are often identified when one is a noncreative expansion of the other. A structure with a rich set of components can be developed by repeatedly expanding a smaller structure.

Let the *minimum structure* be the structure $S_{\min} = (\emptyset, \emptyset)$. The minimum structure has no base domains, but it does have the implicit domain $\mathbb{B}$ and all the domains constructed from $\mathbb{B}$ by the function space, power set, Cartesian product, and Kleene star operations. Notice that $S_{\min} \leq S$ for every structure S.

The following are some common kinds of structures of the form

$$(D, R) = (\{D\}, \{R\}),$$

where R is a binary relation over D, that are defined using the properties of relations given in Subsection 3.2.3:

- (D, R) is a *pre-order* if R is reflexive and transitive.

- (D, E) is an *equivalence relation* if E is reflexive, symmetric, and transitive.

- $(D, \leq)$ is a *weak (or nonstrict) partial order* if $\leq$ is reflexive, antisymmetric, and transitive.

- $(D, \leq)$ is a *weak (or nonstrict) total order* if $(D, \leq)$ is a weak partial order such that $\leq$ is total.

- $(D, <)$ is a *strong (or strict) partial order* if $<$ is irreflexive, asymmetric, and transitive.

- $(D, <)$ is a *strong (or strict) total order* if $(D, <)$ is a strong partial order such that $<$ is trichotomous.

For every weak partial order $(D, \leq)$, there is an associated strong partial order $(D, <)$ where $<$ is defined by $x < y$ iff $x \leq y \wedge x \neq y$, and for every strong partial order $(D, <)$, there is an associated weak partial order $(D, \leq)$ where $\leq$ is defined by $x \leq y$ iff $x < y \vee x = y$. Thus how an order structure is expressed, whether weak or strong, is just a matter of preference.

Let $(D, \leq)$ be a weak partial order and $A \subseteq D$. A *maximal element* [*minimal element*] of A is an $M \in A$ [$m \in A$] such that $\neg(M < x)$ [$\neg(x < m)$] for all $x \in A$. The *maximum element* [*minimum element*] of A, if it exists, is the $M \in A$ [$m \in A$] such that $x \leq M$ [$m \leq x$] for all $x \in A$. Of course, the maximum [minimum] element of A, if it exists, is also a maximal [minimal] element of A, but a maximal [minimal] element of A may not be the maximum [minimum] element of A. A *well-order* is a weak total order $(D, \leq)$ such that every nonempty subset of D has a minimum element.

Since mathematical structures are the building blocks of mathematical models, a big part of the mathematical process is creating and exploring mathematical structures. Thus simple type theory — as a logic ideally suited for reasoning about mathematical structures — has an important role to play in mathematics.

3.6   Examples of Mathematical Structures

The mathematical literature is replete with different kinds of mathematical structures. We will present now a selection of mathematical structures and state (usually without definition) the properties they have.

Example 3.7 $(\mathbb{N}, \leq)$ where $\leq \, \in \mathcal{P}(\mathbb{N} \times \mathbb{N})$ is the usual weak total order on the natural numbers. $(\mathbb{N}, \leq)$ is a very simple structure; it has just one base domain $\mathbb{N}$ and one distinguished value $\leq$. $(\mathbb{N}, \leq)$ is a description of the structure $(\{\mathbb{N}\}, \{\leq\})$ as a tuple. As a weak total order, $\leq$ is reflexive, antisymmetric, transitive, and total. Every nonempty set $A \subseteq \mathbb{N}$ has a minimum member with respect to $\leq$, and so $(\mathbb{N}, \leq)$ is also a well-order.   □

Example 3.8 $(\mathbb{N}, 0, 1, +, *)$ where $0, 1 \in \mathbb{N}$ are the natural numbers 0 and 1, $+ : \mathbb{N} \times \mathbb{N} \to \mathbb{N}$ is the addition function, and $* : \mathbb{N} \times \mathbb{N} \to \mathbb{N}$ is the multiplication function. By virtue of containing only distinguished values that are members of $\mathbb{N}$ or functions over $\mathbb{N}$, $(\mathbb{N}, 0, 1, +, *)$ is an example of an *algebraic structure*. More precisely, $(\mathbb{N}, 0, 1, +, *)$ is a commutative semiring. It exhibits the structure of natural number arithmetic.   □

Example 3.9 $(\mathbb{N}, \mid, \mathsf{lcm}, \gcd, 1, 0)$ where $\mid \, \in \mathcal{P}(\mathbb{N} \times \mathbb{N})$ is the binary relation such that $m \mid n$ iff m divides n; $\mathsf{lcm} : \mathbb{N} \times \mathbb{N} \to \mathbb{N}$ is the function such that $\mathsf{lcm}(m, n)$ is the least common multiple of m and n; $\gcd : \mathbb{N} \times \mathbb{N} \to \mathbb{N}$ is the function such that $\gcd(m, n)$ is the greatest common divisor of m and n; and $0, 1 \in \mathbb{N}$. $(\mathbb{N}, \mid, \mathsf{lcm}, \gcd, 1, 0)$ is a complete bounded lattice in which $\mid$ is the lattice's weak partial order, lcm is the lattice's join, $\gcd$ is the lattice's meet, and 1 and 0 are the lattice's bottom and top, respectively.   □

Example 3.10 $(\mathbb{R}, 0, 1, +, *, -, \cdot^{-1})$ where $\mathbb{R}$ is the set of the real numbers; 0 and 1 are the real numbers 0 and 1; and $+, *, -, \cdot^{-1}$ are the usual addition, multiplication, negation, and reciprocal functions on $\mathbb{R}$. $(\mathbb{R}, 0, 1, +, *, -, \cdot^{-1})$

can be expanded to a complete ordered field.[3] It exhibits the interrelated algebraic, order, topological structures of the real numbers. □

Example 3.11 $(\mathbb{R}, \mathbb{C}, 0_\mathbb{R}, 1_\mathbb{R}, +_\mathbb{R}, *_\mathbb{R}, -_\mathbb{R}, \cdot^{-1}_\mathbb{R}, 0_\mathbb{C}, 1_\mathbb{C}, +_\mathbb{C}, *_\mathbb{C}, -_\mathbb{C}, \cdot^{-1}_\mathbb{C}, \otimes)$ where $\mathbb{R}$ and $\mathbb{C}$ are base domains of the real and complex numbers, $(\mathbb{R}, 0_\mathbb{R}, 1_\mathbb{R}, +_\mathbb{R}, *_\mathbb{R}, -_\mathbb{R}, \cdot^{-1}_\mathbb{R})$ is the field of the real numbers, $(\mathbb{C}, 0_\mathbb{C}, 1_\mathbb{C}, +_\mathbb{C}, *_\mathbb{C}, -_\mathbb{C}, \cdot^{-1}_\mathbb{C})$ is the field of the complex numbers, and $\otimes : \mathbb{R}, \mathbb{C} \to \mathbb{C}$ is scalar multiplication. Note that this is our first example of a structure with more than one base domain. This structure is the complex numbers structured as an algebra over a field. □

Example 3.12 $(\mathbb{R}, O)$ where $O \in \mathcal{P}(\mathcal{P}(\mathbb{R}))$ is the set of open subsets of $\mathbb{R}$. Notice that $\mathbb{R}$ is a base domain, while O is a distinguished value. $(\mathbb{R}, O)$ is a topological space. O is the first example we have seen of a distinguished value that is not an element of one of the base domains, a function whose inputs and outputs reside in the base domains, or a relation (or predicate) over the base domains. □

Example 3.13 $(\mathbb{R}, d)$ where $d : \mathbb{R} \times \mathbb{R} \to \mathbb{R}$ is the function such that $d(x, y) = |x - y|$. d is a metric on $\mathbb{R}$, and so $(\mathbb{R}, d)$ is a metric space. It can be used to define the set of open subsets of $\mathbb{R}$, so $(\mathbb{R}, d)$ can be considered as a topological space. □

Example 3.14 $(\mathbb{B}, \text{F}, \text{T}, \vee, \wedge, \neg)$ where $\mathbb{B} = \{\text{F}, \text{T}\}$ and $\vee, \wedge, \neg$ are the usual Boolean operators. $(\mathbb{B}, \text{F}, \text{T}, \vee, \wedge, \neg)$ is a Boolean algebra. □

Example 3.15 $(\mathcal{P}(A), \emptyset, A, \cup, \cap, ^-)$ where A is a nonempty set, $\mathcal{P}(A)$ is the power set of A, $\emptyset$ is the empty set, and $\cup, \cap, ^-$ are the usual set operators. $(\mathcal{P}(A), \emptyset, A, \cup, \cap, ^-)$ is a Boolean algebra for every nonempty set A. □

Example 3.16 (A, B, pair) where A and B are base domains and $\mathsf{pair} : A \to (B \to (A \times B))$ is the function such that $\mathsf{pair}(a)(b) = (a, b)$. (A, B, pair) is a structure of pairs constructed from members of A and B. □

Example 3.17 $(A, [], \mathsf{cons})$ where A is a base domain, $[] \in A^*$ is the empty list, and $\mathsf{cons} : A \to (A^* \to A^*)$ is the function such that $\mathsf{cons}(a)([b_1, \ldots, b_n]) = a :: [b_1, \ldots, b_n]$. (Recall A^* is the Kleene star of A, i.e., the set of lists over A.) $(A, [], \mathsf{cons})$ is a structure of lists over the set A. □

---

[3]All complete ordered fields are isomorphic, so $(\mathbb{R}, 0, 1, +, *, -, \cdot^{-1})$ is essentially the unique complete ordered field.

Example 3.18 $(Q, \Sigma, \delta, s, F)$ where Q is a finite set of *states*, Σ is a finite *input alphabet*, $\delta : Q \times \Sigma \to Q$ is a total *transition function*, $s \in Q$ is a *start state*, and $F \in \mathcal{P}(Q)$ is a set of *final states*. $(Q, \Sigma, \delta, s, F)$ is a deterministic finite automaton (DFA).   □

3.7   Conclusions

Mathematics is a process in which mathematical models are created and explored in order to learn about the mathematical aspects of the world. The building blocks of mathematical models are mathematical structures, and the building blocks of mathematical structures are mathematical values.

3.8   Exercises

1. Let $A \,\Delta\, B$ denote the symmetric difference of A and B. $A \,\Delta\, B$ can be defined as either $(A \setminus B) \cup (B \setminus A)$ or $(A \cup B) \setminus (A \cap B)$. Prove

$$(A \setminus B) \cup (B \setminus A) = (A \cup B) \setminus (A \cap B).$$

2. Let $R \subseteq (\mathbb{Z}^+ \times \mathbb{Z}^+) \times (\mathbb{Z}^+ \times \mathbb{Z}^+)$ be a relation such that $(a, b) \, R \, (c, d)$ iff $ad = bc$ where $\mathbb{Z}^+$ is the set of positive integers. Prove that R is an equivalence relation over $\mathbb{Z}^+$. What is the equivalence class of $(1, 2)$? Give a natural interpretation of the equivalence classes of R.

3. Show how a relation $R \subseteq A \times B$ can be transformed into an "equivalent" total function $f_R : A \to \mathcal{P}(B)$, where $\mathcal{P}(B)$ is the power set of B. A function like f_R is called a *many-valued function*. Show how a partial function $f : A \to B$ can be represented by a total many-valued function $g : A \to \mathcal{P}(B)$.

4. Let $f : U \to V$ be a function and $S, T \subseteq U$. Prove that:

    a. $f[S \cup T] = f[S] \cup f[T]$.
    b. $f[S \cap T] \subseteq f[S] \cap f[T]$.

5. Let $R \subseteq A \times A$ and $R \circ R = \{(a, b) \mid a \, R \, c \text{ and } c \, R \, b \text{ for some } c \in A\}$. Prove that R is transitive iff $R \circ R \subseteq R$.

6. Let $f : (\mathbb{R} \times \mathbb{R}) \to \mathbb{R}$ be the function defined by:

$$f(x, y) = \frac{1}{\sqrt{x - y}}.$$

    a. What is the domain of f?

    b. What is the range of f?

    c. Write f in Curryed form using the function abstraction binder.

7. Let $f : A \to B$ and $g : B \to C$ be total, and let $h = f \circ g : A \to C$ be the composition of f and g (i.e., $(f \circ g)(x) = g(f(x))$ for all $x \in A$). Prove that, if f and g are injective, then h is injective, but the converse is false.

8. Let $f : A \to B$ and $g : B \to C$ be total, and let $h = f \circ g : A \to C$ be the composition of f and g. Prove that, if f and g are surjective, then h is surjective, but the converse is false.

9. Prove that a function $f : A \to B$ is bijective iff there is a function $f^{-1} : B \to A$ such that $f \circ f^{-1} = \mathrm{id}_A$ and $f^{-1} \circ f = \mathrm{id}_B$ (where id_X is the identity function on the set X).

10. Determine which of the following functions are bijections from $\mathbb{R}$ to $\mathbb{R}$:

    a. $f(x) = -3x + 4$.

    b. $f(x) = -3x^2 + 7$.

    c. $f(x) = (x+1)/(x+2)$.

    d. $f(x) = x^5 + 1$.

11. Define the square root function $\sqrt{\cdot} : \mathbb{R} \to \mathbb{R}$ from $\leq\, : \mathbb{R}, \mathbb{R} \to \mathbb{B}$ and $* : \mathbb{R}, \mathbb{R} \to \mathbb{R}$ using the function abstraction and definite description binders.

12. Define the division function $/ : \mathbb{R}, \mathbb{R} \to \mathbb{R}$ from $* : \mathbb{R}, \mathbb{R} \to \mathbb{R}$ using the function abstraction and definite description binders.

13. Suppose $(S_1, \leq_1)$ and $(S_2, \leq_2)$ are weak partial orders. Prove that $(S_1 \times S_2, \leq)$ is a weak partial order where $(s_1, s_2) \leq (s_1', s_2')$ iff $s_1 \leq_1 s_1'$ and $s_2 \leq_2 s_2'$.

14. Prove that the structure $(\mathbb{N}, \leq)$ in Example 3.7 is a weak total order and also a well-order.

15. Prove that $(\mathbb{N} \times \mathbb{N}, \leq)$, where $\leq$ is lexicographical order on $\mathbb{N} \times \mathbb{N}$, is a weak total order and also a well-order.

16. Prove that the structure $(\mathbb{N}, 0, 1, +, *)$ in Example 3.8 is a commutative semiring.

17. Prove that the structure $(\mathbb{N}, |, \mathsf{lcm}, \gcd, 1, 0)$ in Example 3.9 is a complete bounded lattice.

18. Prove that the structure $(\mathbb{R}, 0, 1, +, *, -, \cdot^{-1})$ in Example 3.10 is a complete ordered field.

19. Prove that the structure

$$(\mathbb{R}, \mathbb{C}, 0_{\mathbb{R}}, 1_{\mathbb{R}}, +_{\mathbb{R}}, *_{\mathbb{R}}, -_{\mathbb{R}}, \cdot^{-1}_{\mathbb{R}}, 0_{\mathbb{C}}, 1_{\mathbb{C}}, +_{\mathbb{C}}, *_{\mathbb{C}}, -_{\mathbb{C}}, \cdot^{-1}_{\mathbb{C}}, \otimes)$$

in Example 3.11 is an algebra over the field $(\mathbb{R}, 0_{\mathbb{R}}, 1_{\mathbb{R}}, +_{\mathbb{R}}, *_{\mathbb{R}}, -_{\mathbb{R}}, \cdot^{-1}_{\mathbb{R}})$.

20. Prove that the structure $(\mathbb{R}, O)$ in Example 3.12 is a topological space.

21. Prove that the structure $(\mathbb{R}, d)$ in Example 3.13 is a metric space.

22. Prove that the structure $(\mathbb{B}, \mathrm{F}, \mathrm{T}, \vee, \wedge, \neg)$ in Example 3.14 is a Boolean algebra.

23. Prove that the structure $(\mathcal{P}(A), \emptyset, A, \cup, \cap, \bar{\cdot})$ in Example 3.15 is a Boolean algebra for every nonempty set A.

24. A *cofinite* subset of a set S is a subset S' of S such that the complement of S' with respect to S is finite. Let K be the finite and cofinite subsets of $\mathbb{N}$. Prove that $(K, \emptyset, \mathbb{N}, \cup, \cap, \bar{\cdot})$ is a Boolean algebra but is not a complete lattice.

25. Suppose F is the set of partial and total functions from $\mathbb{N}$ to $\mathbb{N}$. Recall that, for $f, g \in F$, f is a *subfunction* of g, written $f \sqsubseteq g$, if $f(x) = g(x)$ for all $x \in \mathsf{dom}(f)$.

   a. Prove that $(F, \sqsubseteq)$ is a weak partial order.

   b. Describe the set of minimal elements of $(F, \sqsubseteq)$.

   c. Describe the set of maximal elements of $(F, \sqsubseteq)$.

   d. Does $(F, \sqsubseteq)$ have a minimum element? If so, what is it?

   e. Does $(F, \sqsubseteq)$ have a maximum element? If so, what is it?

   f. Prove that $(F, \sqsubseteq)$ is a meet-semilattice with a bottom element.

   g. Prove that $(F, \sqsubseteq)$ is a not a lattice.

26. The weak total order $(\mathbb{R}, \leq)$ is both dense and complete. A weak partial order $(S, \leq)$ is *dense* if, for all $x, y \in S$, $x < y$ implies $x < z < y$ for some $z \in S$ and is *complete* if, for all nonempty $S' \subseteq S$, S' has an upper bound in S implies S' has a least upper bound in S. Let $(\mathbb{R} \cup \{-\infty, +\infty\}, \leq)$ be the weak total order that extends $(\mathbb{R}, \leq)$ by adding $-\infty$ and $+\infty$ as minimum and maximum elements. Prove that $(\mathbb{R} \cup \{-\infty, +\infty\}, \leq)$ is dense and complete.

Chapter 4

Syntax

This chapter defines the syntax of Alonzo. We will define "types" that denote nonempty sets of values and "expressions" that either denote values (when they are defined) or denote nothing at all (when they are undefined).

4.1   Notation

There are two notations for types and expressions, a *formal notation* based on a small number of constructors and a *compact notation* that resembles the notation found in mathematical practice. The compact notation is introduced using *notational definitions* (see the definition below) and *notational conventions* (which are listed in Table 4.1). The formal notation is an "internal" syntax intended for machines, while the compact notation is an "external" syntax intended for humans. The compact notation is used extensively in this book, but the formal notation is used hardly at all.

A *notational definition* has the form

> A stands for B,

where A and B are notations that present types or expressions; it defines A to be an alternate — and usually more compact, convenient, or standard — notation for presenting the type or expression that B presents. The meaning of A is the meaning of B. Hence an occurrence of A in the notation for a type or expression can always be replaced with B without changing the meaning of the presented type or expression. Any compact notation can be transformed into an equivalent formal notation by repeatedly unfolding the notational definitions and conventions that have been employed. The notational definitions are given in tables with boxes surrounding the definitions

W. M. Farmer, *Simple Type Theory*, Computer Science Foundations and Applied Logic, https://doi.org/10.1007/978-3-031-21112-6_4

Number	Kind	Subject	Page number
1	type	dropped parentheses	38
2	type	dropped parentheses	39
3	expression	dropped parentheses	40
4	expression	dropped parentheses	41
5	expression	dropped types from constants	41
6	expression	dropped types from variables	41
7	expression	dropped types from variables	41
8	expression	dropped types from variables	41
9	expression	pseudoconstant names	66
10	expression	weak order symbol families	67
11	expression	quantifier notation	68
12	expression	quantifier notation	68
13	expression	abbreviation names	70

Table 4.1: Notational Conventions

so that the reader can easily find them, and the notational conventions are assigned names of the form "Notational Convention n".

A separate document entitled "LaTeX for Alonzo" describes a set of La-TeX macros and environments for Alonzo.[1] The macros are for presenting Alonzo types and expressions in both the formal and compact notations. The environments are for presenting Alonzo *mathematical knowledge modules* (*modules* for short) — which define theories, developments, theory and development translations, and definition and theorem transportations — that will be introduced in subsequent chapters. All the Alonzo types, expressions, and modules presented in this work have been written using the LaTeX macros and environments. Alonzo modules are printed in brown color.

4.2 Symbols

Let $\mathcal{S}_{bt}, \mathcal{S}_{var}, \mathcal{S}_{con}$ be fixed countably infinite sets of symbols that will serve as names of base types, variables, and constants, respectively.[2] We assume that $\mathcal{S}_{bt}$ contains the symbols $A, B, C \ldots, X, Y, Z$, etc., $\mathcal{S}_{var}$ contains the symbols

---

[1]The document "LaTeX for Alonzo" is available at http://imps.mcmaster.ca/doc/latex-for-alonzo.pdf, and a LaTeX source file containing just the macros and environments for Alonzo is available at http://imps.mcmaster.ca/doc/alonzo-notation.tex.

[2]Some examples will require the size of $\mathcal{S}_{con}$ to be uncountable.

$a, b, c \ldots, x, y, z$, etc., and $\mathcal{S}_{\mathsf{con}}$ contains the symbols $A, B, C \ldots, X, Y, Z$, etc., nonalphanumeric symbols, and words in lowercase sans sarif font.[3] We will employ the following syntactic variables for these symbols and the syntactic entities defined later in this chapter:

1. $\mathbf{a}, \mathbf{b}$, etc. range over $\mathcal{S}_{\mathsf{bt}}$.

2. $\mathbf{f}, \mathbf{g}, \mathbf{h}, \mathbf{i}, \mathbf{j}, \mathbf{k}, \mathbf{m}, \mathbf{n}, \mathbf{u}, \mathbf{v}, \mathbf{w}, \mathbf{x}, \mathbf{y}, \mathbf{z}$, etc. range over $\mathcal{S}_{\mathsf{var}}$.

3. $\mathbf{c}, \mathbf{d}$, etc. range over $\mathcal{S}_{\mathsf{con}}$.

4. $\alpha, \beta, \gamma, \delta$, etc. range over types.

5. $\mathbf{A}_\alpha, \mathbf{B}_\alpha, \mathbf{C}_\alpha, \ldots, \mathbf{X}_\alpha, \mathbf{Y}_\alpha, \mathbf{Z}_\alpha$, etc. range over expressions of type α.

Thus, for example, the syntactic variable $\mathbf{a}$ denotes an unspecified symbol in $\mathcal{S}_{\mathsf{bt}}$.

4.3   Types

A *type* of Alonzo is an inductive set of strings of symbols defined by the following constructors:

T1. *Type of truth values*: BoolTy is a type.

T2. *Base type*: BaseTy($\mathbf{a}$) is a type.

T3. *Function type*: FunTy(α, β) is a type.

T4. *Product type*: ProdTy(α, β) is a type.

Hence a type is a string constructed from the members of $\mathcal{S}_{\mathsf{bt}}$ via the constructors BoolTy, BaseTy, FunTy, and ProdTy. Let $\mathcal{T}$ denote the set of types of Alonzo. A type is presented in the *formal notation* when it is written as a string according to the definition given above. However, we will usually present types using the *compact notation* defined in Table 4.2 by means of notational definitions. We assume BoolTy, $o \notin \mathcal{S}_{\mathsf{bt}}$.

When there is no loss of meaning, matching pairs of parentheses in the compact notation for types may be omitted (Notational Convention 1). We

---

[3]An expression like "u, v, w, etc." means, here and elsewhere, the set of symbols that includes u, v, and w, and all possible annotated forms of u, v, and w such as u', v_1, and $\widetilde{w}$.

o	stands for	BoolTy.	
$\mathbf{a}$	stands for	BaseTy($\mathbf{a}$).	
$(\alpha \to \beta)$	stands for	FunTy(α, β).	
$(\alpha \times \beta)$	stands for	ProdTy(α, β).	

Table 4.2: Notational Definitions for Types

$(\mathbf{x} : \alpha)$	stands for	Var($\mathbf{x}, \alpha$).	
$\mathbf{c}_\alpha$	stands for	Con($\mathbf{c}, \alpha$).	
$(\mathbf{A}_\alpha = \mathbf{B}_\alpha)$	stands for	Eq($\mathbf{A}_\alpha, \mathbf{B}_\alpha$).	
$(\mathbf{F}_{\alpha \to \beta} \mathbf{A}_\alpha)$	stands for	FunApp($\mathbf{F}_{\alpha \to \beta}, \mathbf{A}_\alpha$).	
$(\lambda \mathbf{x} : \alpha . \mathbf{B}_\beta)$	stands for	FunAbs(Var($\mathbf{x}, \alpha$), $\mathbf{B}_\beta$).	
$(\mathbf{I} \mathbf{x} : \alpha . \mathbf{A}_o)$	stands for	DefDes(Var($\mathbf{x}, \alpha$), $\mathbf{A}_o$)	where $\alpha \neq o$.
$(\mathbf{A}_\alpha, \mathbf{B}_\beta)$	stands for	OrdPair($\mathbf{A}_\alpha, \mathbf{B}_\beta$).	

Table 4.3: Notational Definitions for Expressions

assume that function type formation associates to the right so that, e.g., a type of the form

$$(\alpha \to (\beta \to \gamma))$$

may be written more simply as $\alpha \to \beta \to \gamma$ (Notational Convention 2).

A type α denotes a nonempty set D_α of values. o denotes the set $D_o = \mathbb{B}$ of the Boolean (truth) values F and T. $(\alpha \to \beta)$ denotes some set $D_{\alpha \to \beta}$ of (partial and total) functions from D_α to D_β. $(\alpha \times \beta)$ denotes the Cartesian product $D_{\alpha \times \beta} = D_\alpha \times D_\beta$. We will use base types to denote the base domains of structures.

Example 4.1 (Type) Consider the type of Alonzo presented in formal notation as

FunTy(ProdTy(BaseTy(A), BaseTy(B)), BoolTy)

where $A, B \in \mathcal{S}_{\mathsf{bt}}$ and in compact notation as

$$(A \times B) \to o$$

using Notational Convention 1. If the base types A and B denote domains D_A and D_B, then this type denotes the set of predicates on $D_A \times D_B$. $\square$

4.4   Expressions

An *expression of type* α of Alonzo is an inductive set of strings of symbols defined by the following constructors:

E1. *Variable*: $\mathsf{Var}(\mathbf{x}, \alpha)$ is an expression of type α.

E2. *Constant*: $\mathsf{Con}(\mathbf{c}, \alpha)$ is an expression of type α.

E3. *Equality*: $\mathsf{Eq}(\mathbf{A}_\alpha, \mathbf{B}_\alpha)$ is an expression of type BoolTy.

E4. *Function application*: $\mathsf{FunApp}(\mathbf{F}_{\mathsf{FunTy}(\alpha,\beta)}, \mathbf{A}_\alpha)$ is an expression of type β.

E5. *Function abstraction*: $\mathsf{FunAbs}(\mathsf{Var}(\mathbf{x}, \alpha), \mathbf{B}_\beta)$ is an expression of type $\mathsf{FunTy}(\alpha, \beta)$.

E6. *Definite description*: $\mathsf{DefDes}(\mathsf{Var}(\mathbf{x}, \alpha), \mathbf{A}_{\mathsf{BoolTy}})$ is an expression of type α where $\alpha \neq \mathsf{BoolTy}$.[4]

E7. *Ordered pair*: $\mathsf{OrdPair}(\mathbf{A}_\alpha, \mathbf{B}_\beta)$ is an expression of type $\mathsf{ProdTy}(\alpha, \beta)$.

Let $\mathcal{E}$ denote the set of expressions of Alonzo. A *formula* is an expression of type BoolTy. $\mathbf{A}_\alpha \equiv \mathbf{B}_\alpha$ means the expressions denoted by $\mathbf{A}_\alpha$ and $\mathbf{B}_\alpha$ are identical. An expression is presented in the *formal notation* when it is written as a string according to the definition given above. However, we will usually present expressions using the *compact notation* defined in Table 4.3 by means of notational definitions.

Note that not all possible constructions are expressions of Alonzo. For example, $\mathsf{Eq}(\mathbf{A}_\alpha, \mathbf{B}_\beta)$ is not an expression of Alonzo whenever $\alpha \neq \beta$. Note also that, if α and β are distinct, then $(\mathbf{x} : \alpha)$ and $(\mathbf{x} : \beta)$ are distinct expressions as well as $\mathbf{c}_\alpha$ and $\mathbf{c}_\beta$. As shown in Table 4.4, there are four kinds of expressions of type o (i.e., formulas); four kinds of expressions of a base type; five kinds of expressions of a function type; and five kinds of expressions of a product type.

When there is no loss of meaning, matching pairs of parentheses in expressions may be omitted (Notational Convention 3). We assume that function application formation associates to the left so that, e.g., an expression of the form $((\mathbf{G}_{\alpha \to \beta \to \gamma} \, \mathbf{A}_\alpha) \, \mathbf{B}_\beta)$ may be written more simply as

---

[4]Constructions of the form $\mathsf{DefDes}(\mathsf{Var}(\mathbf{x}, \mathsf{BoolTy}), \mathbf{A}_{\mathsf{BoolTy}})$ are not expressions of Alonzo since they would not be useful and would have a different semantics than definite descriptions of the form $\mathsf{DefDes}(\mathsf{Var}(\mathbf{x}, \alpha), \mathbf{A}_{\mathsf{BoolTy}})$ where $\alpha \neq \mathsf{BoolTy}$.

Expression \ Type	o	a	$\alpha \to \beta$	$\alpha \times \beta$
Variable	yes	yes	yes	yes
Constant	yes	yes	yes	yes
Equality	yes	no	no	no
Function application	yes	yes	yes	yes
Function abstraction	no	no	yes	no
Definite description	no	yes	yes	yes
Ordered pair	no	no	no	yes

Table 4.4: Kinds of Expressions

$\mathbf{G}_{\alpha \to \beta \to \gamma} \mathbf{A}_\alpha \mathbf{B}_\beta$ (Notational Convention 4). When the type α of a constant $\mathbf{c}_\alpha$ is known from the context of the constant, we will very often write the constant as simply $\mathbf{c}$ (Notational Convention 5). A variable $(\mathbf{x} : \alpha)$ occurring in the body $\mathbf{B}_\beta$ of $\lambda \mathbf{x} : \alpha \,.\, \mathbf{B}_\beta$ or in the body $\mathbf{A}_o$ of $\mathrm{I} \mathbf{x} : \alpha \,.\, \mathbf{A}_o$ may be written as just $\mathbf{x}$ if there is no resulting ambiguity (Notational Convention 6). So, for example, $\lambda \mathbf{x} : \alpha \,.\, (\mathbf{x} : \alpha)$ may be written more simply as $\lambda \mathbf{x} : \alpha \,.\, \mathbf{x}$. We will employ this convention for the other variable binders of Alonzo introduced later by notational definitions (Notational Convention 7). A variable $(\mathbf{x} : \alpha)$ occurring in $\mathbf{B}_\beta$ may be written as just $\mathbf{x}$ if the type α is known from the context of the occurrence of $(\mathbf{x} : \alpha)$ in $\mathbf{B}_\beta$ (Notational Convention 8). For example, $\mathbf{A}_\alpha = (\mathbf{x} : \alpha)$ may be written as $\mathbf{A}_\alpha = \mathbf{x}$.

An expression of type α is always defined if $\alpha = o$ and may be either defined or undefined if $\alpha \neq o$. If defined, it denotes a value in D_α, the denotation of α. If undefined, it denotes nothing at all. We will use constants to denote the distinguished values of structures.

Example 4.2 (Expression) Consider the expression of Alonzo presented in formal notation as

$$\mathsf{FunAbs}(\mathsf{Var}(x, \mathsf{BaseTy}(A)), \mathsf{OrdPair}(\mathsf{Var}(x, \mathsf{BaseTy}(A)), \mathsf{Var}(x, \mathsf{BaseTy}(A))))$$

where $A \in \mathcal{S}_{\mathsf{bt}}$ and $x \in \mathcal{S}_{\mathsf{var}}$ and in compact notation as

$$\lambda x : A \,.\, (x, x)$$

using Notational Conventions 3 and 6. If the base type A denotes the domain D_A, then this expression denotes the function that maps $d \in D_A$ to $(d, d) \in D_A \times D_A$. □

4.5   Bound and Free Variables

An occurrence of a variable $(\mathbf{x} : \alpha)$ in $\mathbf{B}_\beta$ is *bound* [*free*] if it is [not] within a subexpression of $\mathbf{B}_\beta$ of either the form $\lambda \mathbf{x} : \alpha \,.\, \mathbf{C}_\gamma$ or the form $I\mathbf{x} : \alpha \,.\, \mathbf{C}_o$. A variable $(\mathbf{x} : \alpha)$ is *bound* [*free*] *in* $\mathbf{B}_\beta$ if there is a bound [free] occurrence of $(\mathbf{x} : \alpha)$ in $\mathbf{B}_\beta$. An expression is *closed* if it contains no free variables. A *sentence* is a closed formula.

Example 4.3 (Free and Bound Variables) Let $\mathbf{A}_{\alpha \to o}$ be

$$\lambda \mathbf{x} : \alpha \,.\, (\mathbf{x} = (\mathbf{y} : \alpha))$$

and $\mathbf{B}_o$ be

$$(\lambda \mathbf{x} : \alpha \,.\, \mathbf{x}) = \mathbf{c}_{\alpha \to \alpha \to \alpha}\, \mathbf{x}.$$

$\mathbf{A}_{\alpha \to o}$ is written using Notational Conventions 3 and 6, while $\mathbf{B}_o$ is written using Notational Conventions 3, 6, and 8. The two occurrences of $(\mathbf{x} : \alpha)$ in $\mathbf{A}_{\alpha \to o}$ are bound, while the single occurrence of $(\mathbf{y} : \alpha)$ in $\mathbf{A}_{\alpha \to o}$ is free. Thus $(\mathbf{x} : \alpha)$ is bound and $(\mathbf{y} : \alpha)$ is free in $\mathbf{A}_{\alpha \to o}$. The first two occurrences of $(\mathbf{x} : \alpha)$ in $\mathbf{B}_o$ are bound, while the third is free. Thus $(\mathbf{x} : \alpha)$ is both bound and free in $\mathbf{B}_o$. □

$\mathbf{A}_\alpha$ is *free for* $(\mathbf{x} : \alpha)$ *in* $\mathbf{B}_\beta$ if no free occurrence of $(\mathbf{x} : \alpha)$ in $\mathbf{B}_\beta$ is within a subexpression of $\mathbf{B}_\beta$ of either the form $\lambda \mathbf{y} : \gamma \,.\, \mathbf{C}_\delta$ or the form $I\mathbf{y} : \gamma \,.\, \mathbf{C}_o$ where $(\mathbf{y} : \gamma)$ is free in $\mathbf{A}_\alpha$.

Example 4.4 (Free For In) Let $\mathbf{C}_{\alpha \to \alpha}$ be

$$\lambda \mathbf{x} : \alpha \,.\, (\mathbf{y} : \alpha).$$

Then $(\mathbf{z} : \alpha)$ is free for $(\mathbf{y} : \alpha)$ in $\mathbf{C}_{\alpha \to \alpha}$, but $(\mathbf{x} : \alpha)$ is not free for $(\mathbf{y} : \alpha)$ in $\mathbf{C}_{\alpha \to \alpha}$. □

We will see the need for this free-for-in notion in Chapter 7.

4.6   Substitution

The *substitution of* $\mathbf{A}_\alpha$ *for* $(\mathbf{x} : \alpha)$ *in* $\mathbf{B}_\beta$, written $\mathbf{B}_\beta[(\mathbf{x} : \alpha) \mapsto \mathbf{A}_\alpha]$, is the result of replacing each free occurrence of $(\mathbf{x} : \alpha)$ in $\mathbf{B}_\beta$ with $\mathbf{A}_\alpha$. Notice that $\mathbf{B}_\beta[(\mathbf{x} : \alpha) \mapsto \mathbf{A}_\alpha] \in \mathcal{E}$ since the free occurrences of $(\mathbf{x} : \alpha)$ in $\mathbf{B}_\beta$ are replaced with $\mathbf{A}_\alpha$, an expression of the same type as the type of $(\mathbf{x} : \alpha)$. This operation on expressions is defined using recursion and pattern matching by the following identities:

S1. $(\mathbf{x} : \alpha)[(\mathbf{x} : \alpha) \mapsto \mathbf{A}_\alpha] \equiv \mathbf{A}_\alpha$.

S2. $(\mathbf{y} : \beta)[(\mathbf{x} : \alpha) \mapsto \mathbf{A}_\alpha] \equiv (\mathbf{y} : \beta)$

where $(\mathbf{x} : \alpha)$ and $(\mathbf{y} : \beta)$ are distinct.

S3. $\mathbf{c}_\beta[(\mathbf{x} : \alpha) \mapsto \mathbf{A}_\alpha] \equiv \mathbf{c}_\beta$.

S4. $(\mathbf{B}_\beta = \mathbf{C}_\beta)[(\mathbf{x} : \alpha) \mapsto \mathbf{A}_\alpha] \equiv (\mathbf{B}_\beta[(\mathbf{x} : \alpha) \mapsto \mathbf{A}_\alpha] = \mathbf{C}_\beta[(\mathbf{x} : \alpha) \mapsto \mathbf{A}_\alpha])$.

S5. $(\mathbf{F}_{\beta \to \gamma} \mathbf{B}_\beta)[(\mathbf{x} : \alpha) \mapsto \mathbf{A}_\alpha] \equiv$
$(\mathbf{F}_{\beta \to \gamma}[(\mathbf{x} : \alpha) \mapsto \mathbf{A}_\alpha] \mathbf{B}_\beta[(\mathbf{x} : \alpha) \mapsto \mathbf{A}_\alpha])$.

S6. $(\lambda \mathbf{x} : \alpha \,.\, \mathbf{B}_\beta)[(\mathbf{x} : \alpha) \mapsto \mathbf{A}_\alpha] \equiv (\lambda \mathbf{x} : \alpha \,.\, \mathbf{B}_\beta)$.

S7. $(\lambda \mathbf{y} : \gamma \,.\, \mathbf{B}_\beta)[(\mathbf{x} : \alpha) \mapsto \mathbf{A}_\alpha] \equiv (\lambda \mathbf{y} : \gamma \,.\, \mathbf{B}_\beta[(\mathbf{x} : \alpha) \mapsto \mathbf{A}_\alpha])$

where $(\mathbf{x} : \alpha)$ and $(\mathbf{y} : \gamma)$ are distinct.

S8. $(\mathrm{I}\,\mathbf{x} : \alpha \,.\, \mathbf{B}_o)[(\mathbf{x} : \alpha) \mapsto \mathbf{A}_\alpha] \equiv (\mathrm{I}\,\mathbf{x} : \alpha \,.\, \mathbf{B}_o)$.

S9. $(\mathrm{I}\,\mathbf{y} : \gamma \,.\, \mathbf{B}_o)[(\mathbf{x} : \alpha) \mapsto \mathbf{A}_\alpha] \equiv (\mathrm{I}\,\mathbf{y} : \gamma \,.\, \mathbf{B}_o[(\mathbf{x} : \alpha) \mapsto \mathbf{A}_\alpha])$

where $(\mathbf{x} : \alpha)$ and $(\mathbf{y} : \gamma)$ are distinct.

S10. $(\mathbf{B}_\beta, \mathbf{C}_\gamma)[(\mathbf{x} : \alpha) \mapsto \mathbf{A}_\alpha] \equiv (\mathbf{B}_\beta[(\mathbf{x} : \alpha) \mapsto \mathbf{A}_\alpha], \mathbf{C}_\gamma[(\mathbf{x} : \alpha) \mapsto \mathbf{A}_\alpha])$.

Example 4.5 (Substitution) The following are examples of substitution (using the expressions we saw in Examples 4.3 and 4.4):

1. $(\lambda \mathbf{x} : \alpha \,.\, (\mathbf{x} = (\mathbf{y} : \alpha)))[(\mathbf{x} : \alpha) \mapsto \mathbf{D}_\alpha] \equiv \lambda \mathbf{x} : \alpha \,.\, (\mathbf{x} = (\mathbf{y} : \alpha))$.

2. $(\lambda \mathbf{x} : \alpha \,.\, (\mathbf{x} = (\mathbf{y} : \alpha)))[(\mathbf{y} : \alpha) \mapsto \mathbf{D}_\alpha] \equiv \lambda \mathbf{x} : \alpha \,.\, (\mathbf{x} = \mathbf{D}_\alpha)$.

3. $((\lambda \mathbf{x} : \alpha \,.\, \mathbf{x}) = \mathbf{c}_{\alpha \to \alpha \to \alpha} \,\mathbf{x})[(\mathbf{x} : \alpha) \mapsto \mathbf{D}_\alpha] \equiv$
$(\lambda \mathbf{x} : \alpha \,.\, \mathbf{x}) = \mathbf{c}_{\alpha \to \alpha \to \alpha} \mathbf{D}_\alpha$.

4. $(\lambda \mathbf{x} : \alpha \,.\, (\mathbf{y} : \alpha))[(\mathbf{y} : \alpha) \mapsto (\mathbf{x} : \alpha)] \equiv \lambda \mathbf{x} : \alpha \,.\, \mathbf{x}$.

Notice in the last case, since $(\mathbf{x} : \alpha)$ is not free for $(\mathbf{y} : \alpha)$ in $\lambda \mathbf{x} : \alpha \,.\, (\mathbf{y} : \alpha)$, the free occurrence of $(\mathbf{x} : \alpha)$ in $(\mathbf{x} : \alpha)$ is "captured" by the binder λ applied to $(\mathbf{x} : \alpha)$ and becomes a bound occurrence. A *variable capture* also occurs in the second case if $(\mathbf{x} : \alpha)$ is free in $\mathbf{D}_\alpha$. We will see in Example 7.5 in Chapter 7 that, when a variable is captured in a substitution, the expression produced by the substitution may not be a valid result. So, as a rule, we will only consider a substitution $\mathbf{B}_\beta[(\mathbf{x} : \alpha) \mapsto \mathbf{A}_\alpha]$ when $\mathbf{A}_\alpha$ is free for $(\mathbf{x} : \alpha)$ in $\mathbf{B}_\beta$. □

4.7   Languages

A *language* (or *signature*) of Alonzo is a pair $L = (\mathcal{B}, \mathcal{C})$ where $\mathcal{B}$ is a finite set of base types and $\mathcal{C}$ is a set of constants $\mathbf{c}_\alpha$ where each base type occurring in α is a member of $\mathcal{B}$. A type α is a *type of L* if all the base types occurring in α are members of $\mathcal{B}$, and an expression $\mathbf{A}_\alpha$ is an *expression of L* if all the base types occurring in $\mathbf{A}_\alpha$ are members of $\mathcal{B}$ and all the constants occurring in $\mathbf{A}_\alpha$ are members of $\mathcal{C}$. Let $\mathcal{T}(L) \subseteq \mathcal{T}$ denote the set of types of L and $\mathcal{E}(L) \subseteq \mathcal{E}$ denote the set of expressions of L. Notice that $\mathcal{B}$ and $\mathcal{C}$ may be empty, but $\mathcal{T}(L)$ and $\mathcal{E}(L)$ are always nonempty since $o \in \mathcal{T}(L)$. The *minimum language* is the language $L_{\mathsf{min}} = (\emptyset, \emptyset)$.

The base types and constants of a language are used to represent, respectively, the base domains and distinguished values of a structure. So it is sometimes convenient, when the set of constants is finite, to write a language

$$L = (\{\mathbf{a}_1, \dots, \mathbf{a}_m\}, \{\mathbf{c}^1_{\beta_1}, \dots, \mathbf{c}^n_{\beta_n}\})$$

as the tuple

$$(\mathbf{a}_1, \dots, \mathbf{a}_m, \mathbf{c}^1_{\beta_1}, \dots, \mathbf{c}^n_{\beta_n})$$

in the same way a structure can be written as a tuple.

Let $L_i = (\mathcal{B}_i, \mathcal{C}_i)$ be a language for $i \in \{1, 2\}$. L_2 is an *extension* of L_1 (or L_1 is a *sublanguage* of L_2), written $L_1 \leq L_2$, if $\mathcal{B}_1 \subseteq \mathcal{B}_2$ and $\mathcal{C}_1 \subseteq \mathcal{C}_2$. Notice that $L_{\mathsf{min}} \leq L$ for every language L. Recall that the *cardinality* of a set S, denoted $|S|$, is the cardinal number κ such that there is a bijection $f : \kappa \to S$. The *power* of a language $L = (\mathcal{B}, \mathcal{C})$, written $\|L\|$, is $|\mathcal{E}(L)|$. In the usual case, when $\mathcal{C}$ is countable (i.e., finite or countably infinite), $\|L\| = \omega$. When $\mathcal{C}$ is uncountable, $\|L\| = |\mathcal{C}|$. Of course, languages of uncountable power — i.e., languages having an uncountable number of constants — are only of use for theoretical purposes.

4.8   Conclusions

Alonzo has two kinds of syntactic entities: types and expressions. There are four kinds of types: the type of truth values, base types, function types, and product types. Types denote nonempty sets of values. There are seven kinds of expressions: variables, constants, equalities, function applications, function abstractions, definite descriptions, and ordered pairs. Each expression is assigned a type on the basis of its syntax. An expression is either defined and denotes a value in the set denoted by its type or undefined and denotes nothing at all.

4.9 Exercises

1. Write the following expressions of Alonzo in formal notation:

   a. $(x : A) = c_A$.

   b. $c_{A \to B \to C} \, d_A \, e_B$.

   c. $c_{(A \times B) \to o} \left((u : A), (v : B) \right)$.

   d. $\lambda x : A \,.\, \lambda y : A \,.\, x = y$.

   e. $(\lambda x : A \,.\, x) = (f : A \to A)$.

   f. $I x : A \,.\, (x = c_{A \to A} \, x)$.

   What is the type of each of these expressions?

2. Explain why each of the following expressions is or is not a well-formed expression of Alonzo:

   a. $c_A \, d_B$.

   b. $c_{A \to B} \, d_B$.

   c. $x_{A \to B} \, x_A$.

   d. $\lambda x : A \,.\, \lambda y : B \,.\, x = y$.

   e. $\lambda x : A \,.\, \lambda x : A \,.\, x = x$.

   f. $\lambda x : A \,.\, \lambda y : B \,.\, (x, y) = (y, x)$.

   g. $\lambda x : A \,.\, \lambda y : B \,.\, c_{A \to B \to C} \, (x \, y)$.

   h. $\lambda x : A \,.\, \lambda y : B \,.\, c_{A \to B \to C} \, x \, y$.

   i. $\lambda x : A \,.\, \lambda x : B \,.\, c_{A \to B \to C} \, x \, x$.

   j. $\lambda x : A \,.\, \lambda x : A \,.\, c_{A \to A \to B} \, x \, x$.

   k. $\lambda x : A \,.\, \lambda y : B \,.\, \lambda z : C \,.\, x = (y, z)$.

   l. $(\lambda x : A \,.\, \lambda y : B \,.\, c_{A \to B \to C} \, x \, y) = $
      $(\lambda u : B \,.\, \lambda v : A \,.\, d_{B \to A \to C} \, u \, v)$.

   m. $I x : o \,.\, x = c_{o \to o} \, x$.

   n. $I x : A \,.\, c_{A \to B} \, x$.

   o. $\mathsf{FunApp}(\mathsf{Con}(c, \mathsf{BaseTy}(A)), \mathsf{Con}(d, \mathsf{BaseTy}(B)))$.

   p. $\mathsf{FunAbs}(\mathsf{Con}(c, \mathsf{BaseTy}(A)), \mathsf{Con}(d, \mathsf{BaseTy}(B)))$.

3. Let $\mathbf{A}_B$ be the expression

$$c_{(A\times A)\rightarrow(A\rightarrow(A\times A))\rightarrow B}$$
$$((x:A),(y:A))$$
$$(\lambda x:A \,.\, ((x:A),(y:A))).$$

Compute the following substitutions using the identities S1–S10:

    a. $\mathbf{A}_B[(x:A)\mapsto(y:A)]$.

    b. $\mathbf{A}_B[(x:A)\mapsto(z:A)]$.

    c. $\mathbf{A}_B[(y:A)\mapsto(x:A)]$.

    d. $\mathbf{A}_B[(y:A)\mapsto(z:A)]$.

Which of these substitutions cause a variable capture?

4. Prove

$$(\mathbf{A}_\alpha[(x:\beta)\mapsto(y:\beta)])[(y:\beta)\mapsto\mathbf{B}_\beta] \equiv \mathbf{A}_\alpha[(x:\beta)\mapsto\mathbf{B}_\beta]$$

provided $(y:\beta)$ does not occur in $\mathbf{A}_\alpha$. (Exercise X5100 in [5].)

5. Prove

$$(\mathbf{A}_\alpha[(x:\beta)\mapsto(y:\beta)])[(y:\beta)\mapsto\mathbf{B}_\beta] \equiv \mathbf{A}_\alpha[(x:\beta)\mapsto\mathbf{B}_\beta]$$

provided $(y:\beta)$ is not free in $\mathbf{A}_\alpha$ and $(y:\beta)$ is free for $(x:\beta)$ in $\mathbf{A}_\alpha$. (Exercise X5101 in [5].)

6. Prove $\mathbf{A}_\alpha[(\mathbf{x}:\beta)\mapsto(\mathbf{x}:\beta)] \equiv \mathbf{A}_\alpha$ using equations S1–S10.

7. Prove $\mathbf{A}_\alpha[(\mathbf{x}:\beta)\mapsto\mathbf{B}_\beta] \equiv \mathbf{A}_\alpha$ provided $(\mathbf{x}:\beta)$ is not free in $\mathbf{A}_\alpha$ using equations S1–S10.

8. Construct a language $L = (\mathcal{B},\mathcal{C})$ of Alonzo such that $\mathcal{B}$ is empty but $\mathcal{C}$ is nonempty.

9. Construct a language of Alonzo that has uncountable power. What needs to be assumed about $\mathcal{S}_{con}$?

10. Let $\mathbb{L}$ be the set of languages of Alonzo and $\leq$ be the sublanguage relation over $\mathbb{L}$. Prove that $(\mathbb{L},\leq)$ is a lattice with a bottom element.

Chapter 5

Semantics

Let $L = (\mathcal{B}, \mathcal{C})$ be a language of Alonzo. In this chapter we will define the semantics of L.

5.1 Interpretations

A *frame* for L is a collection $\mathcal{D} = \{D_\alpha \mid \alpha \in \mathcal{T}(L)\}$ of nonempty domains (sets) of values such that:

F1. *Domain of truth values*: $D_o = \mathbb{B} = \{\text{F}, \text{T}\}$.

F2. *Predicate domain*: $D_{\alpha \to o}$ is a set of *some* total functions from D_α to D_o for $\alpha \in \mathcal{T}(L)$.

F3. *Function domain*: $D_{\alpha \to \beta}$ is a set of *some* partial and total functions from D_α to D_β for $\alpha, \beta \in \mathcal{T}(L)$ with $\beta \neq o$.

F4. *Product domain*: $D_{\alpha \times \beta} = D_\alpha \times D_\beta$ for $\alpha, \beta \in \mathcal{T}(L)$.

A predicate domain $D_{\alpha \to o}$ is *full* if it is the set of *all* total functions from D_α to D_o, and a function domain $D_{\alpha \to \beta}$ with $\beta \neq o$ is *full* if it is the set of *all* partial and total functions from D_α to D_β. The frame is *full* if $D_{\alpha \to \beta}$ is full for all $\alpha, \beta \in \mathcal{T}(L)$. Notice that the only restriction on a *base domain*, i.e., $D_{\mathbf{a}}$ for some $\mathbf{a} \in \mathcal{B}$, is that it is nonempty and that the frame is completely determined by its base domains when the frame is full.

An *interpretation* of L is a pair $M = (\mathcal{D}, I)$ where $\mathcal{D} = \{D_\alpha \mid \alpha \in \mathcal{T}(L)\}$ is a frame for L and I is an *interpretation function* that maps each constant in $\mathcal{C}$ of type α to an element of D_α. Notice that

$$(\{D_{\mathbf{a}} \mid \mathbf{a} \in \mathcal{B}\}, \{I(\mathbf{c}_\alpha) \mid \mathbf{c}_\alpha \in \mathcal{C}\})$$

© The Author(s), under exclusive license to Springer Nature Switzerland AG 2023
W. M. Farmer, *Simple Type Theory*, Computer Science Foundations
and Applied Logic, https://doi.org/10.1007/978-3-031-21112-6_5

is a structure. Hence an interpretation of a language *defines* (1) a structure
and (2) a mapping of the base types and constants of the language to the
base domains and distinguished values, respectively, of the structure. This
mapping can be lifted to a mapping of the (formal) expressions of the language to the (informal) expressions constructed from the base domains and
distinguished values. And so the interpretation provides the means to use
the language to make statements about and describe values in the structure
defined by the interpretation.

We will say that a language L is a *language for* a structure S if there
is an interpretation M of L that defines S. The same language can be for
different structures, and different languages can be for the same structure.

Example 5.1 (Interpretation) Let $L = (\{N\}, \{\leq_{(N \times N) \to o}\})$ be a language which we can write as $(N, \leq_{(N \times N) \to o})$; $(\mathbb{N}, \leq)$ be the structure described in Example 3.7 where $\leq : \mathbb{N} \times \mathbb{N} \to \mathbb{B}$ (instead of being a binary
relation over $\mathbb{N}$); and $\mathcal{D} = \{D_\alpha \mid \alpha \in \mathcal{T}(L)\}$ be a frame for L such that
$D_N = \mathbb{N}$ and $\leq \in D_{(N \times N) \to o}$. Then $M = (\mathcal{D}, I)$, where $I(\leq_{(N \times N) \to o}) = \leq$,
is an interpretation of L, and thus L is a language for $(\mathbb{N}, \leq)$. L has many
other possible interpretations, and L is a language for many other structures. For instance, if $(\mathbb{N}, <)$ is the same structure as $(\mathbb{N}, \leq)$ except the
weak total order $\leq : (\mathbb{N} \times \mathbb{N}) \to \mathbb{B}$ is replaced by the strong total order
$< : (\mathbb{N} \times \mathbb{N}) \to \mathbb{B}$, then $M' = (\mathcal{D}', I')$, where $\mathcal{D}'$ is $\mathcal{D}$ with $<$ added to
$D_{(N \times N) \to o}$ (if necessary) and $I'(\leq_{(N \times N) \to o}) = <$, is also an interpretation
of L, and thus L is a language for $(\mathbb{N}, <)$ as well as $(\mathbb{N}, \leq)$.   □

Let $\mathcal{D} = \{D_\alpha \mid \alpha \in \mathcal{T}(L)\}$ be a frame for L. An *assignment into* $\mathcal{D}$ is a
function φ whose domain is the set of variables of L such that $\varphi((\mathbf{x} : \alpha)) \in
D_\alpha$ for each variable $(\mathbf{x} : \alpha)$ of L. Given an assignment φ, a variable $(\mathbf{x} : \alpha)$
of L, and $d \in D_\alpha$, let $\varphi[(\mathbf{x} : \alpha) \mapsto d]$ be the assignment ψ in $\mathcal{D}$ such that
$\psi((\mathbf{x} : \alpha)) = d$ and $\psi((\mathbf{y} : \beta)) = \varphi((\mathbf{y} : \beta))$ for all variables $(\mathbf{y} : \beta)$ of L
distinct from $(\mathbf{x} : \alpha)$. Given an interpretation M of L, let $\mathsf{assign}(M)$ be the
set of assignments into the frame of M. Assignments are used to perform
"semantic substitution" of values for variables.

5.2   General Models

Let $\mathcal{D} = \{D_\alpha \mid \alpha \in \mathcal{T}(L)\}$ be a frame for L and $M = (\mathcal{D}, I)$ be an interpretation of L. M is a *general model* of L if there is a partial binary *valuation
function* V^M such that, for all assignments $\varphi \in \mathsf{assign}(M)$ and expressions

$\mathbf{C}_\gamma$ of L, (1) either $V_\varphi^M(\mathbf{C}_\gamma) \in D_\gamma$ or $V_\varphi^M(\mathbf{C}_\gamma)$ is undefined[1] and (2) each of the following conditions is satisfied:

V1. $V_\varphi^M((\mathbf{x}:\alpha)) = \varphi((\mathbf{x}:\alpha))$.

V2. $V_\varphi^M(\mathbf{c}_\alpha) = I(\mathbf{c}_\alpha)$.

V3. $V_\varphi^M(\mathbf{A}_\alpha = \mathbf{B}_\alpha) = \mathrm{T}$ if $V_\varphi^M(\mathbf{A}_\alpha)$ is defined, $V_\varphi^M(\mathbf{B}_\alpha)$ is defined, and $V_\varphi^M(\mathbf{A}_\alpha) = V_\varphi^M(\mathbf{B}_\alpha)$. Otherwise, $V_\varphi^M(\mathbf{A}_\alpha = \mathbf{B}_\alpha) = \mathrm{F}$.

V4. $V_\varphi^M(\mathbf{F}_{\alpha\to\beta}\,\mathbf{A}_\alpha) = V_\varphi^M(\mathbf{F}_{\alpha\to\beta})(V_\varphi^M(\mathbf{A}_\alpha))$ — i.e., the application of the function $V_\varphi^M(\mathbf{F}_{\alpha\to\beta})$ to the argument $V_\varphi^M(\mathbf{A}_\alpha)$ — if $V_\varphi^M(\mathbf{F}_{\alpha\to\beta})$ is defined, $V_\varphi^M(\mathbf{A}_\alpha)$ is defined, and $V_\varphi^M(\mathbf{F}_{\alpha\to\beta})$ is defined at $V_\varphi^M(\mathbf{A}_\alpha)$. Otherwise, $V_\varphi^M(\mathbf{F}_{\alpha\to\beta}\,\mathbf{A}_\alpha) = \mathrm{F}$ if $\beta = o$ and $V_\varphi^M(\mathbf{F}_{\alpha\to\beta}\,\mathbf{A}_\alpha)$ is undefined if $\beta \neq o$.

V5. $V_\varphi^M(\lambda\mathbf{x}:\alpha\,.\,\mathbf{B}_\beta)$ is the (partial or total) function $f \in D_{\alpha\to\beta}$ such that, for each $d \in D_\alpha$, $f(d) = V_{\varphi[(\mathbf{x}:\alpha)\mapsto d]}^M(\mathbf{B}_\beta)$ if $V_{\varphi[(\mathbf{x}:\alpha)\mapsto d]}^M(\mathbf{B}_\beta)$ is defined and $f(d)$ is undefined if $V_{\varphi[(\mathbf{x}:\alpha)\mapsto d]}^M(\mathbf{B}_\beta)$ is undefined.

V6. $V_\varphi^M(\mathrm{I}\mathbf{x}:\alpha\,.\,\mathbf{A}_o)$ is the $d \in D_\alpha$ such that $V_{\varphi[(\mathbf{x}:\alpha)\mapsto d]}^M(\mathbf{A}_o) = \mathrm{T}$ if there is exactly one such d. Otherwise, $V_\varphi^M(\mathrm{I}\mathbf{x}:\alpha\,.\,\mathbf{A}_o)$ is undefined.

V7. $V_\varphi^M((\mathbf{A}_\alpha,\mathbf{B}_\beta)) = (V_\varphi^M(\mathbf{A}_\alpha), V_\varphi^M(\mathbf{B}_\beta))$ if $V_\varphi^M(\mathbf{A}_\alpha)$ and $V_\varphi^M(\mathbf{B}_\beta)$ are defined. Otherwise, $V_\varphi^M((\mathbf{A}_\alpha,\mathbf{B}_\beta))$ is undefined.

$V_\varphi^M(\mathbf{C}_\gamma)$ is called the *value of* $\mathbf{C}_\gamma$ *in* M *with respect to* φ when $V_\varphi^M(\mathbf{C}_\gamma)$ is defined. $\mathbf{C}_\gamma$ is said to have no value in M with respect to φ when $V_\varphi^M(\mathbf{C}_\gamma)$ is undefined.

Proposition 5.2 *General models for L exist.*

Proof It is easy to construct an interpretation $M = (\mathcal{D}, I)$ of L such that $\mathcal{D}$ is full. It is also easy to define by induction on the syntactic structure of expressions a valuation function V^M so that M is a general model of L. $\square$

Example 5.3 (General Model) Let $S = (\mathbb{N}, 0, 1, +, *)$ be the structure presented in Example 3.8 where $+, * : \mathbb{N} \times \mathbb{N} \to \mathbb{N}$. Define $L = (\mathcal{B}, \mathcal{C})$, $\mathcal{B} = \{N\}$,

$$\mathcal{C} = \{0_N, 1_N, +_{(N\times N)\to N}, *_{(N\times N)\to N}\},$$

---

[1] We write $V_\varphi^M(\mathbf{C}_\gamma)$ instead of $V^M(\varphi, \mathbf{C}_\gamma)$.

$\mathcal{D} = \{D_\alpha \mid \alpha \in \mathcal{T}(L)\}$ to be a full frame for L such that $D_N = \mathbb{N}$, and I to be the function on $\mathcal{C}$ such that $I(0_N) = 0$, $I(1_N) = 1$, $I(+_{(N\times N)\to N}) = +$, and $I(*_{(N\times N)\to N}) = *$. Then $M = (\mathcal{D}, I)$ is a general model of L (by the proof of Proposition 5.2) that defines S.

Let $\mathbf{A}_o$ be the closed expression

$$((\lambda x : N \, . \, ((+_{(N\times N)\to N}\,((x : N), (x : N))) = 1_N))\,0_N)$$

of L and $\varphi \in \mathsf{assign}(M)$. Dropping the unnecessary parentheses and the unnecessary types from the constants and variables (Notational Conventions 3, 5, and 6), we can write $\mathbf{A}_o$ in a more readable form as:

$$(\lambda x : N \, . + (x, x) = 1)\,0.$$

We will "compute" $V_\varphi^M(\mathbf{A}_o)$, the value of $\mathbf{A}_o$ in M with respect to φ, in steps. Let $d \in D_N$, $\psi = \varphi[(x : N) \mapsto d]$, and $\mathbf{B}_N$ be

$$+_{(N\times N)\to N}\,((x : N), (x : N)).$$

1. $V_\psi^M((x : N)) = \psi((x : N)) = d$ by V1 of the definition of a general model.

2. $V_\varphi^M(0_N) = I(0_N) = 0$ by V2.

3. $V_\psi^M(1_N) = I(1_N) = 1$ by V2.

4. $V_\psi^M(+_{(N\times N)\to N}) = I(+_{(N\times N)\to N}) = +$ by V2.

5. $V_\psi^M(((x : N), (x : N))) = (V_\psi^M((x : N)), V_\psi^M((x : N))) = (d, d)$ by V7.

6. $V_\psi^M(\mathbf{B}_N) = V_\psi^M(+_{(N\times N)\to N})(V_\psi^M(((x : N), (x : N)))) = +(d, d) = d + d$ by V4.

7. $V_\psi^M(\mathbf{B}_N = 1_N) = (V_\psi^M(\mathbf{B}_N) = V_\psi^M(1_N)) = (d + d = 1)$ by V3.

8. $V_\varphi^M(\lambda x : N \, . \, \mathbf{B}_N = 1_N)$ is the function f that maps each $d \in D_N$ to $V_\psi^M(\mathbf{B}_N = 1_N) = (d + d = 1)$ by V5.

9. $V_\varphi^M(\mathbf{A}_o) = V_\varphi^M(\lambda x : N \, . \, \mathbf{B}_N = 1_N)(V_\varphi^M(0_N)) = f(0) = (0 + 0 = 1) = (0 = 1) = \mathrm{F}$ by V4.

Therefore, $V_\varphi^M(\mathbf{A}_o) = \mathrm{F}$. Notice that $V_\varphi^M(\mathbf{A}_o)$ does not depend on φ. This is a consequence of $\mathbf{A}_o$ being closed (see part 17 of Lemma 5.4 below). $\square$

We will use "Let M be a general model of L" as shorthand for "Let $M = (\mathcal{D}^M, I^M)$ be a general model of $L = (\mathcal{B}, \mathcal{C})$ where $\mathcal{D}^M = \{D_\alpha^M \mid \alpha \in \mathcal{T}(L)\}$", and "Let M_i be a general model of L_i for $i \in \{1,2\}$" as shorthand for "Let $M_i = (\mathcal{D}_i, I_i)$ be a general model of $L_i = (\mathcal{B}_i, \mathcal{C}_i)$ where $\mathcal{D}_i = \{D_\alpha^i \mid \alpha \in \mathcal{T}(L_i)\}$ for $i \in \{1,2\}$".

Lemma 5.4 (Semantics Lemma 1) *Let M be a general model of L and $\varphi, \psi \in \mathsf{assign}(M)$.*

1. $V_\varphi^M((\mathbf{x} : \alpha))$ *is defined.*

2. $V_\varphi^M(\mathbf{c}_\alpha)$ *is defined.*

3. $V_\varphi^M(\mathbf{A}_\alpha = \mathbf{B}_\alpha)$ *is defined.*

4. $V_\varphi^M(\mathbf{F}_{\alpha \to o} \mathbf{A}_\alpha)$ *is defined.*

5. *If* $V_\varphi^M(\mathbf{F}_{\alpha \to \beta})$, $V_\varphi^M(\mathbf{A}_\alpha)$, *or* $V_\varphi^M(\mathbf{F}_{\alpha \to \beta})(V_\varphi^M(\mathbf{A}_\alpha))$ *is undefined, then* $V_\varphi^M(\mathbf{F}_{\alpha \to \beta} \mathbf{A}_\alpha) = \mathrm{F}$ *if $\beta = o$ and* $V_\varphi^M(\mathbf{F}_{\alpha \to \beta} \mathbf{A}_\alpha)$ *is undefined if $\beta \neq o$.*

6. $V_\varphi^M(\lambda \mathbf{x} : \alpha \,.\, \mathbf{B}_\beta)$ *is defined.*

7. $V_\varphi^M(\mathrm{I}\, \mathbf{x} : \alpha \,.\, \mathbf{A}_o)$ *is defined iff* $V_{\varphi[(\mathbf{x}:\alpha) \mapsto d]}^M(\mathbf{A}_o) = \mathrm{T}$ *for exactly one* $d \in D_\alpha^M$.

8. $V_\varphi^M((\mathbf{A}_\alpha, \mathbf{B}_\alpha))$ *is defined iff* $V_\varphi^M(\mathbf{A}_\alpha)$ *and* $V_\varphi^M(\mathbf{B}_\alpha)$ *are defined.*

9. $V_\varphi^M(\mathbf{A}_o)$ *is defined.*

10. $V_\varphi^M(\mathbf{A}_\alpha = \mathbf{B}_\alpha) = \mathrm{T}$ *iff* $V_\varphi^M(\mathbf{A}_\alpha) = V_\varphi^M(\mathbf{B}_\alpha)$.

11. *If* $V_\varphi^M(\mathbf{A}_\alpha)$ *is defined, then*

$$V_\varphi^M((\lambda \mathbf{x} : \alpha \,.\, \mathbf{B}_\beta) \, \mathbf{A}_\alpha) \simeq V_{\varphi[(\mathbf{x}:\alpha) \mapsto V_\varphi^M(\mathbf{A}_\alpha)]}^M(\mathbf{B}_\beta).$$

12. *If* $V_\varphi^M(\mathrm{I}\, \mathbf{x} : \alpha \,.\, \mathbf{A}_o)$ *is defined, then*

$$V_{\varphi[(\mathbf{x}:\alpha) \mapsto d]}^M(\mathbf{A}_o) = \mathrm{T} \quad \textit{iff} \quad d = V_\varphi^M(\mathrm{I}\, \mathbf{x} : \alpha \,.\, \mathbf{A}_o)$$

for all $d \in D_\alpha^M$.

13. $V_\varphi^M(\lambda \mathbf{x} : \alpha \,.\, \mathbf{B}_\beta) = V_{\varphi[(\mathbf{x}:\alpha) \mapsto d]}^M(\lambda \mathbf{x} : \alpha \,.\, \mathbf{B}_\beta)$ *for all $d \in D_\alpha^M$.*

14. $V_\varphi^M(\mathrm{I}\, \mathbf{x} : \alpha \,.\, \mathbf{A}_o) \simeq V_{\varphi[(\mathbf{x}:\alpha) \mapsto d]}^M(\mathrm{I}\, \mathbf{x} : \alpha \,.\, \mathbf{A}_o)$ *for all $d \in D_\alpha^M$.*

15. *If* $(\mathbf{x}:\alpha)$ *is not free in* $\mathbf{B}_\beta$, *then*

$$V^M_{\varphi[(\mathbf{x}:\alpha)\mapsto d_1]}(\mathbf{B}_\beta) \simeq V^M_{\varphi[(\mathbf{x}:\alpha)\mapsto d_2]}(\mathbf{B}_\beta)$$

for all $d_1, d_2 \in D^M_\alpha$.

16. *If* φ, ψ *agree on all free variables of* $\mathbf{A}_\alpha$, *then* $V^M_\varphi(\mathbf{A}_\alpha) \simeq V^M_\psi(\mathbf{A}_\alpha)$.

17. *If* $\mathbf{A}_\alpha$ *is closed, then* $V^M_\varphi(\mathbf{A}_\alpha)$ *does not depend on* φ.

Proof The proof is left to the reader as an exercise. □

Theorem 5.5 (Traditional Approach to Undefinedness) *Alonzo satisfies the three principles of TATU (given in Section 3.4).*

Proof The first principle is satisfied by parts 1 and 2 of Lemma 5.4. The second is satisfied by parts 5 and 7. And the third is satisfied by parts 3, 4, 5, 9, and 10. □

Let M be a general model of $L = (\mathcal{B}, \mathcal{C})$. The *size* of M, written $|M|$, is the cardinality of $\bigcup_{\mathbf{a}\in\mathcal{B}} D^M_\mathbf{a}$. M is *finite* if its size is finite and is *infinite* otherwise. The *power* of M, written $\|M\|$, is the least cardinal κ such that $|D^M_\alpha| \leq \kappa$ for all $\alpha \in \mathcal{T}(L)$. The power of a model need not exist; whether it exists can depend on the underlying set-theoretic assumptions that one makes. For instance, the power of a model of countably infinite size with a full frame exists if a strongly inaccessible cardinal exists.

5.3   Finite General Models

Proposition 5.6 *Let M be a general model of L.*

1. *M is finite iff D^M_α is finite for all $\alpha \in \mathcal{T}(L)$.*

2. *If M is finite, then $\|M\| = \omega$.*

Proof

Part 1 The $\Leftarrow$ direction is obvious. So assume M is finite. Then D^M_o is finite and $D^M_\mathbf{a}$ is finite for all $\mathbf{a} \in \mathcal{B}$. If D^M_α and D^M_β are finite, then obviously $D^M_{\alpha\to\beta}$ and $D^M_{\alpha\times\beta}$ are finite. From these two facts it easily follows that D^M_α is finite for all $\alpha \in \mathcal{T}(L)$ by induction on the syntactic structure of types.

Part 2 Assume M is finite and $\varphi \in \mathsf{assign}(M)$. Then D_α^M is finite for all $\alpha \in \mathcal{T}(L)$ by part 1 and so $\|M\| \leq \omega$. Suppose $(\star)$ $\|M\| = n$ for some $n \in \mathbb{N}$. ($n > 1$ since $|D_o| = 2$.) Then $D_\alpha^M = \{d_1, \ldots, d_n\}$ for some $\alpha \in \mathcal{T}(L)$.

$$V_{\varphi[(y:\alpha) \mapsto d_1]}^M(\lambda x : \alpha . (y : \alpha)), \ \ldots, \ V_{\varphi[(y:\alpha) \mapsto d_n]}^M(\lambda x : \alpha . (y : \alpha))$$

are members of $D_{\alpha \to \alpha}^M$; these are the n distinct constant functions from D_α^M to D_α^M. $V_\varphi^M(\lambda x : \alpha . x)$ is also a member of $D_{\alpha \to \alpha}^M$; it is the identity function on D_α^M. Hence $|D_{\alpha \to \alpha}^M| \geq n + 1 > n = |D_\alpha^M|$ which contradicts $(\star)$. Therefore, $\|M\| = \omega$. $\qquad\square$

5.4 Standard Models

An interpretation $M = (\mathcal{D}, I)$ of L is a *standard model* for L if $\mathcal{D}$ is full.

Proposition 5.7 *A standard model of L is a general model of L.*

Proof By the proof of Proposition 5.2. $\qquad\square$

A language of Alonzo thus has two different semantics, a *general semantics* based on general models and a *standard semantics* based on standard models. (See Section 5.8 for a comparison of the two semantics.) A general model of L is a *nonstandard model* of L if it is not a standard model.

Proposition 5.8 *Let M be a general model of L that defines a structure S. Then there is a unique standard model of L with the interpretation function I^M that defines S.*

Proof There is clearly a standard model $N = (\mathcal{D}^N, I^M)$ of L that defines S where $\mathcal{D}^N = \{D_\alpha^N \mid \alpha \in \mathcal{T}(L)\}$ and $D_\alpha^M \subseteq D_\alpha^N$ for all $\alpha \in \mathcal{T}(L)$. $\{D_\alpha^N \mid \alpha \in \mathcal{B}\}$ is determined since N defines S and $\{D_\alpha^N \mid \alpha \notin \mathcal{B}\}$ is determined since $\mathcal{D}^N$ is full by definition. Hence there is only one standard model of L with interpretation function I^M that defines S. $\qquad\square$

Theorem 5.9 (Finite General Models are Standard Models) *Every finite general model is a standard model.*

Proof The proof of this theorem is given in Section 6.1 on page 67 after we have introduced notation for the Boolean operators. $\qquad\square$

Figure 5.1 shows pictorially how general models, standard models, and finite general models are related according to Proposition 5.7 and Theorem 5.9.

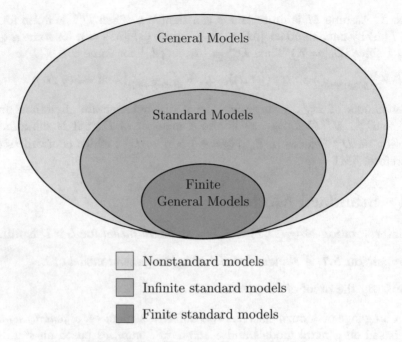

Figure 5.1: Kinds of General Models

5.5   Satisfiability, Validity, and Semantic Consequence

We will say that a general model M *interprets* an expression $\mathbf{A}_\alpha$ if $\mathbf{A}_\alpha$ is an expression of the language of which M is an interpretation and *interprets* a set of expressions if it interprets each member of the set. Let M be a general model that interprets $\mathbf{A}_\alpha$ and $\varphi \in \mathsf{assign}(M)$. The following are basic definitions about the semantics of Alonzo:

1. φ *satisfies* $\mathbf{A}_o$ in M, written $M \vDash_\varphi \mathbf{A}_o$, if $V_\varphi^M(\mathbf{A}_o) = \mathrm{T}$.

2. $\mathbf{A}_o$ is *satisfiable in* M if $M \vDash_\varphi \mathbf{A}_o$ for some $\varphi \in \mathsf{assign}(M)$.

3. $\mathbf{A}_o$ is *satisfiable* if $M \vDash_\varphi \mathbf{A}_o$ for some general model M and some $\varphi \in \mathsf{assign}(M)$.

4. $\mathbf{A}_o$ is *valid in* M (or M is a *model of* $\mathbf{A}_o$), written $M \vDash \mathbf{A}_o$, if $M \vDash_\varphi \mathbf{A}_o$ for all $\varphi \in \mathsf{assign}(M)$.

5. If $\mathbf{A}_o$ is a sentence, then $\mathbf{A}_o$ is *true* [*false*] *in* M if $V_\varphi^M(\mathbf{A}_o) = \mathrm{T}$ [F] (for all $\varphi \in \mathsf{assign}(M)$).

6. $\mathbf{A}_o$ is *valid (in the general sense)*, written $\vDash \mathbf{A}_o$, if $M \vDash \mathbf{A}_o$ for all general models M that interpret $\mathbf{A}_o$.

7. $\mathbf{A}_o$ is *valid in the standard sense*, written $\vDash^s \mathbf{A}_o$, if $M \vDash \mathbf{A}_o$ for all standard models M that interpret $\mathbf{A}_o$.

8. $\mathbf{A}_o$ is *logically equivalent* to $\mathbf{B}_o$ if $V_\varphi^M(\mathbf{A}_o) = V_\varphi^M(\mathbf{B}_o)$ for all general models M that interpret $\mathbf{A}_o$ and $\mathbf{B}_o$ and all $\varphi \in \mathsf{assign}(M)$.

Now let M be a general model that interprets a set Γ of formulas. The following are additional basic definitions about the semantics of Alonzo:

1. φ *satisfies* Γ *in* M, written $M \vDash_\varphi \Gamma$, if $M \vDash_\varphi \mathbf{A}_o$ for all $\mathbf{A}_o \in \Gamma$.

2. Γ is *satisfiable in* M if $M \vDash_\varphi \Gamma$ for some $\varphi \in \mathsf{assign}(M)$.

3. Γ is *satisfiable* if $M \vDash_\varphi \Gamma$ for some general model M for L and some $\varphi \in \mathsf{assign}(M)$.

4. M is a *model of* Γ, written $M \vDash \Gamma$, if $M \vDash \mathbf{A}_o$ for all $\mathbf{A}_o \in \Gamma$.

5. $\mathbf{A}_o$ is a *semantic consequence of* Γ *(in the general sense)*, written $\Gamma \vDash \mathbf{A}_o$, if $M \vDash_\varphi \Gamma$ implies $M \vDash_\varphi \mathbf{A}_o$ for all general models M that interpret $\mathbf{A}_o$ and Γ and all $\varphi \in \mathsf{assign}(M)$.

6. $\mathbf{A}_o$ is a *semantic consequence of* Γ *in the standard sense*, written $\Gamma \vDash^s \mathbf{A}_o$, if $M \vDash_\varphi \Gamma$ implies $M \vDash_\varphi \mathbf{A}_o$ for all standard models M that interpret $\mathbf{A}_o$ and Γ and all $\varphi \in \mathsf{assign}(M)$.

Obviously, $\emptyset \vDash \mathbf{A}_o$ iff $\vDash \mathbf{A}_o$, and $\Gamma \vDash \mathbf{A}_o$ implies $\Gamma \vDash^s \mathbf{A}_o$. Semantic consequence is a form of logical consequence.

Proposition 5.10 *Let* $\mathbf{A}_o$ *be a formula and* Γ *be a set of sentences.*

1. $\Gamma \vDash \mathbf{A}_o$ *iff every model of* Γ *is a model of* $\mathbf{A}_o$.

2. $\Gamma \vDash^s \mathbf{A}_o$ *iff every standard model of* Γ *is a model of* $\mathbf{A}_o$.

Proof

Part 1

($\Rightarrow$): Assume $\Gamma \vDash \mathbf{A}_o$. Then $(\star)$ $M \vDash_\varphi \Gamma$ implies $M \vDash_\varphi \mathbf{A}_o$ for all general models M that interpret $\mathbf{A}_o$ and Γ and all $\varphi \in \mathsf{assign}(M)$. Let M be a model of Γ. By part 17 of Lemma 5.4, $M \vDash_\varphi \Gamma$ for all $\varphi \in \mathsf{assign}(M)$ since Γ is a set of sentences, and so $M \vDash_\varphi \mathbf{A}_o$ for all $\varphi \in \mathsf{assign}(M)$ by $(\star)$. Hence M is a model of $\mathbf{A}_o$.

($\Leftarrow$): Assume $(\star)$ every model of Γ is a model of $\mathbf{A}_o$. Let $M \vDash_\varphi \Gamma$ for a general model M that interprets $\mathbf{A}_o$ and Γ and $\varphi \in \mathsf{assign}(M)$. By part 17 of Lemma 5.4, M is a model of Γ since Γ is a set of sentences, and so M is a model of $\mathbf{A}_o$ by $(\star)$. This implies $M \vDash_\varphi \mathbf{A}_o$. Hence $\Gamma \vDash \mathbf{A}_o$.

Part 2 Similar to part 1.   □

Remark 5.11 (Languages as Specifications) *A language L can be viewed as a specification of the set of structures defined by the general models of L. The structures in this set can contain vastly different values, but they all share the structural form prescribed by L.*

5.6   Isomorphic General Models

Let M_i be a general model of L for $i \in \{1, 2\}$. An *isomorphism* from M_1 to M_2 is a set $\Theta = \{\theta_\alpha \mid \alpha \in \mathcal{T}(L)\}$ of mappings such that:

1. θ_α is a bijection from D_α^1 to D_α^2 for all $\alpha \in \mathcal{T}(L)$.

2. $\theta_o(\mathrm{F}) = \mathrm{F}$ and $\theta_o(\mathrm{T}) = \mathrm{T}$.

3. $\theta_\beta(f(a)) \simeq \theta_{\alpha \to \beta}(f)(\theta_\alpha(a))$ for all $\alpha, \beta \in \mathcal{T}(L)$, $f \in D_{\alpha \to \beta}^1$, and $a \in D_\alpha^1$.

4. $\theta_{\alpha \times \beta}((a, b)) = (\theta_\alpha(a), \theta_\beta(b))$ for all $\alpha, \beta \in \mathcal{T}(L)$, $a \in D_\alpha^1$, and $b \in D_\beta^1$.

5. $\theta_\alpha(I_1(\mathbf{c}_\alpha)) = I_2(\mathbf{c}_\alpha)$ for all $\mathbf{c}_\alpha \in \mathcal{C}$.

M_1 and M_2 are *isomorphic*, written $M_1 \equiv M_2$, if there is an isomorphism from M_1 to M_2.

Proposition 5.12 *Let L be a language, M_1 and M_2 be general models of L, $\Theta = \{\theta_\alpha \mid \alpha \in \mathcal{T}(L)\}$ be an isomorphism from M_1 to M_2, $\varphi_1 \in \mathsf{assign}(M_1)$, $\Theta(\varphi_1) \in \mathsf{assign}(M_2)$ such that $\Theta(\varphi_1)((\mathbf{x} : \alpha)) = \theta_\alpha(\varphi_1((\mathbf{x} : \alpha)))$ for all variables $(\mathbf{x} : \alpha) \in \mathcal{E}(L)$, and $\mathbf{A}_\alpha \in \mathcal{E}(L)$. Then*

$$\theta_\alpha(V_{\varphi_1}^{M_1}(\mathbf{A}_\alpha)) = V_{\Theta(\varphi_1)}^{M_2}(\mathbf{A}_\alpha).$$

Proof The proof, which is by induction on the syntactic structure of expressions, is left to the reader as an exercise.   □

Corollary 5.13 *If M_1 and M_2 are isomorphic general models of L, then $M_1 \vDash \mathbf{A}_o$ iff $M_2 \vDash \mathbf{A}_o$.*

Proof Let $\{\theta_\alpha \mid \alpha \in \mathcal{T}(L)\}$ be an isomorphism from M_1 to M_2.

$$M_1 \vDash \mathbf{A}_o$$

$$\text{iff} \quad V_{\varphi_1}^{M_1}(\mathbf{A}_o) = \mathrm{T} \text{ for all } \varphi_1 \in \mathsf{assign}(M_1) \tag{1}$$

$$\text{iff} \quad \theta_o(V_{\varphi_1}^{M_1}(\mathbf{A}_o)) = \mathrm{T} \text{ for all } \varphi_1 \in \mathsf{assign}(M_1) \tag{2}$$

$$\text{iff} \quad V_{\varphi_2}^{M_2}(\mathbf{A}_o) = \mathrm{T} \text{ for all } \varphi_2 \in \mathsf{assign}(M_2) \tag{3}$$

$$\text{iff} \quad M_2 \vDash \mathbf{A}_o \tag{4}$$

(1) and (4) are by definition of a formula being valid in a general model; (2) follows from θ_o being an identity function; and (3) is by Proposition 5.12 and the fact that each θ_α is a bijection. $\qquad\square$

In summary, two isomorphic general models (of the same language) have exactly the same structure and can be considered to be identical. We will say that a statement is true "up to isomorphism" to express the assertion that the statement holds under the assumption that isomorphic general models are identical.

5.7 Expansion of a General Model

Let M_i be a general model of L_i for $i \in \{1, 2\}$. Assume $L_1 \leq L_2$. M_2 is an *expansion of M_1 to L_2* (or M_1 is a *reduct of M_2 to L_1*), written $M_1 \leq M_2$, if $\mathcal{D}_1 \subseteq \mathcal{D}_2$ and $I_1 \sqsubseteq I_2$. If $L_1 \neq L_2$, then M_1 has many possible expansions to L_2, one for each way of assigning domains to the types in $\mathcal{T}(L_2) \setminus \mathcal{T}(L_1)$ and values to the constants in $\mathcal{C}_2 \setminus \mathcal{C}_1$. However, M_2 has only one reduct to L_1, namely, the general model M_1 where $\mathcal{D}_1 = \{D_\alpha^2 \in \mathcal{D}_2 \mid \alpha \in \mathcal{T}(L_1)\}$ and $I_1 = I_2{\restriction_{\mathcal{C}_1}}$ (i.e., the restriction of I_2 to $\mathcal{C}_1$).

The following lemma says that expanding a general model M of a language L preserves the set of formulas of L that are valid in M:

Lemma 5.14 *Let M_i be a general model of L_i for $i \in \{1, 2\}$ such that $L_1 \leq L_2$ and M_2 is an expansion of M_1 to L_2. If $\mathbf{A}_o$ is a formula of L_1, then $M_1 \vDash \mathbf{A}_o$ iff $M_2 \vDash \mathbf{A}_o$.*

Proof

$$M_1 \vDash \mathbf{A}_o$$

$$\text{iff } V_\varphi^{M_1}(\mathbf{A}_o) = \text{T for all } \varphi \in \mathsf{assign}(M_1) \tag{1}$$

$$\text{iff } V_\varphi^{M_2}(\mathbf{A}_o) = \text{T for all } \varphi \in \mathsf{assign}(M_2) \tag{2}$$

$$\text{iff } M_2 \vDash \mathbf{A}_o \tag{3}$$

(1) and (3) are by the definition of a formula being valid in a general model; and (2) is implied by (a) M_2 is an expansion of M_1 to L_2 and (b) $\mathbf{A}_o$ is a formula of L_1. □

The minimum language L_{min} has a unique standard model $M_{\mathsf{min}} = (\mathcal{D}, I)$, where $\mathcal{D}$ is the unique full frame for L_{min} and I is the empty function, which we will call the *minimum general model*. M_{min} defines S_{min}, the minimum structure. Every model of L_{min} is finite, and so every general model of L_{min} is a standard model by by Theorem 5.9. Hence M_{min} is the unique general model of L_{min} since M_{min} is the unique standard model of L_{min}. Notice that $M_{\mathsf{min}} \leq M$ for every general model M.

5.8   Standard vs. General Semantics

A standard model defines the kind of structure that is commonly found in mathematical practice, while a nonstandard model defines the kind of structure that is much more likely found in logical practice than in mathematical practice. Thus, roughly speaking, the standard semantics is the semantics of mathematical practice and the general semantics is the semantics of logical practice.

A language of Alonzo with the general semantics can be encoded as a language of many-sorted first-order logic with a version of first-order semantics that admits undefined terms. Thus Alonzo with the general semantics can be viewed as a first-order theory presented in a more natural and convenient form. We will see later that, by virtue of the general semantics, Alonzo has analogs to the fundamental theorems of first-order model theory including the completeness, compactness, and the Löwenheim-Skolem theorems.

The distinction between standard and nonstandard models is perfectly clear in Alonzo (and other versions of simple type theory with the general semantics), but there is no precise way to distinguish between the standard and nonstandard models of a first-order theory without a higher-order perspective [79]. The general semantics of Alonzo provides the necessary higher-order perspective to fully understand the distinction.

In summary, both semantics are important. The standard models semantics is needed for the practical application of Alonzo, while the general semantics is needed for the theoretical study of Alonzo.

5.9 Examples of Standard Models

We will now use the machinery of Alonzo to construct several examples of standard models. For each structure S presented in Section 3.6, a language L and a standard model M of L is constructed so that L is a language for S and M defines S.

Example 5.15 Let $S = (\mathbb{N}, \leq)$ be the structure presented in Example 3.7 where $\leq : \mathbb{N} \times \mathbb{N} \to \mathbb{B}$ (instead of being a binary relation over $\mathbb{N}$). Define $L = (\mathcal{B}, \mathcal{C})$, $\mathcal{B} = \{N\}$, $\mathcal{C} = \{\leq_{(N \times N) \to o}\}$, $\mathcal{D} = \{D_\alpha \mid \alpha \in \mathcal{T}(L)\}$ to be the full frame for L such that $D_N = \mathbb{N}$, and I to be the function on $\mathcal{C}$ such that $I(\leq_{(N \times N) \to o}) = \leq$. Then $M = (\mathcal{D}, I)$ is a standard model of L that defines S. □

Example 5.16 This is the same as Example 5.3. Let $S = (\mathbb{N}, 0, 1, +, *)$ be the structure presented in Example 3.8 where $+, * : \mathbb{N} \times \mathbb{N} \to \mathbb{N}$. Define $L = (\mathcal{B}, \mathcal{C})$, $\mathcal{B} = \{N\}$,

$$\mathcal{C} = \{0_N,\ 1_N,\ +_{(N \times N) \to N},\ *_{(N \times N) \to N}\},$$

$\mathcal{D} = \{D_\alpha \mid \alpha \in \mathcal{T}(L)\}$ to be the full frame for L such that $D_N = \mathbb{N}$, and I to be the function on $\mathcal{C}$ such that $I(0_N) = 0$, $I(1_N) = 1$, $I(+_{(N \times N) \to N}) = +$, and $I(*_{(N \times N) \to N}) = *$. Then $M = (\mathcal{D}, I)$ is a standard model of L that defines S. □

Example 5.17 Let $S = (\mathbb{N}, |, \mathsf{lcm}, \mathsf{gcd}, 1, 0)$ be the structure presented in Example 3.9 where $| : \mathbb{N} \times \mathbb{N} \to \mathbb{B}$. Define $L = (\mathcal{B}, \mathcal{C})$, $\mathcal{B} = \{N\}$,

$$\mathcal{C} = \{|_{(N \times N) \to o},\ \mathsf{lcm}_{(N \times N) \to N},\ \mathsf{gcd}_{(N \times N) \to N},\ 1_N,\ 0_N\},$$

$\mathcal{D} = \{D_\alpha \mid \alpha \in \mathcal{T}(L)\}$ to be the full frame for L such that $D_N = \mathbb{N}$, and I to be the function that maps the constants in $\mathcal{C}$ to the corresponding distinguished values in S. Then $M = (\mathcal{D}, I)$ is a standard model of L that defines S. □

Example 5.18 Let $S = (\mathbb{R}, 0, 1, +, *, -, \cdot^{-1})$ be the structure presented in Example 3.10 where $+, * : \mathbb{R} \times \mathbb{R} \to \mathbb{R}$. Define $L = (\mathcal{B}, \mathcal{C})$, $\mathcal{B} = \{R\}$,

$$\mathcal{C} = \{0_R,\ 1_R,\ +_{(R \times R) \to R},\ *_{(R \times R) \to R},\ -_{R \to R},\ \cdot^{-1}{}_{R \to R}\},$$

and $\mathcal{D} = \{D_\alpha \mid \alpha \in \mathcal{T}(L)\}$ to be the full frame for L such that $D_R = \mathbb{R}$, and I to be the function that maps the constants in $\mathcal{C}$ to the corresponding distinguished values in S. Then $M = (\mathcal{D}, I)$ is a standard model of L that defines S. $\square$

Example 5.19 Let

$$S = (\mathbb{R}, \mathbb{C}, 0_\mathbb{R}, 1_\mathbb{R}, +_\mathbb{R}, *_\mathbb{R}, -_\mathbb{R}, \cdot^{-1}_\mathbb{R}, 0_\mathbb{C}, 1_\mathbb{C}, +_\mathbb{C}, *_\mathbb{C}, -_\mathbb{C}, \cdot^{-1}_\mathbb{C}, \otimes)$$

be the structure presented in Example 3.11 where $+_\mathbb{R}, *_\mathbb{R} : \mathbb{R} \times \mathbb{R} \to \mathbb{R}$, $+_\mathbb{C}, *_\mathbb{C} : \mathbb{C} \times \mathbb{C} \to \mathbb{C}$, and $\otimes : \mathbb{R} \times \mathbb{C} \to \mathbb{C}$. Define $L = (\mathcal{B}, \mathcal{C}_1 \cup \mathcal{C}_2 \cup \mathcal{C}_3)$, $\mathcal{B} = \{R, C\}$,

$$\mathcal{C}_1 = \{0_R, \ 1_R, \ +_{(R \times R) \to R}, \ *_{(R \times R) \to R}, \ -_{R \to R}, \ \cdot^{-1}_{R \to R}\},$$

$$\mathcal{C}_2 = \{0_C, \ 1_C, \ +_{(C \times C) \to C}, \ *_{(C \times C) \to C}, \ -_{C \to C}, \ \cdot^{-1}_{C \to C}\},$$

$$\mathcal{C}_3 = \{\otimes_{(R \times C) \to C}\},$$

$\mathcal{D} = \{D_\alpha \mid \alpha \in \mathcal{T}(L)\}$ to be the full frame for L such that $D_R = \mathbb{R}$ and $D_C = \mathbb{C}$, and I to be the function that maps the constants in $\mathcal{C}_1 \cup \mathcal{C}_2 \cup \mathcal{C}_3$ to the corresponding distinguished values in S. Then $M = (\mathcal{D}, I)$ is a standard model of L that defines S. $\square$

Example 5.20 Let $S = (\mathbb{R}, O)$ be the structure presented in Example 3.12 where $O : (\mathbb{R} \to \mathbb{B}) \to \mathbb{B}$. Notice that O is a predicate on predicates on $\mathbb{R}$ that represents a set of sets of members of $\mathbb{R}$. Define $L = (\mathcal{B}, \mathcal{C})$, $\mathcal{B} = \{R\}$, $\mathcal{C} = \{O_{(R \to o) \to o}\}$, $\mathcal{D} = \{D_\alpha \mid \alpha \in \mathcal{T}(L)\}$ to be the full frame for L such that $D_R = \mathbb{R}$, and I to be the function that maps $O_{(R \to o) \to o}$ to O. Then $M = (\mathcal{D}, I)$ is a standard model of L that defines S. $\square$

Example 5.21 Let $S = (\mathbb{R}, d)$ be the structure presented in Example 3.13 where $d : \mathbb{R} \times \mathbb{R} \to \mathbb{R}$. Define $L = (\mathcal{B}, \mathcal{C})$, $\mathcal{B} = \{R\}$, $\mathcal{C} = \{d_{(R \times R) \to R}\}$, $\mathcal{D} = \{D_\alpha \mid \alpha \in \mathcal{T}(L)\}$ to be the full frame for L such that $D_R = \mathbb{R}$, and I to be the function on $\mathcal{C}$ such that $I(d_{(R \times R) \to R}) = d$. Then $M = (\mathcal{D}, I)$ is a standard model of L that defines S. $\square$

Example 5.22 Let $S = (\mathbb{B}, \mathrm{F}, \mathrm{T}, \vee, \wedge, \neg)$ be the structure presented in Example 3.14 where $\vee, \wedge : \mathbb{B} \times \mathbb{B} \to \mathbb{B}$ and $\neg : \mathbb{B} \to \mathbb{B}$. Define $L = (\mathcal{B}, \mathcal{C})$, $\mathcal{B} = \{B\}$,

$$\mathcal{C} = \{0_B, \ 1_B, \ +_{(B \times B) \to B}, \ *_{(B \times B) \to B}, \ \overline{\cdot}_{B \to B}\}$$

$\mathcal{D} = \{D_\alpha \mid \alpha \in \mathcal{T}(L)\}$ to be the full frame for L such that $D_B = \mathbb{B}$, and I to be the function that maps the constants in $\mathcal{C}$ to the corresponding distinguished values in S. Then $M = (\mathcal{D}, I)$ is a standard model of L that defines S. □

Example 5.23 Let $S = (\mathcal{P}(A), \emptyset, A, \cup, \cap, \bar{\cdot})$ be the structure presented in Example 3.15 where $\cup, \cap : \mathcal{P}(A) \times \mathcal{P}(A) \to \mathcal{P}(A)$ and $\bar{\cdot} \to \mathcal{P}(A)$. Define $L = (\mathcal{B}, \mathcal{C})$, $\mathcal{B} = \{B\}$,

$$\mathcal{C} = \{0_B, 1_B, +_{(B \times B) \to B}, *_{(B \times B) \to B}, \bar{\cdot}_{B \to B}\}$$

$\mathcal{D} = \{D_\alpha \mid \alpha \in \mathcal{T}(L)\}$ to be the full frame for L such that $D_B = \mathcal{P}(A)$, and I to be the function that maps the constants in $\mathcal{C}$ to the corresponding distinguished values in S. Then $M = (\mathcal{D}, I)$ is a standard model of L that defines S. □

Example 5.24 Let $S = (A, B, \mathsf{pair})$ be the structure presented in Example 3.16. Define $L = (\mathcal{B}, \mathcal{C})$, $\mathcal{B} = \{A, B\}$,

$$\mathcal{C} = \{\mathsf{pair}_{A \to B \to (A \times B)}\},$$

$\mathcal{D} = \{D_\alpha \mid \alpha \in \mathcal{T}(L)\}$ to be the full frame for L such that $D_A = A$ and $D_B = B$, and I to be the function on $\mathcal{C}$ such that $I(\mathsf{pair}_{A \to B \to (A \times B)}) = \mathsf{pair}$. Then $M = (\mathcal{D}, I)$ is a standard model of L that defines S. □

Example 5.25 Let $S = (A, [], \mathsf{cons})$ be the structure presented in Example 3.17. Define $L = (\mathcal{B}, \mathcal{C})$, $\mathcal{B} = \{A, B\}$,

$$\mathcal{C} = \{[]_B, \mathsf{cons}_{A \to B \to B}\},$$

$\mathcal{D} = \{D_\alpha \mid \alpha \in \mathcal{T}(L)\}$ to be the full frame for L such that $D_A = A$ and $D_B = A^*$, and I to be the function that maps the constants in $\mathcal{C}$ to the corresponding distinguished values in S. Then $M = (\mathcal{D}, I)$ is a standard model of L that defines S. □

Example 5.26 Let $S = (Q, \Sigma, \delta, s, F)$ be the structure presented in Example 3.18 where $F : Q \to \mathbb{B}$. Define $L = (\mathcal{B}, \mathcal{C})$, $\mathcal{B} = \{Q, \Sigma\}$,

$$\mathcal{C} = \{\delta_{(Q \times \Sigma) \to Q}, s_Q, F_{Q \to o}\},$$

$\mathcal{D} = \{D_\alpha \mid \alpha \in \mathcal{T}(L)\}$ to be the full frame for L such that $D_Q = Q$ and $D_\Sigma = \Sigma$, and I to be the function that maps the constants in $\mathcal{C}$ to the corresponding distinguished values in S. Then $M = (\mathcal{D}, I)$ is a standard model of L that defines S. □

5.10   Conclusions

The semantics of Alonzo is based on general models whose function domains need not contain all possible functions. The general models of Alonzo are the basis of a structure

$$(\mathbb{M}, \equiv, \leq, \mathsf{fin}, \mathsf{stand})$$

where:

- $\mathbb{M}$ is the set of general models of Alonzo.

- $\equiv\; : (\mathbb{M} \times \mathbb{M}) \to \mathbb{B}$ such that $M_1 \equiv M_2$ holds iff M_1 and M_2 are isomorphic.

- $\leq\; : (\mathbb{M} \times \mathbb{M}) \to \mathbb{B}$ such that $M_1 \leq M_2$ holds iff M_1 is a reduct of M_2.

- $\mathsf{fin} : \mathbb{M} \to \mathbb{B}$ such that $\mathsf{fin}(M)$ holds iff M is finite.

- $\mathsf{stand} : \mathbb{M} \to \mathbb{B}$ such that $\mathsf{stand}(M)$ holds iff M is standard.

Alonzo has two semantics, the *standard semantics* based on standard models for the practical application of Alonzo and the *general semantics* based on general (standard and nonstandard) models for the theoretical study of Alonzo.

5.11   Exercises

1. Let $S = (K, \emptyset, \mathbb{N}, \cup, \cap, ^-)$ be the structure presented in Exercise 3.24. Construct a language L and a standard model M of L so that the language L is a language for S and M defines S.

2. Let $S = (F, \sqsubseteq)$ be the structure presented in Exercise 3.25. Construct a language L and a standard model M of L so that the language L is a language for S and M defines S.

3. Prove Lemma 5.4.

4. Let M be a general model of L. Prove $|M| \leq \|M\|$ provided $\|M\|$ is defined.

5. Prove Proposition 5.12.

6. Prove that the isomorphic relation $\equiv$ on general models of L is an equivalence relation.

7. Let $\mathbb{M}$ be the set of general models of Alonzo and $\leq$ be the reduct relation over $\mathbb{M}$. Prove that $(\mathbb{M}, \leq)$ is a meet-semilattice with a bottom element.

8. Unlike most versions of predicate logic, Alonzo admits undefined expressions. What needs to be changed in Alonzo to make it a logic in which only defined expressions are admitted?

Chapter 6

Additional Notation

In chapter 4, we defined a convenient compact notation for the four kinds of types and seven kinds of expressions in Alonzo. In this chapter we extend this compact notation to include notation for a variety of useful operators, binders, and abbreviations. The new notation is introduced using notational definitions. Equipped with this expanded compact notation, Alonzo becomes a practical logic in which mathematics ideas can be expressed naturally and succinctly.

To make the notational definitions as readable as possible we have omitted matching parentheses in the right-hand side of the definitions when there is no loss of meaning and it is obvious where they should occur.

6.1   Boolean Operators

In Table 6.1, we present notation for the truth values, the standard Boolean operators, and a conditional expression

$$(\mathbf{A}_o \mapsto \mathbf{B}_\alpha \mid \mathbf{C}_\alpha)$$

that denotes the value of $\mathbf{B}_\alpha$ if $\mathbf{A}_o$ holds and otherwise denotes the value of $\mathbf{C}_\alpha$.

T_o	stands for	$(\lambda x : o . x) = (\lambda x : o . x)$.
F_o	stands for	$(\lambda x : o . T_o) = (\lambda x : o . x)$.
$\wedge_{o \to o \to o}$	stands for	$\lambda x : o . \lambda y : o .$
		$(\lambda g : o \to o \to o . g T_o T_o) =$
		$(\lambda g : o \to o \to o . g x y)$.
$(\mathbf{A}_o \wedge \mathbf{B}_o)$	stands for	$\wedge_{o \to o \to o} \mathbf{A}_o \mathbf{B}_o$.
$\Rightarrow_{o \to o \to o}$	stands for	$\lambda x : o . \lambda y : o . x = (x \wedge y)$.
$(\mathbf{A}_o \Rightarrow \mathbf{B}_o)$	stands for	$\Rightarrow_{o \to o \to o} \mathbf{A}_o \mathbf{B}_o$.
$\neg_{o \to o}$	stands for	$\lambda x : o . x = F_o$.
$(\neg \mathbf{A}_o)$	stands for	$\neg_{o \to o} \mathbf{A}_o$.
$\vee_{o \to o \to o}$	stands for	$\lambda x : o . \lambda y : o . \neg(\neg x \wedge \neg y)$.
$(\mathbf{A}_o \vee \mathbf{B}_o)$	stands for	$\vee_{o \to o \to o} \mathbf{A}_o \mathbf{B}_o$.
$\text{if}_{o \to \alpha \to \alpha \to \alpha}$	stands for	$\lambda b : o . \lambda x : \alpha . \lambda y : \alpha .$
		$\text{I} z : \alpha . (b \Rightarrow z = x) \wedge (\neg b \Rightarrow z = y)$.
$(\mathbf{A}_o \mapsto \mathbf{B}_\alpha \mid \mathbf{C}_\alpha)$	stands for	$\text{if}_{o \to \alpha \to \alpha \to \alpha} \mathbf{A}_o \mathbf{B}_\alpha \mathbf{C}_\alpha$.

Table 6.1: Notational Definitions for Boolean Operators

Example 6.1 (Expression Unfolding) The following derivation of the expression T_o shows how the notational definitions employed in an expression can be repeatedly unfolded until an equivalent formal notation is produced:

$$T_o$$

stands for $((\lambda x : o . (x : o)) = (\lambda x : o . (x : o)))$

stands for $\text{Eq}((\lambda x : o . (x : o)), (\lambda x : o . (x : o)))$

stands for $\text{Eq}(\text{FunAbs}(\text{Var}(x, o), (x : o)), \text{FunAbs}(\text{Var}(x, o), (x : o)))$

stands for $\text{Eq}(\text{FunAbs}(\text{Var}(x, o), \text{Var}(x, o)),$

  $\text{FunAbs}(\text{Var}(x, o), \text{Var}(x, o)))$

stands for $\text{Eq}(\text{FunAbs}(\text{Var}(x, \text{BoolTy}), \text{Var}(x, \text{BoolTy})),$

  $\text{FunAbs}(\text{Var}(x, \text{BoolTy}), \text{Var}(x, \text{BoolTy})))$

This example, as simple as it is, illustrates how unfolding notational definitions can explode the size of an expression's notation. □

The notation $\wedge_{o \to o \to o}$ is an example of a *pseudoconstant*. It is not a real constant of Alonzo, but it stands for an expression $\mathbf{C}_\gamma$ that can be used just like a constant $\mathbf{c}_\gamma$. Unlike a normal constant, $\wedge_{o \to o \to o}$ and most other pseudoconstants can be employed in any language. Thus they serve as logical constants.

The notation $\mathsf{if}_{o \to \alpha \to \alpha \to \alpha}$ is an example of a *parametric pseudoconstant*. It stands for an expression $\mathbf{C}_\gamma$ where $\gamma = o \to \alpha \to \alpha \to \alpha$ is a *parametric type* with the syntactic variable α serving as a parameter that can be freely replaced with any type. Thus $\mathsf{if}_{o \to \alpha \to \alpha \to \alpha}$ is polymorphic in the sense that it can be used with expressions of different types by simply replacing the syntactic variable α with the type that is needed. Alonzo is not a polymorphic logic: it does not include actual type variables and so a constant of Alonzo cannot be used in different contexts involving different types. However, there is a sense in which Alonzo is polymorphic: a parametric pseudoconstant like $\mathsf{if}_{o \to \alpha \to \alpha \to \alpha}$ defined using a notational definition can be used as if it were a polymorphic constant. Thus Alonzo gains much of the benefit of being a polymorphic logic — without the additional complexity of allowing types and expressions to include type variables — by using the syntactic type variables in the definitions of parametric pseudoconstants as if they were actual type variables. So, in summary, Alonzo is a nonpolymorphic logic with a polymorphic notation.

The same symbols that are used to write constants are used to write pseudoconstants and parametric pseudoconstants (Notational Convention 9).

Lemma 6.2 (Semantics Lemma 2) *Let M be a general model and $\varphi \in$* assign(M).

   1. $V_\varphi^M(T_o) = \mathrm{T}$.

   2. $V_\varphi^M(F_o) = \mathrm{F}$.

   3. $V_\varphi^M(\neg \mathbf{A}_o) = \mathrm{T}$ iff $V_\varphi^M(\mathbf{A}_o) = \mathrm{F}$.

   4. $V_\varphi^M(\mathbf{A}_o \vee \mathbf{B}_o) = \mathrm{T}$ iff $V_\varphi^M(\mathbf{A}_o) = \mathrm{T}$ or $V_\varphi^M(\mathbf{B}_o) = \mathrm{T}$.

   5. $V_\varphi^M(\mathbf{A}_o \wedge \mathbf{B}_o) = \mathrm{T}$ iff $V_\varphi^M(\mathbf{A}_o) = \mathrm{T}$ and $V_\varphi^M(\mathbf{B}_o) = \mathrm{T}$.

   6. $V_\varphi^M(\mathbf{A}_o \Rightarrow \mathbf{B}_o) = \mathrm{T}$ iff $V_\varphi^M(\mathbf{A}_o) = \mathrm{F}$ or $V_\varphi^M(\mathbf{B}_o) = \mathrm{T}$.

   7. $V_\varphi^M(\mathbf{A}_o \mapsto \mathbf{B}_\alpha \mid \mathbf{C}_\alpha) \simeq V_\varphi^M(\mathbf{B}_\alpha)$ if $V_\varphi^M(\mathbf{A}_o) = \mathrm{T}$.

   8. $V_\varphi^M(\mathbf{A}_o \mapsto \mathbf{B}_\alpha \mid \mathbf{C}_\alpha) \simeq V_\varphi^M(\mathbf{C}_\alpha)$ if $V_\varphi^M(\mathbf{A}_o) = \mathrm{F}$.

Proof The proof is left to the reader as an exercise.   □

Now we have the notation needed to prove the following theorem stated in Chapter 5:

Proof of Theorem 5.9 *(Every finite general model is a standard model.)*
Let M be a finite general model of L. To show that M is a standard
model, we must show that $\mathcal{D}^M$ is full, i.e., that, for all $\alpha, \beta \in \mathcal{T}(L)$ and
all $f : D_\alpha^M \to D_\beta^M$, $f \in D_{\alpha \to \beta}^M$. Let α, β be arbitrary types in $\mathcal{T}(L)$ and
$f : D_\alpha^M \to D_\beta^M$ be an arbitrary function. By part 1 of Proposition 5.6,
$D_\alpha^M = \{d_1, \ldots, d_n\}$. Then, without loss of generality, for some m with
$0 \le m \le n$ and $d_1', \ldots, d_m' \in D_\beta^M$, $f(d_i) = d_i'$ for all i with $1 \le i \le m$ and
$f(d_i)$ is undefined for all i with $m + 1 \le i \le n$. Choose distinct symbols
$x, y, u_1, \ldots, u_n, v_1, \ldots, v_m$ from $\mathcal{S}_{\mathsf{var}}$ and define $\mathbf{A}_o^i$ to be

$$(x : \alpha) = (u_i : \alpha) \Rightarrow (y : \beta) = (v_i : \beta)$$

for all i with $1 \le i \le m$ and $\mathbf{B}_o^i$ to be

$$(x : \alpha) = (u_i : \alpha) \Rightarrow F_o$$

for all i with $m + 1 \le i \le n$. Let ψ be any assignment

$$\varphi[(u_1 : \alpha) \mapsto d_1] \cdots [(u_n : \alpha) \mapsto d_n] \cdots [(v_1 : \beta) \mapsto d_1'] \cdots [(v_m : \beta) \mapsto d_m']$$

where $\varphi \in \mathsf{assign}(M)$. Then

$$V_\psi^M(\lambda x : \alpha . \mathrm{I} y : \beta . (\mathbf{A}_o^1 \wedge \cdots \wedge \mathbf{A}_o^m \wedge \mathbf{B}_o^{m+1} \wedge \cdots \wedge \mathbf{B}_o^n)) = f$$

and so $f \in D_{\alpha \to \beta}^M$. □

Lemma 6.3 (Semantics Lemma 3) *Let $\mathbf{A}_o$ be a formula and Γ be a set
of formulas. Then $\Gamma \vDash \mathbf{A}_o$ iff $\Gamma \cup \{\neg \mathbf{A}_o\}$ is unsatisfiable.*

Proof The proof is left to the reader as an exercise. □

6.2 Binary Operators

In Table 6.2, we present notation for binary operators which is extensively
used throughout the book. We use the notation $(\mathbf{A}_\alpha \; \mathbf{c} \; \mathbf{B}_\alpha)$ only when
$\mathbf{c}$ stands for a constant $\mathbf{c}_{\alpha \to \alpha \to \beta}$ or $\mathbf{c}_{(\alpha \times \alpha) \to \beta}$ whose type is known from
the context of the constant. We will occasionally use implicit notational
definitions analogous to the notational definitions in Table 6.2 for the infix
operators $<, >$, and $\ge$ corresponding to $\le$ for other weak order operators
such as $\subseteq$ and $\unlhd$ (Notational Convention 10).

$(\mathbf{A}_\alpha\ \mathbf{c}\ \mathbf{B}_\alpha)$	stands for	$\mathbf{c}_{\alpha\to\alpha\to\beta}\,\mathbf{A}_\alpha\,\mathbf{B}_\alpha$ or $\mathbf{c}_{(\alpha\times\alpha)\to\beta}\,\mathbf{A}_\alpha\,\mathbf{B}_\alpha$.
$(\mathbf{A}_o \Leftrightarrow \mathbf{B}_o)$	stands for	$\mathbf{A}_o = \mathbf{B}_o$.
$(\mathbf{A}_\alpha \neq \mathbf{B}_\alpha)$	stands for	$\neg(\mathbf{A}_\alpha = \mathbf{B}_\alpha)$.
$(\mathbf{A}_\alpha < \mathbf{B}_\alpha)$	stands for	$(\leq_{\alpha\to\alpha\to o}\mathbf{A}_\alpha\,\mathbf{B}_\alpha) \wedge (\mathbf{A}_\alpha \neq \mathbf{B}_\alpha)$.
$(\mathbf{A}_\alpha > \mathbf{B}_\alpha)$	stands for	$\mathbf{B}_\alpha < \mathbf{A}_\alpha$.
$(\mathbf{A}_\alpha \geq \mathbf{B}_\alpha)$	stands for	$\mathbf{B}_\alpha \leq \mathbf{A}_\alpha$.
$(\mathbf{A}_\alpha = \mathbf{B}_\alpha = \mathbf{C}_\alpha)$	stands for	$(\mathbf{A}_\alpha = \mathbf{B}_\alpha) \wedge (\mathbf{B}_\alpha = \mathbf{C}_\alpha)$.
$(\mathbf{A}_\alpha\ \mathbf{c}\ \mathbf{B}_\alpha\ \mathbf{d}\ \mathbf{C}_\alpha)$	stands for	$(\mathbf{A}_\alpha\ \mathbf{c}\ \mathbf{B}_\alpha) \wedge (\mathbf{B}_\alpha\ \mathbf{d}\ \mathbf{C}_\alpha)$.

Table 6.2: Notational Definitions for Binary Operators

$(\forall\,\mathbf{x} : \alpha\ .\ \mathbf{A}_o)$	stands for	$(\lambda x : \alpha\ .\ T_o) = (\lambda\,\mathbf{x} : \alpha\ .\ \mathbf{A}_o)$.
$(\exists\,\mathbf{x} : \alpha\ .\ \mathbf{A}_o)$	stands for	$\neg(\forall\,\mathbf{x} : \alpha\ .\ \neg\mathbf{A}_o)$.
$(\exists!\,\mathbf{x} : \alpha\ .\ \mathbf{A}_o)$	stands for	$\exists y : \alpha\ .\ (\lambda\,\mathbf{x} : \alpha\ .\ \mathbf{A}_o) = (\lambda\,\mathbf{x} : \alpha\ .\ \mathbf{x} = y)$
		where y is not free in $(\lambda\,\mathbf{x} : \alpha\ .\ \mathbf{A}_o)$.

Table 6.3: Notational Definitions for Quantifiers

6.3   Quantifiers

In Table 6.3, we present notation for the universal and existential quantifiers. The expression $(\exists!\,\mathbf{x} : \alpha\ .\ \mathbf{A}_o)$ asserts that there is a unique value for $(\mathbf{x} : \alpha)$ that satisfies $\mathbf{A}_o$. A *universal closure* of $\mathbf{C}_o$ is a formula

$$\forall\,\mathbf{x}_1 : \alpha_1\ .\ \cdots\ \forall\,\mathbf{x}_n : \alpha_n\ .\ \mathbf{C}_o$$

such that $\{(\mathbf{x}_1 : \alpha_1), \ldots, (\mathbf{x}_n : \alpha_n)\}$ is the set of variables free in $\mathbf{C}_o$. We will usually write a sequence of universal quantifiers and a sequence of existential quantifiers in a more compact form with a single quantifier (Notational Convention 11). Thus, for example,

$$\forall\,\mathbf{x} : \alpha\ .\ \forall\,\mathbf{y} : \alpha\ .\ \forall\,\mathbf{z} : \beta\ .\ \mathbf{A}_o$$

will be written as

$$\forall\,\mathbf{x}, \mathbf{y} : \alpha,\ \mathbf{z} : \beta\ .\ \mathbf{A}_o.$$

We will also use this form with quasitypes which are introduced in Section 6.9 (Notational Convention 12).

$\perp_o$	stands for	F_o.
$\perp_\alpha$	stands for	$I x : \alpha . x \neq x$ where $\alpha \neq o$.
$\Delta_{\alpha \to \beta}$	stands for	$\lambda x : \alpha . \perp_\beta$ where $\beta \neq o$.
$(\mathbf{A}_\alpha {\downarrow})$	stands for	$\mathbf{A}_\alpha = \mathbf{A}_\alpha$.
$(\mathbf{A}_\alpha {\uparrow})$	stands for	$\neg (\mathbf{A}_\alpha {\downarrow})$.
$(\mathbf{A}_\alpha \simeq \mathbf{B}_\alpha)$	stands for	$(\mathbf{A}_\alpha {\downarrow} \vee \mathbf{B}_\alpha {\downarrow}) \Rightarrow \mathbf{A}_\alpha = \mathbf{B}_\alpha$.
$(\mathbf{A}_\alpha \not\simeq \mathbf{B}_\alpha)$	stands for	$\neg (\mathbf{A}_\alpha \simeq \mathbf{B}_\alpha)$.

Table 6.4: Notational Definitions for Definedness

Lemma 6.4 (Semantics Lemma 4) *Let M be a general model of L and $\varphi \in \mathsf{assign}(M)$.*

1. *$V_\varphi^M (\forall \mathbf{x} : \alpha . \mathbf{A}_o) = \mathrm{T}$ iff $V_{\varphi[(\mathbf{x}:\alpha) \mapsto d]}^M (\mathbf{A}_o) = \mathrm{T}$ for all $d \in D_\alpha^M$.*

2. *$V_\varphi^M (\exists \mathbf{x} : \alpha . \mathbf{A}_o) = \mathrm{T}$ iff $V_{\varphi[(\mathbf{x}:\alpha) \mapsto d]}^M (\mathbf{A}_o) = \mathrm{T}$ for some $d \in D_\alpha^M$.*

3. *$V_\varphi^M (\exists! \mathbf{x} : \alpha . \mathbf{A}_o) = \mathrm{T}$ iff $V_{\varphi[(\mathbf{x}:\alpha) \mapsto d]}^M (\mathbf{A}_o) = \mathrm{T}$ for exactly one $d \in D_\alpha^M$.*

4. *If $\mathbf{B}_o$ is a universal closure of $\mathbf{A}_o$, then $M \vDash \mathbf{A}_o$ iff $M \vDash \mathbf{B}_o$.*

Proof The proof is left to the reader as an exercise. $\square$

6.4 Definedness

In Table 6.4, we present notation for expressions involving definedness. $\perp_o$ is a canonical "undefined" formula. $\perp_\alpha$ is a canonical undefined expression of type $\alpha \neq o$. $\Delta_{\alpha \to \beta}$ is the empty function of type $\alpha \to \beta$ (where $\beta \neq o$). $(\mathbf{A}_\alpha {\downarrow})$ and $(\mathbf{A}_\alpha {\uparrow})$ assert that the expression $\mathbf{A}_\alpha$ is defined and undefined, respectively. $(\mathbf{A}_\alpha \simeq \mathbf{B}_\alpha)$ asserts that the expressions $\mathbf{A}_\alpha$ and $\mathbf{B}_\alpha$ are quasi-equal, i.e., they are both defined and equal or both undefined.

Lemma 6.5 (Semantics Lemma 5) *Let M be a general model and $\varphi \in \mathsf{assign}(M)$.*

1. *$V_\varphi^M (\perp_\alpha) = \mathrm{F}$ if $\alpha = o$, and $V_\varphi^M (\perp_\alpha)$ is undefined if $\alpha \neq o$.*

2. *$V_\varphi^M (\Delta_{\alpha \to \beta} \mathbf{A}_\alpha)$ is undefined.*

3. *$V_\varphi^M (\mathbf{A}_\alpha {\downarrow}) = \mathrm{T}$ iff $V_\varphi^M (\mathbf{A}_\alpha)$ is defined.*

4. $V_\varphi^M(\mathbf{A}_\alpha\!\uparrow) = \mathrm{T}$ *iff* $V_\varphi^M(\mathbf{A}_\alpha)$ *is undefined.*

5. $V_\varphi^M(\mathbf{A}_\alpha \simeq \mathbf{B}_\alpha) = \mathrm{T}$ *iff* $V_\varphi^M(\mathbf{A}_\alpha) \simeq V_\varphi^M(\mathbf{B}_\alpha)$.

6. If $V_\varphi^M(\mathbf{A}_\alpha)$ *or* $V_\varphi^M(\mathbf{B}_\alpha)$ *are defined, then*

$$V_\varphi^M(\mathbf{A}_\alpha \simeq \mathbf{B}_\alpha) = V_\varphi^M(\mathbf{A}_\alpha = \mathbf{B}_\alpha).$$

Proof The proof is left to the reader as an exercise.   □

6.5   Sets

Since we can identify a set $S \subseteq U$ with the predicate $p_S : U \to \mathbb{B}$ such that $a \in S$ iff $p_S(a)$, we will introduce a power set type of α, i.e., a type of the subsets of α, as the type $\alpha \to o$ of predicates on α. The compact notation for $\alpha \to o$ is $\{\alpha\}$. We introduce this notation and compact notation for the common set operators in Table 6.5. $\emptyset_{\{\alpha\}}$ and $U_{\{\alpha\}}$ are parametric pseudo-constants that denote the empty set and the universal set, respectively, of the members in the domain of α.

6.6   Tuples

We introduce notation for product types, tuples, and the accessors for ordered pairs in Table 6.6.

6.7   Functions

Some convenient notation for functions is found in Table 6.7.

6.8   Miscellaneous Notation

In Table 6.8, we introduce various miscellaneous notation.[1] The notational definitions introduced in this table are abbreviations of the form

$$A(\mathbf{B}_{\alpha_1}^1, \ldots, \mathbf{B}_{\alpha_n}^n) \text{ stands for } C$$

where A is a name, $n \geq 0$, and the syntactic variables $\mathbf{B}_{\alpha_1}^1, \ldots, \mathbf{B}_{\alpha_1}^1$ appear in the expression C. A is written in uppercase to distinguish it from the name of a pseudoconstant (Notational Convention 13).

---

[1]The missing parentheses in the definitions of $\mathsf{BIJ}(\mathbf{F}_{\alpha\to\beta})$ and $\mathsf{DISTINCT}(\mathbf{A}_\alpha^1, \ldots, \mathbf{A}_\alpha^n)$ can be added in any way that is appropriate.

$\{\alpha\}$	stands for	$\alpha \to o$.
$(\mathbf{A}_\alpha \in \mathbf{B}_{\{\alpha\}})$	stands for	$\mathbf{B}_{\{\alpha\}}\,\mathbf{A}_\alpha$.
$(\mathbf{A}_\alpha \notin \mathbf{B}_{\{\alpha\}})$	stands for	$\neg(\mathbf{A}_\alpha \in \mathbf{B}_{\{\alpha\}})$.
$\{\mathbf{x} : \alpha \mid \mathbf{A}_o\}$	stands for	$\lambda\,\mathbf{x} : \alpha\,.\,\mathbf{A}_o$.
$\emptyset_{\{\alpha\}}$	stands for	$\lambda\,x : \alpha\,.\,F_o$.
$\{\,\}_{\{\alpha\}}$	stands for	$\emptyset_{\{\alpha\}}$.
$U_{\{\alpha\}}$	stands for	$\lambda\,x : \alpha\,.\,T_o$.
$n\text{-}\alpha\text{-SET}$	stands for	$\lambda\,x_1 : \alpha\,.\,\cdots\,.\,\lambda\,x_n : \alpha\,.\,\lambda\,x : \alpha\,.$
		$\qquad x = x_1 \vee \cdots \vee x = x_n$ where $n \ge 1$.
$\{\mathbf{A}_\alpha^1, \ldots, \mathbf{A}_\alpha^n\}$	stands for	$n\text{-}\alpha\text{-SET}\,\mathbf{A}_\alpha^1 \cdots \mathbf{A}_\alpha^n$ where $n \ge 1$.
$\subseteq_{\{\alpha\}\to\{\alpha\}\to o}$	stands for	$\lambda\,a : \{\alpha\}\,.\,\lambda\,b : \{\alpha\}\,.$
		$\qquad \forall\,x : \alpha\,.\,x \in a \Rightarrow x \in b$.
$\cup_{\{\alpha\}\to\{\alpha\}\to\{\alpha\}}$	stands for	$\lambda\,a : \{\alpha\}\,.\,\lambda\,b : \{\alpha\}\,.$
		$\qquad \{x : \alpha \mid x \in a \vee x \in b\}$.
$\cap_{\{\alpha\}\to\{\alpha\}\to\{\alpha\}}$	stands for	$\lambda\,a : \{\alpha\}\,.\,\lambda\,b : \{\alpha\}\,.$
		$\qquad \{x : \alpha \mid x \in a \wedge x \in b\}$.
$\bar{}_{\{\alpha\}\to\{\alpha\}}$	stands for	$\lambda\,a : \{\alpha\}\,.\,\{x : \alpha \mid x \notin a\}$
$\overline{\mathbf{A}_{\{\alpha\}}}$	stands for	$\bar{}_{\{\alpha\}\to\{\alpha\}}\,\mathbf{A}_{\{\alpha\}}$.

Table 6.5: Notational Definitions for Sets

(α)	stands for	α.
$(\alpha_1 \times \cdots \times \alpha_n)$	stands for	$(\alpha_1 \times (\alpha_2 \times \cdots \times \alpha_n))$ where $n \ge 2$.
$(\mathbf{A}_\alpha)$	stands for	$\mathbf{A}_\alpha$.
$(\mathbf{A}_{\alpha_1}^1, \ldots, \mathbf{A}_{\alpha_n}^n)$	stands for	$(\mathbf{A}_{\alpha_1}^1, (\mathbf{A}_{\alpha_1}^2, \ldots, \mathbf{A}_{\alpha_n}^n))$ where $n \ge 2$.
$\mathsf{fst}_{(\alpha\times\beta)\to\alpha}$	stands for	$\lambda\,p : \alpha \times \beta\,.\,\mathrm{I}\,x : \alpha\,.\,\exists\,y : \beta\,.\,p = (x, y)$.
$\mathsf{snd}_{(\alpha\times\beta)\to\beta}$	stands for	$\lambda\,p : \alpha \times \beta\,.\,\mathrm{I}\,y : \beta\,.\,\exists\,x : \alpha\,.\,p = (x, y)$.

Table 6.6: Notational Definitions for Tuples

$\mathsf{dom}_{(\alpha\to\beta)\to\{\alpha\}}$	stands for	$\lambda f : \alpha \to \beta$. $\{x : \alpha \mid (f\,x)\!\downarrow\}$.
$\mathsf{ran}_{(\alpha\to\beta)\to\{\beta\}}$	stands for	$\lambda f : \alpha \to \beta$. $\{y : \beta \mid \exists x : \alpha . f\,x = y\}$.
$\sqsubseteq_{(\alpha\to\beta)\to(\alpha\to\beta)\to o}$	stands for	$\lambda f : \alpha \to \beta . \lambda g : \alpha \to \beta$. $\forall x : \alpha . x \in \mathsf{dom}_{(\alpha\to\beta)\to\{\alpha\}} f \Rightarrow$ $f\,x = g\,x$.
$\circ_{(\alpha\to\beta)\to(\beta\to\gamma)\to(\alpha\to\gamma)}$	stands for	$\lambda f : \alpha \to \beta . \lambda g : \beta \to \gamma$. $\lambda x : \alpha . g\,(f\,x)$.
$(\mathbf{F}_{\alpha\to\beta} \circ \mathbf{G}_{\beta\to\gamma})$	stands for	$\circ_{(\alpha\to\beta)\to(\beta\to\gamma)\to(\alpha\to\gamma)} \mathbf{F}_{\alpha\to\beta} \, \mathbf{G}_{\beta\to\gamma}$.
$\lceil_{(\alpha\to\beta)\to\{\alpha\}\to(\alpha\to\beta)}$	stands for	$\lambda f : \alpha \to \beta . \lambda s : \{\alpha\}$. $\lambda x : \alpha . x \in s \mapsto f\,x \mid \perp_\beta$.
$(\mathbf{F}_{\alpha\to\beta}\lceil\mathbf{A}_{\{\alpha\}})$	stands for	$\lceil_{(\alpha\to\beta)\to\{\alpha\}\to(\alpha\to\beta)} \mathbf{F}_{\alpha\to\beta} \mathbf{A}_{\{\alpha\}}$.

Table 6.7: Notational Definitions for Functions

6.9   Quasitypes and Dependent Quasitypes

A *quasitype within type* $\alpha \in \mathcal{T}$ is any expression of type $\{\alpha\} = \alpha \to o$. A quasitype $\mathbf{Q}_{\{\alpha\}}$ denotes a subset of the domain denoted by α. Thus quasitypes represent subtypes and are useful for specifying subdomains of a domain. Unlike a type, a quasitype may denote an empty domain. Notice that an expression $\mathbf{A}_{\alpha\to o}$ is simultaneously an expression of type $\alpha \to o$, an expression of type of $\{\alpha\}$, and a quasitype within type α. So $\mathbf{A}_{\alpha\to o}$ (or $\mathbf{A}_{\{\alpha\}}$) can be used as a function, as a set, and like a type as shown below.

In Table 6.9, we introduce various notations for using quasitypes in place of types. Quasitypes can be used to restrict the range of a variable bound by a binder. For example, $(\lambda x : \mathbf{Q}_{\{\alpha\}} . \mathbf{B}_\beta)$ denotes the function denoted by $\lambda x : \alpha . \mathbf{B}_\beta$ restricted to the domain denoted by $\mathbf{Q}_{\{\alpha\}}$. Quasitypes can also be used to sharpen definedness statements. For example, $(\mathbf{A}_\alpha \downarrow \mathbf{Q}_{\{\alpha\}})$, read as $\mathbf{A}_\alpha$ *is defined in* $\mathbf{Q}_{\{\alpha\}}$, asserts that the value of $\mathbf{A}_\alpha$ is defined and is a member of the set denoted by $\mathbf{Q}_{\{\alpha\}}$. $(\mathbf{Q}_{\{\alpha\}} \to \mathbf{R}_{\{\beta\}})$ is a quasitype within $\alpha \to \beta$ that denotes the function space from the denotation of $\mathbf{Q}_{\{\alpha\}}$ to the denotation of $\mathbf{R}_{\{\beta\}}$, and $(\mathbf{Q}_{\{\alpha\}} \times \mathbf{R}_{\{\beta\}})$ is a quasitype within $\alpha \times \beta$ that denotes the Cartesian product of the denotation of $\mathbf{Q}_{\{\alpha\}}$ and the denotation of $\mathbf{R}_{\{\beta\}}$. $\mathsf{INF}(\mathbf{Q}_{\{\alpha\}})$, $\mathsf{FIN}(\mathbf{Q}_{\{\alpha\}})$, and $\mathsf{COUNT}(\mathbf{Q}_{\{\alpha\}})$ assert that the set denoted by the quasitype $\mathbf{Q}_{\{\alpha\}}$ is infinite, finite, and countable, respectively.

Let $(x : \alpha)$, $(x' : \alpha)$, $(y : \beta)$, and $(z : \gamma)$ not be free in $\mathbf{F}_{\alpha \to \beta}$ and $\mathbf{F}_{\alpha \to \beta \to \gamma}$.

TOTAL($\mathbf{F}_{\alpha \to \beta}$)	stands for	$\forall x : \alpha . (\mathbf{F}_{\alpha \to \beta} \, x) \!\downarrow$.
TOTAL2($\mathbf{F}_{\alpha \to \beta \to \gamma}$)	stands for	$\forall x : \alpha, y : \beta . (\mathbf{F}_{\alpha \to \beta \to \gamma} \, x \, y) \!\downarrow$.
SURJ($\mathbf{F}_{\alpha \to \beta}$)	stands for	$\forall y : \beta . \exists x : \alpha . \mathbf{F}_{\alpha \to \beta} \, x = y$.
SURJ2($\mathbf{F}_{\alpha \to \beta \to \gamma}$)	stands for	$\forall z : \gamma . \exists x : \alpha, y : \beta . \mathbf{F}_{\alpha \to \beta \to \gamma} \, x \, y = z$.
INJ($\mathbf{F}_{\alpha \to \beta}$)	stands for	$\forall x, x' : \alpha . \mathbf{F}_{\alpha \to \beta} \, x = \mathbf{F}_{\alpha \to \beta} \, x' \Rightarrow x = x'$.
BIJ($\mathbf{F}_{\alpha \to \beta}$)	stands for	TOTAL($\mathbf{F}_{\alpha \to \beta}$) $\wedge$ SURJ($\mathbf{F}_{\alpha \to \beta}$) $\wedge$ INJ($\mathbf{F}_{\alpha \to \beta}$).
DISTINCT($\mathbf{A}_\alpha^1, \mathbf{A}_\alpha^2$)	stands for	$(\mathbf{A}_\alpha^1 \neq \mathbf{A}_\alpha^2)$.
DISTINCT($\mathbf{A}_\alpha^1, \ldots, \mathbf{A}_\alpha^n$)	stands for	$(\mathbf{A}_\alpha^1 \neq \mathbf{A}_\alpha^2) \wedge \cdots \wedge (\mathbf{A}_\alpha^1 \neq \mathbf{A}_\alpha^n) \wedge$ DISTINCT($\mathbf{A}_\alpha^2, \ldots, \mathbf{A}_\alpha^n$) where $n \geq 3$.

Table 6.8: Miscellaneous Notational Definitions

Example 6.6 (Number Quasitypes) Let

$$L = (\{C\}, \{N_{\{C\}}, Z_{\{C\}}, Q_{\{C\}}, R_{\{C\}}\})$$

be a language of Alonzo. Then, if the type C denotes the domain $\mathbb{C}$ of complex numbers, the quasitypes $N_{\{C\}}, Z_{\{C\}}, Q_{\{C\}}, R_{\{C\}}$ can denote the subdomains $\mathbb{N}, \mathbb{Z}, \mathbb{Q}, \mathbb{R}$ of $\mathbb{C}$, respectively. And a quasitype like $Z_{\{C\}} \to R_{\{C\}}$ can denote the domain $\mathbb{Z} \to \mathbb{R}$ of (partial and total) functions from the integers to the real numbers (which is a subdomain of $\mathbb{C} \to \mathbb{C}$). $\square$

A *dependent quasitype constructor within type* $\alpha_1 \to \cdots \to \alpha_n \to \beta \in \mathcal{T}$ with $n \geq 1$ is any expression of type

$$\alpha_1 \to \cdots \to \alpha_n \to \{\beta\} = \alpha_1 \to \cdots \to \alpha_n \to (\beta \to o).$$

A *dependent quasitype* is a quasitype that results from the application of a dependent quasitype constructor. The dependent quasitype $\mathbf{C}_{\alpha \to \{\beta\}} \, \mathbf{A}_\alpha$, resulting from the application of the dependent quasitype constructor $\mathbf{C}_{\alpha \to \{\beta\}}$ to the expression $\mathbf{A}_\alpha$, denotes a subset of the domain denoted by β that depends on the value of $\mathbf{A}_\alpha$. Thus dependent quasitypes represent dependent subtypes and are useful for specifying subdomains of a domain that depend on values. For example, if $[\alpha]$ is a type of lists over α and $[\alpha]^n$ is the dependent subtype of $[\alpha]$ that denotes the lists over α of length n, then the function mapping values of n to $[\alpha]^n$ can be represented by a dependent quasitype constructor as shown in Example 10.3. $\to_{\{\alpha\} \to \{\beta\} \to \{\alpha \to \beta\}}$ and $\times_{\{\alpha\} \to \{\beta\} \to \{\alpha \times \beta\}}$ are two other examples of dependent quasitype constructors.

Let $(x : \alpha)$, $(x' : \alpha)$, $(y : \beta)$, $(z : \gamma)$, and $(u : \{\alpha\})$ not be free in
$\mathbf{F}_{\alpha \to \beta}$, $\mathbf{F}_{\alpha \to \beta \to \gamma}$, $\mathbf{Q}_{\{\alpha\}}$, $\mathbf{R}_{\{\beta\}}$, and $\mathbf{S}_{\{\gamma\}}$.

$(\lambda \mathbf{x} : \mathbf{Q}_{\{\alpha\}} \,.\, \mathbf{B}_\beta)$	stands for	$\lambda \mathbf{x} : \alpha \,.\, (\mathbf{x} \in \mathbf{Q}_{\{\alpha\}} \mapsto \mathbf{B}_\beta \mid \bot_\beta)$.
$(\forall \mathbf{x} : \mathbf{Q}_{\{\alpha\}} \,.\, \mathbf{B}_o)$	stands for	$\forall \mathbf{x} : \alpha \,.\, (\mathbf{x} \in \mathbf{Q}_{\{\alpha\}} \Rightarrow \mathbf{B}_o)$.
$(\exists \mathbf{x} : \mathbf{Q}_{\{\alpha\}} \,.\, \mathbf{B}_o)$	stands for	$\exists \mathbf{x} : \alpha \,.\, (\mathbf{x} \in \mathbf{Q}_{\{\alpha\}} \wedge \mathbf{B}_o)$.
$(\mathrm{I} \mathbf{x} : \mathbf{Q}_{\{\alpha\}} \,.\, \mathbf{B}_o)$	stands for	$\mathrm{I} \mathbf{x} : \alpha \,.\, (\mathbf{x} \in \mathbf{Q}_{\{\alpha\}} \wedge \mathbf{B}_o)$.
$(\mathbf{A}_\alpha \downarrow \mathbf{Q}_{\{\alpha\}})$	stands for	$\mathbf{A}_\alpha \downarrow \wedge\, \mathbf{A}_\alpha \in \mathbf{Q}_{\{\alpha\}}$.
$(\mathbf{A}_\alpha \uparrow \mathbf{Q}_{\{\alpha\}})$	stands for	$\neg(\mathbf{A}_\alpha \downarrow \mathbf{Q}_{\{\alpha\}})$.
$\to_{\{\alpha\} \to \{\beta\} \to \{\alpha \to \beta\}}$	stands for	$\lambda s : \{\alpha\} \,.\, \lambda t : \{\beta\} \,.$
		$\{f : \alpha \to \beta \mid \forall x : \alpha \,.$
		$(f\,x)\downarrow \,\Rightarrow (x \in s \wedge f\,x \in t)\}$.
$\times_{\{\alpha\} \to \{\beta\} \to \{\alpha \times \beta\}}$	stands for	$\lambda s : \{\alpha\} \,.\, \lambda t : \{\beta\} \,.$
		$\{p : \alpha \times \beta \mid$
		$\mathsf{fst}_{(\alpha \times \beta) \to \alpha}\, p \in s \,\wedge$
		$\mathsf{snd}_{(\alpha \times \beta) \to \beta}\, p \in t\}$
$(\mathbf{Q}_{\{\alpha\}} \to \mathbf{R}_{\{\beta\}})$	stands for	$\to_{\{\alpha\} \to \{\beta\} \to \{\alpha \to \beta\}} \mathbf{Q}_{\{\alpha\}}\, \mathbf{R}_{\{\beta\}}$.
$(\alpha \to \mathbf{R}_{\{\beta\}})$	stands for	$(U_{\{\alpha\}} \to \mathbf{R}_{\{\beta\}})$.
$(\mathbf{Q}_{\{\alpha\}} \to \beta)$	stands for	$(\mathbf{Q}_{\{\alpha\}} \to U_{\{\beta\}})$.
$(\mathbf{Q}_{\{\alpha\}} \times \mathbf{R}_{\{\beta\}})$	stands for	$\times_{\{\alpha\} \to \{\beta\} \to \{\alpha \times \beta\}} \mathbf{Q}_{\{\alpha\}}\, \mathbf{R}_{\{\beta\}}$.
$(\alpha \times \mathbf{R}_{\{\beta\}})$	stands for	$(U_{\{\alpha\}} \times \mathbf{R}_{\{\beta\}})$.
$(\mathbf{Q}_{\{\alpha\}} \times \beta)$	stands for	$(\mathbf{Q}_{\{\alpha\}} \times U_{\{\beta\}})$.
TOTAL-ON$(\mathbf{F}_{\alpha \to \beta}, \mathbf{Q}_{\{\alpha\}}, \mathbf{R}_{\{\beta\}})$	stands for	$\forall x : \mathbf{Q}_{\{\alpha\}} \,.\, (\mathbf{F}_{\alpha \to \beta}\, x) \downarrow \mathbf{R}_{\{\beta\}}$.
TOTAL-ON2$(\mathbf{F}_{\alpha \to \beta \to \gamma},$	stands for	$\forall x : \mathbf{Q}_{\{\alpha\}},\, y : \mathbf{R}_{\{\beta\}} \,.$
$\quad \mathbf{Q}_{\{\alpha\}}, \mathbf{R}_{\{\beta\}}, \mathbf{S}_{\{\gamma\}})$		$(\mathbf{F}_{\alpha \to \beta \to \gamma}\, x\, y) \downarrow \mathbf{S}_{\{\gamma\}}$.
SURJ-ON$(\mathbf{F}_{\alpha \to \beta}, \mathbf{Q}_{\{\alpha\}}, \mathbf{R}_{\{\beta\}})$	stands for	$\forall y : \mathbf{R}_{\{\beta\}} \,.\, \exists x : \mathbf{Q}_{\{\alpha\}} \,.$
		$\mathbf{F}_{\alpha \to \beta}\, x = y$.
SURJ-ON2$(\mathbf{F}_{\alpha \to \beta \to \gamma},$	stands for	$\forall z : \mathbf{S}_{\{\gamma\}} \,.\, \exists x : \mathbf{Q}_{\{\alpha\}},\, y : \mathbf{R}_{\{\beta\}} \,.$
$\quad \mathbf{Q}_{\{\alpha\}}, \mathbf{R}_{\{\beta\}}, \mathbf{S}_{\{\gamma\}})$		$\mathbf{F}_{\alpha \to \beta \to \gamma}\, x\, y = z$.
INJ-ON$(\mathbf{F}_{\alpha \to \beta}, \mathbf{Q}_{\{\alpha\}})$	stands for	$\forall x, x' : \mathbf{Q}_{\{\alpha\}} \,.$
		$\mathbf{F}_{\alpha \to \beta}\, x = \mathbf{F}_{\alpha \to \beta}\, x' \Rightarrow x = x'$.
BIJ-ON$(\mathbf{F}_{\alpha \to \beta}, \mathbf{Q}_{\{\alpha\}}, \mathbf{R}_{\{\beta\}})$	stands for	TOTAL-ON$(\mathbf{F}_{\alpha \to \beta}, \mathbf{Q}_{\{\alpha\}}, \mathbf{R}_{\{\beta\}}) \wedge$
		SURJ-ON$(\mathbf{F}_{\alpha \to \beta}, \mathbf{Q}_{\{\alpha\}}, \mathbf{R}_{\{\beta\}}) \wedge$
		INJ-ON$(\mathbf{F}_{\alpha \to \beta}, \mathbf{Q}_{\{\alpha\}})$.
INF$(\mathbf{Q}_{\{\alpha\}})$	stands for	$\exists u : \{\alpha\} \,.\, u \subset \mathbf{Q}_{\{\alpha\}} \wedge$
		$\exists f : \alpha \to \alpha \,.\, \text{BIJ-ON}(f, u, \mathbf{Q}_{\{\alpha\}})$.
FIN$(\mathbf{Q}_{\{\alpha\}})$	stands for	$\neg \text{INF}(\mathbf{Q}_{\{\alpha\}})$.
COUNT$(\mathbf{Q}_{\{\alpha\}})$	stands for	$\forall u : \{\alpha\} \,.\, (u \subseteq \mathbf{Q}_{\{\alpha\}} \wedge \text{INF}(u)) \Rightarrow$
		$\exists f : \alpha \to \alpha \,.\, \text{BIJ-ON}(f, u, \mathbf{Q}_{\{\alpha\}})$.

Table 6.9: Notational Definitions for Quasitypes

Let $\gamma = \{\alpha\} \to (\alpha \to \{\beta\}) \to \{\alpha \to \beta\}$ and $\delta = \{\alpha\} \to (\alpha \to \{\beta\}) \to \{\alpha \times \beta\}$.		
Π_γ	stands for	$\lambda s : \{\alpha\} \, . \, \lambda t : \alpha \to \{\beta\} \, .$ $\{f : \alpha \to \beta \mid \forall x : \alpha \, .$ $(f\,x){\downarrow} \Rightarrow (x \in s \wedge f\,x \in t\,x)\}.$
Σ_δ	stands for	$\lambda s : \{\alpha\} \, . \, \lambda t : \alpha \to \{\beta\} \, .$ $\{p : \alpha \times \beta \mid$ $\mathsf{fst}_{(\alpha \times \beta) \to \alpha}\,p \in s \wedge$ $\mathsf{snd}_{(\alpha \times \beta) \to \beta}\,p \in t\,(\mathsf{fst}_{(\alpha \times \beta) \to \alpha}\,p)\}$
$(\Pi\,\mathbf{x} : \mathbf{Q}_{\{\alpha\}} \, . \, \mathbf{R}_{\{\beta\}})$	stands for	$\Pi_\gamma \, \mathbf{Q}_{\{\alpha\}}(\lambda\,\mathbf{x} : \alpha \, . \, \mathbf{R}_{\{\beta\}}).$
$(\Sigma\,\mathbf{x} : \mathbf{Q}_{\{\alpha\}} \, . \, \mathbf{R}_{\{\beta\}})$	stands for	$\Sigma_\delta \, \mathbf{Q}_{\{\alpha\}}(\lambda\,\mathbf{x} : \alpha \, . \, \mathbf{R}_{\{\beta\}}).$

Table 6.10: Notational Definitions for Dependent Quasitypes

In Table 6.10, we introduce notation involving dependent quasitypes. $(\Pi\,x : \mathbf{Q}_{\{\alpha\}} \, . \, \mathbf{R}_{\{\beta\}})$ and $(\Sigma\,x : \mathbf{Q}_{\{\alpha\}} \, . \, \mathbf{R}_{\{\beta\}})$ are quasitypes within type $\alpha \to \beta$ and $\alpha \times \beta$, respectively. $(\Pi\,x : \mathbf{Q}_{\{\alpha\}} \, . \, \mathbf{R}_{\{\beta\}})$ represents a *dependent function type*, while $(\Sigma\,x : \mathbf{Q}_{\{\alpha\}} \, . \, \mathbf{R}_{\{\beta\}})$ represents a *dependent pair type*.

6.10 Conclusions

Alonzo has an extensive set of notation introduced by notational definitions that includes notation for Boolean operators, binary operators, quantifiers, definedness notions, sets, tuples, quasitypes, and dependent quasitypes.

6.11 Exercises

1. Prove Lemma 6.2.

2. Prove Lemma 6.3.

3. Prove Lemma 6.4.

4. Prove $\mathbf{A}_o \Leftrightarrow \mathbf{B}_o$ is valid iff $\mathbf{A}_o$ and $\mathbf{B}_o$ are logically equivalent.

5. Prove that every formula of the form

$$\mathbf{A}_\alpha = \mathbf{B}_\alpha \Rightarrow (\lambda\,y : \beta \, . \, \mathbf{A}_\alpha) = (\lambda\,y : \beta \, . \, \mathbf{B}_\alpha)$$

is satisfiable, but not necessarily valid, in every general model that interprets it. (Exercises X5401–2 in [5].)

6. Let M be a general model, $\varphi \in \mathsf{assign}(M)$, and $\mathbf{A}_o$ be a formula interpreted by M. Prove $M \vDash_\varphi \exists x : \alpha \,.\, \mathbf{A}_o$ iff $M \vDash_\psi \mathbf{A}_o$ for some $\psi \in \mathsf{assign}(M)$ that agrees with φ off $(x : \alpha)$. (Exercise X5403 in [5].)

7. Prove Lemma 6.5.

8. Prove that the following equalities are valid:

    a. $(\emptyset_{\{\alpha\}} \to \mathbf{R}_{\{\beta\}}) = \{\Delta_{\alpha \to \beta}\}$.
    b. $(\mathbf{Q}_{\{\alpha\}} \to \emptyset_{\{\beta\}}) = \{\Delta_{\alpha \to \beta}\}$.
    c. $(\emptyset_{\{\alpha\}} \times \mathbf{R}_{\{\beta\}}) = \emptyset_{\{\alpha \times \beta\}}$.
    d. $(\mathbf{Q}_{\{\alpha\}} \times \emptyset_{\{\beta\}}) = \emptyset_{\{\alpha \times \beta\}}$.

9. Let $L = (\{A\}, \emptyset)$ be a language of Alonzo. Write an expression of L that denotes the function that maps a binary relation over A to its reflexive closure.

10. Let $L = (\{A\}, \emptyset)$ be a language of Alonzo. Write an expression of L that denotes the function that maps a binary relation over A to its transitive closure.

11. Write a formula that expresses the axiom of choice for the type α.

12. Write a formula that expresses Cantor's theorem for the type α.

Chapter 7

Beta-Reduction and Substitution

With the compact notation defined in the previous chapter, we can now discuss beta-reduction, substitution, and the need for the free-for-in notion.

7.1 Beta-Reduction

The lambda-expression application

$$(\lambda \mathbf{x} : \alpha \, . \, \mathbf{B}_\beta) \, \mathbf{A}_\alpha$$

beta-reduces to

$$\mathbf{B}_\beta[(\mathbf{x} : \alpha) \mapsto \mathbf{A}_\alpha].$$

The following theorem states that, if the free-for-in condition holds and the argument is defined, the lambda-expression application beta-reduces to an expression to which it is quasi-equal:

Theorem 7.1 (Beta-Reduction) *If $\mathbf{A}_\alpha$ is free for $(\mathbf{x} : \alpha)$ in $\mathbf{B}_\beta$, then*

$$\mathbf{A}_\alpha{\downarrow} \Rightarrow (\lambda \mathbf{x} : \alpha \, . \, \mathbf{B}_\beta) \, \mathbf{A}_\alpha \simeq \mathbf{B}_\beta[(\mathbf{x} : \alpha) \mapsto \mathbf{A}_\alpha]$$

is valid.

Beta-reduction thus provides a way, in certain cases, to compute the value of a lambda-expression application using the equations S1–S10 that define substitution in Section 4.6 as rewrite rules. Before proving this theorem, we will prove the following lemma that connects semantic substitution to syntactic substitution:

W. M. Farmer, *Simple Type Theory*, Computer Science Foundations and Applied Logic, https://doi.org/10.1007/978-3-031-21112-6_7

Lemma 7.2 (Semantic and Syntactic Substitution) *Let M be a general model of L that interprets $\mathbf{A}_\alpha$ and $\mathbf{B}_\beta$, and let $\varphi \in \mathsf{assign}(M)$. If $\mathbf{A}_\alpha$ is free for $(\mathbf{x} : \alpha)$ in $\mathbf{B}_\beta$ and $V_\varphi^M(\mathbf{A}_\alpha)$ is defined, then*

$$V_{\varphi[(\mathbf{x}:\alpha)\mapsto V_\varphi^M(\mathbf{A}_\alpha)]}^M(\mathbf{B}_\beta) \simeq V_\varphi^M(\mathbf{B}_\beta[(\mathbf{x}:\alpha) \mapsto \mathbf{A}_\alpha]).$$

Proof Fix M, $(\mathbf{x} : \alpha)$, and $\mathbf{A}_\alpha$, and assume M interprets $\mathbf{A}_\alpha$ and $V_\varphi^M(\mathbf{A}_\alpha)$ is defined. Let $P(\mathbf{B}_\beta)$ be the statement that, if M interprets $\mathbf{B}_\beta$ and $\mathbf{A}_\alpha$ is free for $(\mathbf{x} : \alpha)$ in $\mathbf{B}_\beta$, then

$$V_\psi^M(\mathbf{B}_\beta) \simeq V_\varphi^M(\mathbf{B}_\beta[(\mathbf{x} : \alpha) \mapsto \mathbf{A}_\alpha])$$

for all $\varphi \in \mathsf{assign}(M)$, where

$$\psi = \varphi[(\mathbf{x} : \alpha) \mapsto V_\varphi^M(\mathbf{A}_\alpha)].$$

We will prove, by induction on the syntactic structure of $\mathbf{B}_\beta$, that $P(\mathbf{B}_\beta)$ holds for all $\mathbf{B}_\beta$ — which immediately implies the lemma. Assume (a) M interprets $\mathbf{B}_\beta$ and (b) $\mathbf{A}_\alpha$ is free for $(\mathbf{x} : \alpha)$ in $\mathbf{B}_\beta$.

Case 1: $\mathbf{B}_\beta$ is $(\mathbf{x} : \alpha)$.

$$\begin{aligned} &V_\psi^M(\mathbf{B}_\beta) \\ \simeq\ &V_\varphi^M(\mathbf{A}_\alpha) &&(1) \\ \simeq\ &V_\varphi^M(\mathbf{B}_\beta[(\mathbf{x} : \alpha) \mapsto \mathbf{A}_\alpha]) &&(2) \end{aligned}$$

(1) is by condition V1 of the definition of a general model; and (2) is by equation S1 of the definition of substitution. Therefore, $P(\mathbf{B}_\beta)$ holds.

Case 2: $\mathbf{B}_\beta$ is a constant or a variable that differs from $(\mathbf{x} : \alpha)$.

$$\begin{aligned} &V_\psi^M(\mathbf{B}_\beta) \\ \simeq\ &V_\varphi^M(\mathbf{B}_\beta) &&(1) \\ \simeq\ &V_\varphi^M(\mathbf{B}_\beta[(\mathbf{x} : \alpha) \mapsto \mathbf{A}_\alpha]) &&(2) \end{aligned}$$

(1) is by conditions V1 and V2 of the definition of a general model; and (2) is by equations S2 and S3 of the definition of substitution. Therefore, $P(\mathbf{B}_\beta)$ holds.

Case 3: $\mathbf{B}_\beta$ is $(\mathbf{C}_\gamma = \mathbf{D}_\gamma)$. Assume $P(\mathbf{C}_\gamma)$ and $P(\mathbf{D}_\gamma)$. (a) implies (c) M interprets $\mathbf{C}_\gamma$ and $\mathbf{D}_\gamma$, and (b) implies (d) $\mathbf{A}_\alpha$ is free for $(\mathbf{x} : \alpha)$ in $\mathbf{C}_\gamma$ and $\mathbf{D}_\gamma$.

$$V_\psi^M(\mathbf{B}_\beta)$$
$$\simeq V_\psi^M(\mathbf{C}_\gamma) = V_\psi^M(\mathbf{D}_\gamma) \tag{1}$$
$$\simeq V_\varphi^M(\mathbf{C}_\gamma[(\mathbf{x} : \alpha) \mapsto \mathbf{A}_\alpha]) = V_\varphi^M(\mathbf{D}_\gamma[(\mathbf{x} : \alpha) \mapsto \mathbf{A}_\alpha]) \tag{2}$$
$$\simeq V_\varphi^M(\mathbf{C}_\gamma[(\mathbf{x} : \alpha) \mapsto \mathbf{A}_\alpha] = \mathbf{D}_\gamma[(\mathbf{x} : \alpha) \mapsto \mathbf{A}_\alpha]) \tag{3}$$
$$\simeq V_\varphi^M(\mathbf{B}_\beta[(\mathbf{x} : \alpha) \mapsto \mathbf{A}_\alpha]) \tag{4}$$

(1) and (3) are by condition V3 of the definition of a general model; (2) is by (c), (d), and the induction hypotheses $P(\mathbf{C}_\gamma)$ and $P(\mathbf{D}_\gamma)$; and (4) is by equation S4 of the definition of substitution. Therefore, $P(\mathbf{B}_\beta)$ holds.

Case 4: $\mathbf{B}_\beta$ is $(\mathbf{F}_{\gamma \to \delta}\, \mathbf{A}_\gamma)$. The proof is similar to Case 3.

Case 5: $\mathbf{B}_\beta$ is $(\lambda\, \mathbf{x} : \alpha\, .\, \mathbf{C}_\gamma)$.

$$V_\psi^M(\mathbf{B}_\beta)$$
$$\simeq V_\varphi^M(\mathbf{B}_\beta) \tag{1}$$
$$\simeq V_\varphi^M(\mathbf{B}_\beta[(\mathbf{x} : \alpha) \mapsto \mathbf{A}_\alpha]) \tag{2}$$

(1) is by part 16 of Lemma 5.4; and (2) is by equation S6 of the definition of substitution. Therefore, $P(\mathbf{B}_\beta)$ holds.

Case 6: $\mathbf{B}_\beta$ is $(\lambda\, \mathbf{y} : \gamma\, .\, \mathbf{C}_\delta)$ where $(\mathbf{y} : \gamma)$ differs from $(\mathbf{x} : \alpha)$. Assume $P(\mathbf{C}_\delta)$. (a) implies (c) M interprets $\mathbf{C}_\delta$, and (b) implies (d) $\mathbf{A}_\alpha$ is free for $(\mathbf{x} : \alpha)$ in $\mathbf{C}_\delta$. Let $d \in D_\gamma^M$, the domain in M for γ.

$$V_\psi^M(\mathbf{B}_\beta)(d)$$
$$\simeq V_{\psi[(\mathbf{y}:\gamma)\mapsto d]}^M(\mathbf{C}_\delta) \tag{1}$$
$$\simeq V_{\varphi[(\mathbf{y}:\gamma)\mapsto d][(\mathbf{x}:\alpha)\mapsto V_\varphi^M(\mathbf{A}_\alpha)]}^M(\mathbf{C}_\delta) \tag{2}$$
$$\simeq V_{\varphi[(\mathbf{y}:\gamma)\mapsto d]}^M(\mathbf{C}_\delta[(\mathbf{x} : \alpha) \mapsto \mathbf{A}_\alpha]) \tag{3}$$
$$\simeq V_\varphi^M(\lambda\, \mathbf{y} : \gamma\, .\, \mathbf{C}_\delta[(\mathbf{x} : \alpha) \mapsto \mathbf{A}_\alpha])(d) \tag{4}$$
$$\simeq V_\varphi^M(\mathbf{B}_\beta[(\mathbf{x} : \alpha) \mapsto \mathbf{A}_\alpha])(d) \tag{5}$$

(1) and (4) are by condition V5 of the definition of a general model; (2) is by $(\mathbf{x} : \alpha)$ and $(\mathbf{y} : \gamma)$ being distinct; (3) is by (c), (d), and the induction hypothesis $P(\mathbf{C}_\delta)$; and (5) is by equation S7 the definition of substitution.

Therefore, $P(\mathbf{B}_\beta)$ holds by part 5 of Lemma 5.4 and the extensionality of functions.

Case 7: $\mathbf{B}_\beta$ is $(\mathbf{I}\,\mathbf{x} : \alpha \,.\, \mathbf{C}_o)$. The proof is similar to Case 5.

Case 8: $\mathbf{B}_\beta$ is $(\mathbf{I}\,\mathbf{y} : \gamma \,.\, \mathbf{C}_o)$ where $(\mathbf{y} : \gamma)$ differs from $(\mathbf{x} : \alpha)$. The proof is similar to Case 6.

Case 9: $\mathbf{B}_\beta$ is $(\mathbf{C}_\gamma, \mathbf{D}_\delta)$. The proof is similar to Case 3.   □

Proof of Theorem 7.1   Let $\mathbf{X}_o$ be

$$(\lambda\mathbf{x} : \alpha \,.\, \mathbf{B}_\beta)\,\mathbf{A}_\alpha \simeq \mathbf{B}_\beta[(\mathbf{x} : \alpha) \mapsto \mathbf{A}_\alpha],$$

M be a general model that interprets $\mathbf{X}_o$, and $\varphi \in \mathsf{assign}(M)$. Assume (a) $\mathbf{A}_\alpha$ is free for $(\mathbf{x} : \alpha)$ in $\mathbf{B}_\beta$ and (b) $V_\varphi^M(\mathbf{A}_\alpha)$ is defined. By part 6 of Lemma 6.2 and part 3 of Lemma 6.5, it suffices to show (c) $V_\varphi^M(\mathbf{X}_o) = \mathrm{T}$. By part 5 of Lemma 6.5, (c) is equivalent to

$$\text{(d) } V_\varphi^M((\lambda\mathbf{x} : \alpha \,.\, \mathbf{B}_\beta)\,\mathbf{A}_\alpha) \simeq V_\varphi^M(\mathbf{B}_\beta[(\mathbf{x} : \alpha) \mapsto \mathbf{A}_\alpha]),$$

and by (b) and part 11 of Lemma 5.4, (d) is equivalent to

$$\text{(e) } V_{\varphi[(\mathbf{x}:\alpha)\mapsto V_\varphi^M(\mathbf{A}_\alpha)]}^M(\mathbf{B}_\beta) \simeq V_\varphi^M(\mathbf{B}_\beta[(\mathbf{x} : \alpha) \mapsto \mathbf{A}_\alpha]).$$

(a) and (b) imply (e) by Lemma 7.2.   □

Proposition 7.3 (Improper Lambda-Expression Application)
The formula

$$\mathbf{A}_\alpha{\uparrow} \Rightarrow (\lambda\mathbf{x} : \alpha \,.\, \mathbf{B}_\beta)\,\mathbf{A}_\alpha \simeq {\perp}_\beta$$

is valid.

Proof   Follows easily from part 6 of Lemma 6.2, parts 1, 4, and 5 of Lemma 6.5, and condition V4 of the definition of a general model.   □

7.2   Universal Instantiation

Corollary 7.4 (Universal Instantiation) *If $\mathbf{A}_\alpha$ is free for $(\mathbf{x} : \alpha)$ in $\mathbf{B}_o$, then*

$$((\forall\mathbf{x} : \alpha \,.\, \mathbf{B}_o) \wedge \mathbf{A}_\alpha{\downarrow}) \Rightarrow \mathbf{B}_o[(\mathbf{x} : \alpha) \mapsto \mathbf{A}_\alpha]$$

is valid.

Proof   Let $\mathbf{X}_o$ be $\mathbf{B}_o[(\mathbf{x} : \alpha) \mapsto \mathbf{A}_\alpha]$, M be a general model that interprets $\mathbf{A}_\alpha$ and $\mathbf{B}_o$, and $\varphi \in \mathsf{assign}(M)$. Assume (a) $\mathbf{A}_\alpha$ is free for $(\mathbf{x} : \alpha)$ in $\mathbf{B}_o$,

(b) $V_\varphi^M(\forall \mathbf{x} : \alpha . \mathbf{B}_o) = \mathrm{T}$,

and (c) $V_\varphi^M(\mathbf{A}_\alpha)$ is defined. By parts 5 and 6 of Lemma 6.2 and part 3 of Lemma 6.5, it suffices to show (d) $V_\varphi^M(\mathbf{X}_o) = \mathrm{T}$. (b) implies

(e) $V_{\varphi[(\mathbf{x}:\alpha)\mapsto V_\varphi^M(\mathbf{A}_\alpha)]}^M(\mathbf{B}_o) = \mathrm{T}$

by (c) and part 1 of Lemma 6.4. (a), (c), and (e) imply (d) by Lemma 7.2. $\square$

7.3   Invalid Beta-Reduction

Example 7.5 (Invalid Beta-Reduction) Let

$$\mathbf{F}_{\alpha\to\alpha\to\alpha} \equiv \lambda \mathbf{x} : \alpha . \lambda \mathbf{y} : \alpha . (\mathbf{x} : \alpha).$$

Then $\mathbf{F}_{\alpha\to\alpha\to\alpha} \mathbf{A}_\alpha$ should denote a constant function if $\mathbf{A}_\alpha$ is defined and should be undefined if $\mathbf{A}_\alpha$ is undefined.

1. $\mathbf{F}_{\alpha\to\alpha\to\alpha} (\mathbf{y} : \alpha)$ beta-reduces to:

   $$(\lambda \mathbf{y} : \alpha . (\mathbf{x} : \alpha))[(\mathbf{x} : \alpha) \mapsto (\mathbf{y} : \alpha)] \equiv \lambda \mathbf{y} : \alpha . (\mathbf{y} : \alpha).$$

   Hence, although $(\mathbf{y} : \alpha)$ is defined, $\mathbf{F}_{\alpha\to\alpha\to\alpha} (\mathbf{y} : \alpha)$ beta-reduces to an identity function, not a constant function as expected. This happens because $(\mathbf{y} : \alpha)$ is not free for $(\mathbf{x} : \alpha)$ in $\lambda \mathbf{y} : \alpha . (\mathbf{x} : \alpha)$.

2. $\mathbf{F}_{\alpha\to\alpha\to\alpha} \perp_\alpha$ beta-reduces to:

   $$(\lambda \mathbf{y} : \alpha . (\mathbf{x} : \alpha))[(\mathbf{x} : \alpha) \mapsto \perp_\alpha] \equiv \lambda \mathbf{y} : \alpha . \perp_\alpha.$$

   Hence $\mathbf{F}_{\alpha\to\alpha\to\alpha} \perp_\alpha$ beta-reduces to the empty function and is defined, not undefined as expected. This happens because $\perp_\alpha$ is undefined. $\square$

7.4   Alpha-Conversion

We can compute a lambda-expression application $(\lambda \mathbf{x} : \alpha \, . \, \mathbf{B}_\beta) \, \mathbf{A}_\alpha$ when $\mathbf{A}_\alpha$ is defined but $\mathbf{A}_\alpha$ is *not* free for $(\mathbf{x} : \alpha)$ in $\mathbf{B}_\beta$ if we judiciously rename the bound variables in $\mathbf{B}_\beta$. The lambda-expression

$$\lambda \mathbf{x} : \alpha \, . \, \mathbf{B}_\beta$$

alpha-converts to

$$\lambda \mathbf{y} : \alpha \, . \, \mathbf{B}_\beta[(\mathbf{x} : \alpha) \mapsto (\mathbf{y} : \alpha)]$$

provided (1) $(\mathbf{y} : \alpha)$ is not free in $\mathbf{B}_\beta$ and (2) $(\mathbf{y} : \alpha)$ is free for $(\mathbf{x} : \alpha)$ in $\mathbf{B}_\beta$. The following theorem states that a lambda-expression is equal to an expression to which it alpha-converts if these two conditions hold:

Theorem 7.6 (Alpha-Conversion) *If $(\mathbf{y} : \alpha)$ is not free in $\mathbf{B}_\beta$ and $(\mathbf{y} : \alpha)$ is free for $(\mathbf{x} : \alpha)$ in $\mathbf{B}_\beta$, then*

$$(\lambda \mathbf{x} : \alpha \, . \, \mathbf{B}_\beta) = (\lambda \mathbf{y} : \alpha \, . \, \mathbf{B}_\beta[(\mathbf{x} : \alpha) \mapsto (\mathbf{y} : \alpha)])$$

is valid.

Proof Let $\mathbf{X}_o$ be

$$(\lambda \mathbf{x} : \alpha \, . \, \mathbf{B}_\beta) = (\lambda \mathbf{y} : \alpha \, . \, \mathbf{B}_\beta[(\mathbf{x} : \alpha) \mapsto (\mathbf{y} : \alpha)]),$$

M be a general model that interprets $\mathbf{X}_o$, and $\varphi \in \mathsf{assign}(M)$. Assume (a) $(\mathbf{y} : \alpha)$ is not free in $\mathbf{B}_\beta$ and (b) $(\mathbf{y} : \alpha)$ is free for $(\mathbf{x} : \alpha)$ in $\mathbf{B}_\beta$. We must show (c) $V_\varphi^M(\mathbf{X}_o) = \mathrm{T}$. Let $d \in D_\alpha^M$, the domain in M for α.

$$V_\varphi^M(\lambda \mathbf{x} : \alpha \, . \, \mathbf{B}_\beta)(d)$$

$$\simeq V_{\varphi[(\mathbf{x}:\alpha)\mapsto d]}^M(\mathbf{B}_\beta) \tag{1}$$

$$\simeq V_{\varphi[(\mathbf{x}:\alpha)\mapsto V_{\varphi[(\mathbf{y}:\alpha)\mapsto d]}^M(\mathbf{y}:\alpha)]}^M(\mathbf{B}_\beta) \tag{2}$$

$$\simeq V_{\varphi[(\mathbf{y}:\alpha)\mapsto d][(\mathbf{x}:\alpha)\mapsto V_{\varphi[(\mathbf{y}:\alpha)\mapsto d]}^M(\mathbf{y}:\alpha)]}^M(\mathbf{B}_\beta) \tag{3}$$

$$\simeq V_{\varphi[(\mathbf{y}:\alpha)\mapsto d]}^M(\mathbf{B}_\beta[(\mathbf{x} : \alpha) \mapsto (\mathbf{y} : \alpha)]) \tag{4}$$

$$\simeq V_\varphi^M(\lambda \mathbf{y} : \alpha \, . \, \mathbf{B}_\beta[(\mathbf{x} : \alpha) \mapsto (\mathbf{y} : \alpha)])(d) \tag{5}$$

(1) and (5) are by condition V5 of the definition of a general model; (2) is by condition V1 of the definition of a general model; (3) is by (a) and part 15 of Lemma 5.4; and (4) is by (b), part 1 of Lemma 5.4, and Lemma 7.2. Therefore, (c) holds by part 5 of Lemma 5.4 and the extensionality of functions. $\qquad\square$

Corollary 7.7 (Alpha-Conversion) *If* $(\mathbf{y} : \alpha)$ *does not occur in* $\mathbf{B}_\beta$, *then*

$$(\lambda \mathbf{x} : \alpha . \mathbf{B}_\beta) = (\lambda \mathbf{y} : \alpha . \mathbf{B}_\beta[(\mathbf{x} : \alpha) \mapsto (\mathbf{y} : \alpha)])$$

is valid.

Proof If $(\mathbf{y} : \alpha)$ does not occur in $\mathbf{B}_\beta$, then clearly $(\mathbf{y} : \alpha)$ is not free in $\mathbf{B}_\beta$ and $(\mathbf{y} : \alpha)$ is free for $(\mathbf{x} : \alpha)$ in $\mathbf{B}_\beta$ □

When $\mathbf{A}_\alpha$ is defined but $\mathbf{A}_\alpha$ is *not* free for $(\mathbf{x} : \alpha)$ in $\mathbf{B}_\beta$, we can compute the value of $(\lambda \mathbf{x} : \alpha . \mathbf{B}_\beta) \mathbf{A}_\alpha$ using Theorem 7.6 (or Corollary 7.7) as follows. Repeatedly apply alpha-conversion to the lambda-expressions in $\mathbf{B}_\beta$ until an expression $\mathbf{B}'_\beta$ is obtained such that $\mathbf{A}_\alpha$ is free for $(\mathbf{x} : \alpha)$ in $\mathbf{B}'_\beta$. Then $(\lambda \mathbf{x} : \alpha . \mathbf{B}'_\beta) \mathbf{A}_\alpha$ is quasi-equal to

$$(\lambda \mathbf{x} : \alpha . \mathbf{B}_\beta) \mathbf{A}_\alpha$$

by Theorem 7.6, and so we can compute the value of $(\lambda \mathbf{x} : \alpha . \mathbf{B}_\beta) \mathbf{A}_\alpha$ by beta-reducing $(\lambda \mathbf{x} : \alpha . \mathbf{B}'_\beta) \mathbf{A}_\alpha$.

7.5 Conclusions

Beta-reduction provides a way to compute the value of a lambda-expression application $(\lambda \mathbf{x} : \alpha . \mathbf{B}_\beta) \mathbf{A}_\alpha$ when $\mathbf{A}_\alpha$ is free for $(\mathbf{x} : \alpha)$ in $\mathbf{B}_\beta$ and $\mathbf{A}_\alpha$ is defined. Alpha-conversion provides a way to use beta-reduction to compute the value of the lambda-expression application when $\mathbf{A}_\alpha$ is defined but $\mathbf{A}_\alpha$ is *not* free for $(\mathbf{x} : \alpha)$ in $\mathbf{B}_\beta$.

7.6 Exercises

1. Prove that $(\lambda \mathbf{x} : \alpha . \mathbf{B}_\beta) (\mathbf{x} : \alpha) \simeq \mathbf{B}_\beta$ is valid.

2. Prove that $(\lambda \mathbf{x} : \alpha . \mathbf{x}) \mathbf{A}_\alpha \simeq \mathbf{A}_\alpha$ is valid.

3. Prove that *alpha-conversion for definite descriptions* is valid: If $(\mathbf{y} : \alpha)$ is not free in $\mathbf{A}_o$ and $(\mathbf{y} : \alpha)$ is free for $(\mathbf{x} : \alpha)$ in $\mathbf{A}_o$, then

$$(\mathbf{I}\,\mathbf{x} : \alpha . \mathbf{A}_o) \simeq (\mathbf{I}\,\mathbf{y} : \alpha . \mathbf{A}_o[(\mathbf{x} : \alpha) \mapsto (\mathbf{y} : \alpha)])$$

   is valid.

4. Prove that *eta-reduction* is valid: If $(\mathbf{x} : \alpha)$ is not free in $\mathbf{A}_{\alpha \to \beta}$, then

$$\mathbf{A}_{\alpha \to \beta}\!\downarrow\;\Rightarrow (\lambda \mathbf{x} : \alpha . \mathbf{A}_{\alpha \to \beta}\,\mathbf{x}) = \mathbf{A}_{\alpha \to \beta}$$

   is valid.

Chapter 8

Proof Systems

This chapter presents a proof system for Alonzo that is sound and complete with respect to the general semantics defined in Chapter 5.

8.1 Background

A *proof system* $\mathfrak{P}$ (P in fraktur font) for Alonzo consists of a decidable[1] set of *axioms* and *rules of inference*. Each axiom is a formula of Alonzo and each rule of inference has the following form: From the formulas $\mathbf{A}_o^1, \ldots, \mathbf{A}_o^n$, infer the formula $\mathbf{B}_o$ (possibly subject to certain constraints). A *proof of $\mathbf{A}_o$ in $\mathfrak{P}$* is a finite sequence Π of formulas of Alonzo ending with $\mathbf{A}_o$ such that every formula in Π is an axiom of $\mathfrak{P}$ or inferred from previous formulas in Π by one of the rules of inference of $\mathfrak{P}$. We write $\vdash_{\mathfrak{P}} \mathbf{A}_o$ to assert that there is a proof of $\mathbf{A}_o$ in $\mathfrak{P}$. A *theorem* of $\mathfrak{P}$ is a formula that has a proof in $\mathfrak{P}$.

Now let Γ be a set of formulas of Alonzo. A *proof of $\mathbf{A}_o$ from Γ in $\mathfrak{P}$* is a pair (Π_1, Π_2) of finite sequences of formulas of Alonzo ending with $\mathbf{A}_o$ such that Π_1 is a proof in $\mathfrak{P}$, Π_2 ends with $\mathbf{A}_o$, and every formula in Π_2 is a member of Γ, a member of Π_1 (and thus a theorem of $\mathfrak{P}$), or inferred from previous formulas in Π_2 by one of the rules of inference of $\mathfrak{P}$ modified, if necessary, so that the free variables in members of Γ are treated as constants instead of as universally quantified variables as they are in axioms. The formulas in Γ are the *premises* of the proof and $\mathbf{A}_o$ is the *conclusion* of the proof. We write $\Gamma \vdash_{\mathfrak{P}} \mathbf{A}_o$ to assert that there is a proof of $\mathbf{A}_o$ from Γ in $\mathfrak{P}$.

---

[1] A set $A \subseteq U$ is *decidable* if there is an algorithm that can check whether or not a given value of U belongs to A.

© The Author(s), under exclusive license to Springer Nature Switzerland AG 2023
W. M. Farmer, *Simple Type Theory*, Computer Science Foundations
and Applied Logic, https://doi.org/10.1007/978-3-031-21112-6_8

$\mathfrak{P}$ is *sound* if

$\Gamma \vdash_{\mathfrak{P}} \mathbf{A}_o$ implies $\Gamma \vDash \mathbf{A}_o$,

where Γ is a set of formulas. $\mathfrak{P}$ is *complete (in the general sense)* if

$\Gamma \vDash \mathbf{A}_o$ implies $\Gamma \vdash_{\mathfrak{P}} \mathbf{A}_o$

and *complete in the standard sense* if

$\Gamma \vDash^s \mathbf{A}_o$ implies $\Gamma \vdash_{\mathfrak{P}} \mathbf{A}_o$,

where Γ is a set of sentences. Obviously, if $\mathfrak{P}$ is complete in the standard sense, then $\mathfrak{P}$ is complete (in the general sense). A set Γ of formulas is *consistent in* $\mathfrak{P}$ if not $\Gamma \vdash_{\mathfrak{P}} F_o$.

Proposition 8.1 *Let $\mathfrak{P}$ be sound and complete proof system for Alonzo and Γ be a set of sentences that is inconsistent in $\mathfrak{P}$. Then $\Gamma \vdash_{\mathfrak{P}} \mathbf{A}_o$ for all formulas $\mathbf{A}_o$.*

Proof Let $\mathbf{A}_o$ be any formula. Γ inconsistent in $\mathfrak{P}$ means $\Gamma \vdash_{\mathfrak{P}} F_o$. This implies (a) $\Gamma \vDash F_o$ by the soundness of $\mathfrak{P}$. (b) $\Gamma \vDash F_o \Rightarrow \mathbf{A}_o$ holds by parts 2 and 6 of Lemma 6.2. (a) and (b) then imply $\Gamma \vDash \mathbf{A}_o$. This implies $\Gamma \vdash_{\mathfrak{P}} \mathbf{A}_o$ by the completeness of $\mathfrak{P}$. $\square$

Theorem 8.2 (Incompleteness of Alonzo in the Standard Sense)
There is no sound proof system for Alonzo that is complete in the standard sense.

Proof The proof of this theorem is given in Section 9.5 on page 119. $\square$

The following theorem, proved by Leon Henkin [61] in 1950, is the *arch theorem* of simple type theory:

Theorem 8.3 (Henkin's Soundness and Completeness Theorem)
There is a proof system for Church's type theory that is sound and complete with respect to the semantics of general models.

On the basis of this theorem we would expect that there is a proof system for Alonzo that is sound and complete. In the next section we will present just such a proof system.

8.2 A Proof System for Alonzo

In this section we present the axioms and rules of inference of a proof system for Alonzo named $\mathfrak{A}$ (A in fraktur font). It is based on the proof system for $\mathcal{Q}_0$, the version of Church's type theory Peter Andrews presents in [5]. The proof system is named $\mathfrak{A}$ in honor of Andrews.

8.2.1 Axioms

The single instance of Axiom A1 is presented by a formula, while the instances of Axioms A2–A7 are presented by one or more formula schemas. $x, y, f, g, h \in \mathcal{S}_{\mathsf{var}}$ are the actual names of the free variables in A1, A2, and A3.

A1 (Truth Values)

$$((g : o \to o) \, T_o \wedge (g : o \to o) \, F_o) \Leftrightarrow \forall x : o \, . \, (g : o \to o) \, x.$$

A2 (Leibniz's Law[2])

$$(x : \alpha) = (y : \alpha) \Rightarrow ((h : \alpha \to o) \, (x : \alpha) \Leftrightarrow (h : \alpha \to o) \, (y : \alpha)).$$

A3 (Extensionality)

$$(f : \alpha \to \beta) = (g : \alpha \to \beta) \Leftrightarrow \forall x : \alpha \, . \, (f : \alpha \to \beta) \, x \simeq (g : \alpha \to \beta) \, x.$$

A4 (Beta-Reduction)

$\mathbf{A}_\alpha{\downarrow} \Rightarrow (\lambda \mathbf{x} : \alpha \, . \, \mathbf{B}_\beta) \, \mathbf{A}_\alpha \simeq \mathbf{B}_\beta[(\mathbf{x} : \alpha) \mapsto \mathbf{A}_\alpha]$
provided $\mathbf{A}_\alpha$ is free for $(\mathbf{x} : \alpha)$ in $\mathbf{B}_\beta$.

A5 (Definedness)

1. $(\mathbf{x} : \alpha){\downarrow}.$

2. $\mathbf{c}_\alpha{\downarrow}.$[3]

3. $(\mathbf{A}_\alpha = \mathbf{B}_\alpha){\downarrow}.$

---

[2] Also called the *indiscernibility of identicals*.

[3] Since each instance of the parametric pseudoconstant $\perp_\alpha$ with $\alpha \neq o$ is undefined in every general model that interprets the instance, we cannot consider the instances of $\perp_\alpha$ with $\alpha \neq o$ to be constants. However, we can consider all the nonparametric pseudoconstants we have defined and the instances of all the other parametric pseudoconstants we have defined to be constants.

4. $(\mathbf{A}_\alpha = \mathbf{B}_\alpha) \Rightarrow \mathbf{A}_\alpha{\downarrow}$.

5. $(\mathbf{A}_\alpha = \mathbf{B}_\alpha) \Rightarrow \mathbf{B}_\alpha{\downarrow}$.

6. $(\mathbf{F}_{\alpha \to o}\, \mathbf{A}_\alpha){\downarrow}$.

7. $\mathbf{F}_{\alpha \to o}\, \mathbf{A}_\alpha \Rightarrow \mathbf{F}_{\alpha \to o}{\downarrow}$.

8. $\mathbf{F}_{\alpha \to o}\, \mathbf{A}_\alpha \Rightarrow \mathbf{A}_\alpha{\downarrow}$.

9. $(\mathbf{F}_{\alpha \to \beta}\, \mathbf{A}_\alpha){\downarrow} \Rightarrow \mathbf{F}_{\alpha \to \beta}{\downarrow}$ where $\beta \neq o$.

10. $(\mathbf{F}_{\alpha \to \beta}\, \mathbf{A}_\alpha){\downarrow} \Rightarrow \mathbf{A}_\alpha{\downarrow}$ where $\beta \neq o$.

11. $(\lambda\mathbf{x} : \alpha\,.\,\mathbf{B}_\beta){\downarrow}$.

12. $\mathbf{A}_\alpha{\downarrow} \Rightarrow (\mathbf{A}_\alpha \simeq \mathbf{B}_\alpha) \simeq (\mathbf{A}_\alpha = \mathbf{B}_\alpha)$.

13. $\mathbf{B}_\alpha{\downarrow} \Rightarrow (\mathbf{A}_\alpha \simeq \mathbf{B}_\alpha) \simeq (\mathbf{A}_\alpha = \mathbf{B}_\alpha)$.

A6 (Definite Description)

1. $(\exists!\, x : \alpha\,.\,\mathbf{A}_o) \Rightarrow (\mathrm{I}\, x : \alpha\,.\,\mathbf{A}_o) \in \{x : \alpha \mid \mathbf{A}_o\}$ (where $\alpha \neq o$).

2. $\neg(\exists!\, x : \alpha\,.\,\mathbf{A}_o) \Rightarrow (\mathrm{I}\, x : \alpha\,.\,\mathbf{A}_o){\uparrow}$ (where $\alpha \neq o$).

A7 (Pairs)

1. $\forall x : \alpha,\, y : \beta\,.\,(x, y){\downarrow}$.

2. $(\mathbf{A}_\alpha, \mathbf{B}_\beta){\downarrow} \Rightarrow \mathbf{A}_\alpha{\downarrow}$.

3. $(\mathbf{A}_\alpha, \mathbf{B}_\beta){\downarrow} \Rightarrow \mathbf{B}_\alpha{\downarrow}$.

4. $\forall p : \alpha \times \beta\,.\,p = (\mathsf{fst}_{(\alpha \times \beta) \to \alpha}\, p, \mathsf{snd}_{(\alpha \times \beta) \to \beta}\, p)$.

5. $\forall x, x' : \alpha,\, y, y' : \beta\,.\,(x, y) = (x', y') \Rightarrow (x = x' \wedge y = y')$.

Remark 8.4 (Axioms A5.12 and A5.13) *Axioms A5.1–A5.11 give the basic properties of definedness. Axioms A5.12–A5.13 serve as scaffolding; they state needed properties that are implied by Axioms A5.1–A5.11 and the definition of quasi-equal but are not immediately provable.*

8.2.2   Rules of Inference

R1 (Modus Ponens)   From $\mathbf{A}_o$ and $\mathbf{A}_o \Rightarrow \mathbf{B}_o$ infer $\mathbf{B}_o$.

R2 (Quasi-Equality Substitution)   From $\mathbf{A}_\alpha \simeq \mathbf{B}_\alpha$ and $\mathbf{C}_o$ infer the result of replacing one occurrence of $\mathbf{A}_\alpha$ in $\mathbf{C}_o$ by an occurrence of $\mathbf{B}_\alpha$, provided that the occurrence of $\mathbf{A}_\alpha$ in $\mathbf{C}_o$ is not the first argument of a function abstraction or a definite description.

Remark 8.5 (Proof System for $\mathcal{Q}_0$) *The proof system for $\mathcal{Q}_0$ given in [5] has just one rule of inference, equality substitution, in which equality plays the role of quasi-equality in Rule R2. Modus ponens is a derived rule of inference. In the proof system $\mathfrak{A}$, modus ponens needs to be a primitive rule of inference so that Axioms A4, A5.4, A5.5, A5.12, and A5.13 can be immediately utilized (see Appendix A for details).*

8.2.3   Proofs

Let $\mathbf{A}_o$ be a formula and Γ be a set of formulas of Alonzo. A *proof of* $\mathbf{A}_o$ *in* $\mathfrak{A}$ is a finite sequence Π of formulas of Alonzo ending with $\mathbf{A}_o$ such that every formula in Π is an instance of one of the schemas in Axioms A1–A7 or inferred from previous formulas in Π by Rule R1 or R2. A *proof of* $\mathbf{A}_o$ *from* Γ *in* $\mathfrak{A}$ is a pair (Π_1, Π_2) of finite sequences of formulas of Alonzo such that Π_1 is a proof in $\mathfrak{A}$, Π_2 ends with $\mathbf{A}_o$, and every formula $\mathbf{D}_o$ in Π_2 satisfies one of the following conditions:

1. $\mathbf{D}_o$ is a member of Γ.

2. $\mathbf{D}_o$ is in Π_1 (and thus a theorem of $\mathfrak{A}$).

3. $\mathbf{D}_o$ is inferred from two previous formulas in Π_2 by R1.

4. $\mathbf{D}_o$ is obtain from previous formulas $\mathbf{A}_\alpha \simeq \mathbf{B}_\alpha$ and $\mathbf{C}_o$ in Π_2 by replacing one occurrence of $\mathbf{A}_\alpha$ in $\mathbf{C}_o$ by an occurrence of $\mathbf{B}_\alpha$, provided that the occurrence of $\mathbf{A}_\alpha$ in $\mathbf{C}_o$ is not the first argument of a function abstraction or a definite description and not in the second argument of a function abstraction $\lambda \mathbf{x} : \beta \,.\, \mathbf{E}_\gamma$ or a definite description $\mathrm{I}\,\mathbf{x} : \beta \,.\, \mathbf{E}_o$ where $(\mathbf{x} : \beta)$ is free in a member of Γ and free in $\mathbf{A}_\alpha \simeq \mathbf{B}_\alpha$.

It is easy to verify that $\emptyset \vdash_{\mathfrak{A}} \mathbf{A}_o$ ($\mathbf{A}_o$ is provable from the empty set of premises in $\mathfrak{A}$) iff $\vdash_{\mathfrak{A}} \mathbf{A}_o$ ($\mathbf{A}_o$ is a theorem of $\mathfrak{A}$). It is also easy to see that $\Gamma \vdash_{\mathfrak{A}} \mathbf{A}_o$ and $\Gamma \subseteq \Gamma'$ implies $\Gamma' \vdash_{\mathfrak{A}} \mathbf{A}_o$. The basic metatheorems of $\mathfrak{A}$ — theorems about what can be proved in $\mathfrak{A}$ — are given in Appendix A.

8.3 Soundness

Theorem 8.6 (Soundness Theorem) $\mathfrak{A}$ *is sound.*

Proof This theorem is part 2 of Theorem B.11 proved in Appendix B. $\square$

Corollary 8.7 (Consistency Theorem) *If a set of formulas is satisfiable, then it is consistent in* $\mathfrak{A}$.

Proof Suppose a satisfiable set Γ of formulas were inconsistent in $\mathfrak{A}$. Then $\Gamma \vdash_{\mathfrak{A}} F_o$, and so by the Soundness Theorem, $\Gamma \vDash F_o$, which contradicts that Γ is satisfiable by part 2 of Lemma 6.2. Therefore, Γ must be consistent in $\mathfrak{A}$. $\square$

8.4 Frugal General Models

A general model M of L is *frugal* if $\|M\| \leq \|L\|$. Thus, in the usual case where L is a language with a countable number of constants, the power of a frugal general model of L is countable.

Proposition 8.8 *Every finite general model is frugal.*

Proof Let M be a finite general model of L. Then $\|M\| = \omega$ by part 2 of Proposition 5.6. $\omega \leq \|L\|$ for every language L. Hence $\|M\| \leq \|L\|$ and so M is frugal. $\square$

Lemma 8.9 *Let M be a frugal general model of $L = (\mathcal{B}, \mathcal{C})$. If $|D_\alpha^M| = \|L\|$ for some $\alpha \in \mathcal{T}(L)$, then M is a nonstandard model.*

Proof Suppose M were a standard model. Then $D_{\alpha \to o}^M$ represents the set of all subsets of D_α^M, and so $|D_{\alpha \to o}^M| > |D_\alpha^M|$ by Cantor's theorem. Hence $|D_{\alpha \to o}^M| > \|L\|$ by the hypothesis, which contradicts the fact that M is frugal. Therefore, M must be nonstandard. $\square$

Corollary 8.10 *Every frugal infinite general model of a language of power ω is a nonstandard model.*

Proof Let $L = (\mathcal{B}, \mathcal{C})$ be a language of power ω and M be a frugal infinite general model of L. Since M is frugal and infinite, $\omega \leq |D_\mathbf{a}^M| \leq \|L\|$ for some $\mathbf{a} \in \mathcal{B}$. Since L is a language of power ω, $\|L\| = \omega$. Hence $|D_\mathbf{a}^M| = \|L\|$, and so by Lemma 8.9, M is a nonstandard model. $\square$

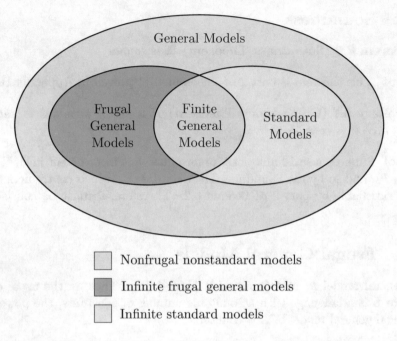

Nonfrugal nonstandard models

Infinite frugal general models

Infinite standard models

Figure 8.1: Kinds of General Models of Languages of Power ω

Figure 8.1 shows pictorially how general models, frugal general models, standard models, and finite general models of languages of power ω are related according to Propositions 5.7 and 8.8, Theorem 5.9, and Corollary 8.10.

Theorem 8.11 (Henkin's Theorem) *Every set of sentences in Alonzo that is consistent in* $\mathfrak{A}$ *has a frugal general model.*

Proof This theorem is the same as Theorem C.3 proved in Appendix C. □

8.5   Completeness

Theorem 8.12 (Provability equals Validity) *Let* $\mathbf{A}_o$ *be a formula and* Γ *be a set of sentences. Then the following are equivalent:*

1. $\Gamma \vdash_{\mathfrak{A}} \mathbf{A}_o$.

2. $\Gamma \vDash \mathbf{A}_o$.

3. Every frugal general model of Γ is a model of $\mathbf{A}_o$.

Proof

($1 \Rightarrow 2$): By the Soundness Theorem (Theorem 8.6).

($2 \Rightarrow 3$): By part 1 of Proposition 5.10 and the fact that Γ is a set of sentences.

($3 \Rightarrow 1$): Assume 3. Let $\mathbf{B}_o$ be a universal closure of $\mathbf{A}_o$. Then $(\star)$ every frugal general model of Γ is a model of $\mathbf{B}_o$ by part 4 of Lemma 6.4. Suppose $\Gamma \cup \{\neg \mathbf{B}_o\}$ is consistent in $\mathfrak{A}$. Then, by Henkin's Theorem (Theorem 8.11), there is a frugal general model M of $\Gamma \cup \{\neg \mathbf{B}_o\}$. This implies $M \vDash \neg \mathbf{B}_o$, and $(\star)$ implies $M \vDash \mathbf{B}_o$, which is a contradiction. Hence $\Gamma \cup \{\neg \mathbf{B}_o\}$ must be inconsistent in $\mathfrak{A}$, i.e., $\Gamma \cup \{\neg \mathbf{B}_o\} \vdash_{\mathfrak{A}} F_o$. So $\Gamma \vdash_{\mathfrak{A}} \neg \mathbf{B}_o \Rightarrow F_o$ by the Deduction Theorem (Theorem A.50) proved in Appendix A. This implies $\Gamma \vdash_{\mathfrak{A}} \mathbf{B}_o$ by the Tautology Rule (Corollary A.46) proved in Appendix A. And finally, this implies $\Gamma \vdash_{\mathfrak{A}} \mathbf{A}_o$ by Axiom A5.1 and Universal Instantiation (Theorem A.14) proved in Appendix A. $\square$

This is the Alonzo analog of the completeness theorem for first-order logic:

Corollary 8.13 (Henkin's Soundness and Completeness Theorem) $\mathfrak{A}$ *is sound and complete.*

Proof By Theorems 8.6 and 8.12. $\square$

Corollary 8.14 (Gödel's Completeness Theorem) $\mathbf{A}_o$ *is a theorem of* $\mathfrak{A}$ *iff* $\mathbf{A}_o$ *is valid.*

Proof By Theorem 8.12. $\square$

Corollary 8.15 (Consistency equals Satisfiability) *Let* Γ *be a set of sentences. Then* Γ *is consistent in* $\mathfrak{A}$ *iff* Γ *is satisfiable.*

Proof

$\quad\quad\quad \Gamma$ is inconsistent in $\mathfrak{A}$

$\quad\quad$ iff $\Gamma \vdash_{\mathfrak{A}} F_o$ $\hfill (1)$

$\quad\quad$ iff $\Gamma \vDash F_o$ $\hfill (2)$

$\quad\quad$ iff Γ is unsatisfiable $\hfill (3)$

(1) is by the definition of consistency; (2) is by Theorem 8.12; and (3) is by part 2 of Lemma 6.2. Therefore, Γ is consistent in $\mathfrak{A}$ iff Γ is satisfiable. $\square$

This is the Alonzo analog of the compactness theorem for first-order logic:

Corollary 8.16 (Compactness Theorem) *Let Γ be a set of sentences. Then Γ is satisfiable iff every finite subset of Γ is satisfiable.*

Proof

$\qquad\qquad$ Γ is satisfiable

$\quad$ iff $\ \Gamma$ is consistent in $\mathfrak{A}$ $\hfill$ (1)

$\quad$ iff $\ $not $\Gamma \vdash_{\mathfrak{A}} F_o$ $\hfill$ (2)

$\quad$ iff $\ $not $\Gamma' \vdash_{\mathfrak{A}} F_o$ for all finite subsets Γ' of Γ $\hfill$ (3)

$\quad$ iff $\ \Gamma'$ is consistent in $\mathfrak{A}$ for all finite subsets Γ' of Γ $\hfill$ (4)

$\quad$ iff $\ \Gamma'$ is satisfiable for all finite subsets Γ' of Γ $\hfill$ (5)

(1) and (5) are by Corollary 8.15; (2) and (4) are by the definition of consistency; and (3) follows from the fact that every proof of a formula A_o from Γ contains only finitely many members of Γ. Therefore, Γ is satisfiable iff every finite subset of Γ is satisfiable. $\hfill\square$

Theorem 8.17 (Unspecifiability of Finite General Models) *Let $L = (\mathcal{B}, \mathcal{C})$ be a language where $\mathcal{B} \neq \emptyset$. Then there is no sentence $\mathbf{X}_o$ of L such that $M \vDash \mathbf{X}_o$ iff M is a finite general model of L.*

Proof   Without loss of generality, we may assume that $B \in \mathcal{B}$. Suppose that there is a sentence $\mathbf{X}_o$ of L such that $M \vDash \mathbf{X}_o$ iff M is a finite general model of L. Let

$$\Sigma = \{\mathbf{X}_o\} \cup \{C_B^m \neq C_B^n \mid m, n \in \mathbb{N} \wedge m < n\}$$

where $\mathcal{C}$ and $\{C_B^n \mid n \in \mathbb{N}\}$ are disjoint sets of constants. For each finite subset Σ' of Σ, there is clearly a model of $\mathbf{X}_o$ for which there is an expansion that satisfies Σ'. Hence, by the Compactness Theorem above, there is a model of Σ, and this model must be infinite by the construction of Σ. This contradicts our assumption about $\mathbf{X}_o$, and therefore, there is no such $\mathbf{X}_o$. $\hfill\square$

8.6 Conclusions

Like Church's type theory, there is no sound proof system for Alonzo that is complete in the standard sense, but there is a sound proof system for Alonzo that is complete (in the general sense). $\mathfrak{A}$ is a proof system for Alonzo that is sound and complete. It is derived from Peter Andrews' elegant proof system for $\mathcal{Q}_0$. Unlike the latter, it has machinery for reasoning about definedness and pairs. For sets of sentences in Alonzo, provability in $\mathfrak{A}$ equals validity, and consistency in $\mathfrak{A}$ equals satisfiability. Alonzo has a compactness theorem (for general models).

8.7 Exercises

1. Let Γ is a set of formulas (which may not necessarily be sentences). Prove that $\Gamma \vDash \mathbf{A}_o$ implies $\Gamma \vdash_{\mathfrak{A}} \mathbf{A}_o$.

2. Let $L = (\mathcal{B}, \mathcal{C})$ be a language where $\mathcal{B} = \emptyset$. Prove that there is a sentence $\mathbf{X}_o$ of L such that $M \vDash \mathbf{X}_o$ iff M is a standard model of L.

3. Let $L = (\mathcal{B}, \mathcal{C})$ be a language where $\mathcal{B} \neq \emptyset$. Prove that there is no sentence $\mathbf{X}_o$ of L such that $M \vDash \mathbf{X}_o$ iff M is a standard model of L. Hint: Use Theorem 8.17. (Exercise X5503 in [5].)

Chapter 9

Theories

This chapter introduces the central notion of a theory of Alonzo as well as several related notions.

9.1 Axiomatic Theories

An *axiomatic theory* (*theory* for short) of Alonzo is a pair $T = (L, \Gamma)$ where L is a language of Alonzo and Γ is a set of sentences of L. L is the *language* of T and the members of Γ are the *axioms* of T. The notion of a theory is one of the most important and useful ideas in logic. We will see that theories specify collections of structures.

A formula $\mathbf{A}_o$ of L is a *valid in T (in the general sense)*, written $T \vDash \mathbf{A}_o$, if $\Gamma \vDash \mathbf{A}_o$. A formula $\mathbf{A}_o$ of L is *valid in T in the standard sense*, written $T \vDash^s \mathbf{A}_o$, if $\Gamma \vDash^s \mathbf{A}_o$. A *theorem of T* is a sentence valid in T. A general model M of L is a *model of T*, written $M \vDash T$, if M is a model of Γ. T is *satisfiable* if Γ is satisfiable.

Let $\mathfrak{P}$ be a proof system for Alonzo. A formula $\mathbf{A}_o$ of L is *provable from T in $\mathfrak{P}$*, written $T \vdash_{\mathfrak{P}} \mathbf{A}_o$, if $\Gamma \vdash_{\mathfrak{P}} \mathbf{A}_o$. T is *consistent* in $\mathfrak{P}$ if Γ is consistent in $\mathfrak{P}$. By Theorem 8.12, $T \vdash_{\mathfrak{A}} \mathbf{A}_o$ iff $T \vDash \mathbf{A}_o$ where $\mathfrak{A}$ is the proof system for Alonzo presented in Section 8.2.

The *closure of a theory T*, written $\overline{T}$, is the set of theorems of T.[1] $\overline{T}$ is the smallest set of sentences of L that contains the valid sentences of L and the members of Γ and is closed under semantic consequence. Two theories T_1 and T_2 having the same language are *equivalent*, written $T_1 \equiv T_2$, if $\overline{T_1} = \overline{T_2}$. (A more general notion of equivalent theories in which the languages of the theory may differ is given in Chapter 13.) A theory T_1 is *included in a*

---

[1]In some nomenclatures, $\overline{T}$ is called a *theory* and T is called a *theory presentation*.

© The Author(s), under exclusive license to Springer Nature Switzerland AG 2023
W. M. Farmer, *Simple Type Theory*, Computer Science Foundations
and Applied Logic, https://doi.org/10.1007/978-3-031-21112-6_9

theory T_2, written $T_1 \preceq T_2$, if $\overline{T_1} \subseteq \overline{T_2}$. The *minimum theory* is the theory $T_{\min} = (L_{\min}, \emptyset)$. $T_{\min}$ has just one model, the minimum general model $M_{\min}$.

An *axiomatization* of T is any theory $T' = (L, \Gamma')$ that is equivalent to T. (A more general notion of the axiomatization of a theory in which the language of the axiomatization may differ from the language of the theory is given in Chapter 13.) T is *finitely axiomatizable* if there is an axiomatization $T' = (L, \Gamma')$ of T such that Γ' is finite. T is *recursively axiomatizable* if there is an axiomatization $T' = (L, \Gamma')$ of T such that Γ' is decidable.

Remark 9.1 (Theories as Specifications) *Recall that in Remark 5.11 we said that a language L can be viewed as a specification of the set $\mathcal{S}_L$ of structures defined by the general models of L. A theory $T = (L, \Gamma)$ can be viewed as a specification of the set $\mathcal{S}_T \subseteq \mathcal{S}_L$ of structures defined by the general models of T, i.e., the general models of L that satisfy the properties expressed by the axioms of T. T can also be viewed as a specification of the set $\mathcal{S}_T^s \subseteq \mathcal{S}_T$ of structures defined by the standard models of T, i.e., the standard models of L that satisfy the properties expressed by the axioms of T. We say that T is a specification in the general sense in the first case and a specification in the standard sense in the second case. The standard models of a theory are usually considered to be the intended models of the theory, so a theory is usually viewed foremost as a specification in the standard sense.*

A *theory definition module* is used to present a theory having finitely many constants and axioms. It has the following form:

Theory Definition X.Y

  Name: Name.

  Base types: $\mathbf{a}_1, \ldots, \mathbf{a}_m$.

  Constants: $\mathbf{c}_{\alpha_1}^1, \ldots, \mathbf{c}_{\alpha_n}^1$.

  Axioms:

  1. $\mathbf{A}_o^1$ (description of $\mathbf{A}_o^1$).

    $\vdots$

  p. $\mathbf{A}_o^p$ (description of $\mathbf{A}_o^p$).

Module kind	Page number
Theory definition	95
Theory extension	101
Development definition	132
Development extension	133
Theory translation definition	161
Theory translation extension	161
Development translation definition	180
Development translation extension	181
Definition transportation	184
Theorem transportation	184

Table 9.1: Kinds of Mathematical Knowledge Modules

The theory definition module is well-formed if

$$L = (\{\mathbf{a}_1, \ldots, \mathbf{a}_m\}, \{\mathbf{c}^1_{\alpha_1}, \ldots, \mathbf{c}^1_{\alpha_n}\})$$

is a language and

$$T = (L, \{\mathbf{A}^1_o, \ldots, \mathbf{A}^p_o\})$$

is a theory. Name is a name for the theory T. X is the number of the chapter in which the module resides, and Y is the number of the module in the list of items (i.e., theorems, examples, remarks, and modules) in the chapter.

The theory definition module is the first kind of *mathematical knowledge module* (*module* for short) of Alonzo that we will employ. Other kinds of modules are introduced later in this chapter and in Chapters 11 and 13; the different kinds of modules are listed in Table 9.1. Modules are printed in brown color. We will see in Chapter 13 how libraries of mathematical knowledge are constructed from modules.

We will give several examples of theories in this chapter starting with the following four:

Example 9.2 (Theory of Monoids) A *monoid* is a structure with an associative binary operation that has an identity element. It is one of the basic algebraic structures. $(\mathbb{N}, +, 0)$, $(\mathbb{N}, *, 1)$, and $(A^*, ++, [\,])$, where A is any set, are three of the many examples of monoids that are seen in mathematics and computing. As our first example of a theory, let us define the following *theory of monoids*:

Theory Definition 9.3 (Monoids)

Name: MON.

Base types: S.

Constants: $\cdot_{S \to S \to S}, e_S$.

Axioms:

1. $\forall x, y, z : S . x \cdot (y \cdot z) = (x \cdot y) \cdot z$ (associativity).

2. $\forall x : S . e \cdot x = x$ (left identity).

3. $\forall x : S . x \cdot e = x$ (right identity).

The name of the theory is MON (for "monoid"). Its language is

$$(\{S\}, \{\cdot_{S \to S \to S}, e_S\})$$

and its axioms are the three sentences given above. The theory definition module is thus well-formed. Notice that the binary operation of a monoid is represented as function in Curryed form and that the types of the constants $\cdot_{S \to S \to S}$ and e_S have been dropped in the axioms for the sake of brevity. Since the three axioms of MON express the three properties that define a monoid, the models of MON define all possible monoids and the theorems of MON are the facts that are true in all monoids. □

Example 9.4 (Theory of Weak Partial Orders) A *weak partial order* is a structure $(S, \leq)$ such that $\leq$ is a binary relation over S that is reflexive, antisymmetric, and transitive. It is the most basic kind of order structure. Our second example of a theory is the following *theory of weak partial orders*:

Theory Definition 9.5 (Weak Partial Orders)

Name: WPO.

Base types: S.

Constants: $\leq_{S \to S \to o}$.

Axioms:

1. $\forall x : S . x \leq x$ (reflexivity).

2. $\forall x, y : S . (x \leq y \wedge y \leq x) \Rightarrow x = y$ (antisymmetry).

3. $\forall\, x, y, z : S \,.\, (x \le y \wedge y \le z) \Rightarrow x \le z$ (transitivity).

Notice that the binary relation of a weak partial order is represented in WPO as a predicate in Curryed form. In the same way that the axioms of MON express the properties that define a monoid, the axioms of WPO express the properties that define a weak partial order. As a result, the models of WPO define all possible weak partial orders and the theorems of WPO are the facts that are true in all weak partial orders.   □

Example 9.6 (Peano Arithmetic) The natural numbers is one of the most basic of mathematical structures. In his 1889 booklet [99], Giuseppe Peano presented a characterization of the natural numbers based on 0 and the successor function S. We express his characterization in Alonzo as the following theory PA called *Peano Arithmetic*[2]:

Theory Definition 9.7 (Peano Arithmetic)

> **Name:** PA.
>
> **Base types:** N.
>
> **Constants:** 0_N, $S_{N \to N}$.
>
> **Axioms:**
>
> 1. $\mathsf{TOTAL}(S)$ (S is total).
>
> 2. $\forall\, x : N \,.\, 0 \ne S\,x$ (0 has no predecessor).
>
> 3. $\forall\, x, y : N \,.\, S\,x = S\,y \Rightarrow x = y.$ (S is injective).
>
> 4. $\forall\, p : N \to o \,.\, (p\,0 \wedge \forall\, x : N \,.\, (p\,x \Rightarrow p\,(S\,x))) \Rightarrow \forall\, x : N \,.\, p\,x$
> (induction principle).

The theory PA specifies the natural numbers as an inductive set of values defined by the constructors denoted by 0 and S. The expressions $0, S\,0, S\,(S\,0), \ldots$ denote the members of the set. Axiom 4 says that there is "no junk": each member of the set is denoted by an expression constructed from 0 and S. Axioms 2 and 3 say that there is "no confusion": each member of the set is denoted by exactly one expression constructed from 0 and S.

---

[2]There are two versions of Peano arithmetic. This version with the induction principle is the higher-order version. The version with the induction schema and axioms for addition and multiplication is the first-order version.

PA is a very important and interesting theory. For example, PA is undecidable, i.e., the decision problem of whether a given formula in the language of PA is valid in PA is undecidable. We will come back to PA several times in this chapter and in subsequent chapters.   □

The construction of PA gives us the following proposition:

Proposition 9.8 (The Natural Numbers is a Model of PA) *There is a standard model of* PA *that defines the structure* $(\mathbb{N}, 0, S)$ *where* $\mathbb{N}$ *is the set of natural numbers,* $0 \in \mathbb{N}$ *is the natural number* 0, *and* $S : \mathbb{N} \to \mathbb{N}$ *is the successor function on* $\mathbb{N}$.

Let M be a model of PA. A member of D_N^M is a *finite element of M* if it is a member of

$$\{V_\varphi^M(S^m\, 0) \mid m \in \mathbb{N} \text{ and } \varphi \in \mathsf{assign}(M)\},$$

where $S^m\, 0$ is the expression formed by applying $S_{N \to N}$ to 0_N $m \geq 0$ times. A member of D_N^M is an *infinite element of M* if it is not a finite element. The finite elements of M represent the standard natural numbers, while the infinite elements of M (if there are any) represent the "infinite natural numbers" in M.

Theorem 9.9 (Standard Models of Peano Arithmetic) *A standard model of* PA *has no infinite elements.*

Proof Let M be a standard model of PA. Suppose D_N^M contains infinite elements. Let $q : D_N^M \to \mathbb{B}$ be the predicate such that $q(x)$ holds iff x is a finite element. Since M is standard, $q \in D_{N \to o}^M$. Axiom 4 of PA (the induction principle) is true in M, and so

$$V_{\varphi[(p:N \to o) \mapsto q]}^M((p\, 0 \wedge \forall x : N\,.\,(p\, x \Rightarrow p\,(S\, x))) \Rightarrow \forall x : N\,.\,p\, x) = \mathrm{T}$$

for any $\varphi \in \mathsf{assign}(M)$ by part 1 of Lemma 6.4. This implies

$$V_{\varphi[(p:N \to o) \mapsto q]}^M(\forall x : N\,.\,p\, x) = \mathrm{T}$$

by the definition of q, and so $q(x)$ holds for all $x \in D_N^M$, which contradicts our hypothesis. Therefore, M must not have any infinite elements.   □

Example 9.10 (Robinson Arithmetic) Raphael Robinson discovered a first-order theory of natural number arithmetic called Q that is undecidable like Peano Arithmetic but much weaker than Peano Arithmetic since it does not contain the induction principle [107]. We express his theory in Alonzo as the following theory RA called *Robinson Arithmetic*:

Theory Definition 9.11 (Robinson Arithmetic)

Name: RA.

Base types: N.

Constants: 0_N, $S_{N \to N}$, $+_{N \to N \to N}$, $*_{N \to N \to N}$.

Axioms:

1. TOTAL(S)   (S is total).

2. TOTAL2($+$)   ($+$ is total).

3. TOTAL2($*$)   ($*$ is total).

4. $\forall x : N . 0 \neq S x$ (0 has no predecessor).

5. $\forall x, y : N . S x = S y \Rightarrow x = y$.   ($S$ is injective).

6. $\forall x : N . (x = 0 \vee (\exists y : N . S y = x))$

   (every element is 0 or a successor).

7. $\forall x : N . x + 0 = x$ (definition of addition).

8. $\forall x, y : N . x + S y = S (x + y)$ (definition of addition).

9. $\forall x : N . x * 0 = 0$ (definition of multiplication).

10. $\forall x, y : N . x * S y = (x * y) + x$ (definition of multiplication).

$\square$

We have seen that a theory can be used to specify (1) a collection of structures (i.e., the structures defined by its models), (2) a collection of sentences (i.e., the sentences in its closure). Accordingly, a theory can be a "maximal specification" in two ways. First, it can be a "categorical theory" that specifies a single structure (up to isomorphism) as discussed in Section 9.4. Second, it can be a "complete theory" that specifies a maximally satisfiable set of sentences as discussed in Section 9.5. We will see that there are significant limitations associated with both kinds of maximal theories.

9.2   Theory Extensions

Let $L_i = (\mathcal{B}_i, \mathcal{C}_i)$ be a language and $T_i = (L_i, \Gamma_i)$ be a theory for $i \in \{1, 2\}$. T_2 is an *extension* of T_1 (or T_1 is a *subtheory* of T_2), written $T_1 \leq T_2$, if $L_1 \leq L_2$ and $\Gamma_1 \subseteq \Gamma_2$. We will define several kinds of theory extensions in this chapter; the different kinds are listed in Table 9.2. Notice that $T_{\mathsf{min}} \leq T$ for every theory T.

Proposition 9.12 $T_1 \leq T_2$ *implies* $T_1 \preceq T_2$.

Proof $T_1 \leq T_2$ implies $\Gamma_1 \subseteq \Gamma_2$, which implies $T_1 \preceq T_2$ $\qquad\qquad\square$

Proposition 9.13 *Let* $T_1 \leq T_2$. T_2 *is satisfiable implies* T_1 *is satisfiable.*

Proof Assume $T_1 \leq T_2$ and M_2 is a model of T_2. Then the reduct M_1 of M_2 to L_1 is obviously a model of T_1. $\qquad\qquad\square$

A *theory extension module* is used to present a theory extension having finitely many new constants and axioms. It has the following form:

Theory Extension X.Y

  Name: Name.

  Extends Name-of-subtheory.

  New base types: $\mathbf{a}_1, \ldots, \mathbf{a}_m$.

  New constants: $\mathbf{c}_{\alpha_1}^1, \ldots, \mathbf{c}_{\alpha_n}^1$.

  New axioms:

  p. $\mathbf{A}_o^p$ (description of $\mathbf{A}_o^p$).

  $\vdots$

  q. $\mathbf{A}_o^q$ (description of $\mathbf{A}_o^q$).

The theory extension module is well-formed if Name-of-subtheory is the name of a theory $T_0 = ((\mathcal{B}, \mathcal{C}), \Gamma)$,

$$L = (\mathcal{B} \cup \{\mathbf{a}_1, \ldots, \mathbf{a}_m\}, \mathcal{C} \cup \{\mathbf{c}_{\alpha_1}^1, \ldots, \mathbf{c}_{\alpha_n}^1\})$$

is a language, and

$$T = (L, \Gamma \cup \{\mathbf{A}_o^p, \ldots, \mathbf{A}_o^q\})$$

is a theory. Thus $T_0 \leq T$. Name is a name for the theory T.

Theory extension kind	Notation
Extension	$T_1 \leq T_2$
Conservative extension	$T_1 \trianglelefteq T_2$
Model conservative extension	$T_1 \trianglelefteq_m T_2$
Simple definitional extension	$T_1 \trianglelefteq_{sd} T_2$
Definitional extension	$T_1 \trianglelefteq_d T_2$
Simple specificational extension	$T_1 \trianglelefteq_{ss} T_2$
Specificational extension	$T_1 \trianglelefteq_s T_2$

Table 9.2: Kinds of Theory Extensions

Example 9.14 (Theory of Groups) A *group* is a monoid with an inverse operation. In Example 9.2, we defined a theory MON of monoids. Let us define the following *theory of groups* by adding an inverse operation to MON:

Theory Extension 9.15 (Groups)

> **Name:** GRP.
>
> **Extends** MON.
>
> **New base types:** (none).
>
> **New constants:** $\cdot^{-1}{}_{S \to S}$.
>
> **New axioms:**
>
> 4. $\forall x : S \, . \, x^{-1} \cdot x = e$ (left inverse).
>
> 5. $\forall x : S \, . \, x \cdot x^{-1} = e$ (right inverse).

The name of the theory is GRP for "group". It includes the base types, constants, and axioms of MON as well as the new constant $\cdot^{-1}{}_{S \to S}$ and two new axioms (in which $\cdot^{-1}{}_{S \to S}\, x$ is written as simply x^{-1}). The theory extension module is thus well-formed, and so MON $\leq$ GRP.   □

Example 9.16 (Theory of Weak Total Orders) A *weak total order* is a weak partial order $(S, \leq)$ such that $\leq$ is total. In Example 9.4, we defined a theory WPO of weak partial orders. Let us define the following *theory of weak total orders* by adding a totality axiom to WPO:

Theory Extension 9.17 (Weak Total Orders)

　Name: WTO.

　Extends WPO.

　New base types: (none).

　New constants: (none).

　New axioms:

　　4. $\forall x, y : S . x \leq y \vee y \leq x$ (totality).

Thus WPO $\leq$ WTO. □

Example 9.18 (Dense Weak Total Orders without Endpoints)
A weak total order with more than one element is *dense* if, for every two
distinct members of the order, there is a member that is strictly between
them. A weak total order is *without endpoints* if it has neither a minimum
nor a maximum element. $(\mathbb{Q}, \leq)$ and $(\mathbb{R}, \leq)$ are examples of a dense weak
total order without endpoints. In Example 9.16, we defined a theory WTO
of weak total orders. Let us define the following *theory of dense weak total
orders without endpoints* as an extension of WTO:

Theory Extension 9.19 (Dense WTOs without Endpoints)

　Name: DWTOWE.

　Extends WTO.

　New base types: (none).

　New constants: (none).

　New axioms:

　　5. $\exists x, y : S . x \neq y$ (more than one element).

　　6. $\forall x, y : S . x < y \Rightarrow (\exists z : S . x < z < y)$ (density).

　　7. $\forall x : S . \exists y : S . y < x$ (no minimum).

　　8. $\forall x : S . \exists y : S . x < y$ (no maximum).

(Recall that $(\mathbf{A}_S < \mathbf{B}_S)$ stands for $(\leq_{S \to S \to o} \mathbf{A}_S \mathbf{B}_S \wedge \mathbf{A}_S \neq \mathbf{B}_S)$ as defined
in Table 6.2.) Thus WTO $\leq$ DWTOWE. □

9.3   Conservative Theory Extensions

Let $T_1 \leq T_2$. T_2 is a *conservative extension* of T_1, written $T_1 \trianglelefteq T_2$, if, for all formulas $\mathbf{A}_o$ of L_1, $T_2 \vDash \mathbf{A}_o$ implies $T_1 \vDash \mathbf{A}_o$. That is, T_2 preserves the set of the formulas that are valid in T_1. T_2 is a *model conservative extension* of T_1, written $T_1 \trianglelefteq_m T_2$, if every model of T_1 expands to a model of T_2 (i.e., if, for every model M_1 of T_1, there is an expansion M_2 of M_1 to L_2 such that $M_2 \vDash T_2$). That is, T_2 preserves the models of T_1.

Proposition 9.20 *If $T_1 = (L_1, \Gamma_1)$, $T_1 \vDash \mathbf{A}_o$, and $T_2 = (L_1, \Gamma_1 \cup \{\mathbf{A}_o\})$, then $T_1 \trianglelefteq T_2$.*

Proof Since $T_1 \vDash \mathbf{A}_o$, $\overline{T_1} = \overline{T_2}$, and so $T_2 \vDash \mathbf{B}_o$ implies $T_1 \vDash \mathbf{B}_o$ for all formulas $\mathbf{B}_o$ of L_1. Hence $T_1 \trianglelefteq T_2$.   □

Proposition 9.21 (Conservativity implies Equisatisfiability)
Let $T_1 \trianglelefteq T_2$. Then T_1 is satisfiable iff T_2 is satisfiable.

Proof

$$T_1 \text{ is satisfiable}$$
$$\text{implies}\quad T_1 \vDash F_o \text{ does not hold} \tag{1}$$
$$\text{implies}\quad T_2 \vDash F_o \text{ does not hold} \tag{2}$$
$$\text{implies}\quad T_2 \text{ is satisfiable} \tag{3}$$

(1) and (3) hold since a satisfiable theory has a model and F_o is not valid in any general model; and (2) is by $T_1 \trianglelefteq T_2$ since F_o is a formula of L_1. This proves the $\Rightarrow$ direction.

The $\Leftarrow$ direction is by Proposition 9.13.   □

Lemma 9.22 $T_1 \trianglelefteq_m T_2$ *implies* $T_1 \trianglelefteq T_2$.

Proof Assume $T_1 \trianglelefteq_m T_2$ and let $\mathbf{A}_o$ be a formula of L_1.

$$T_2 \vDash \mathbf{A}_o$$
$$\text{implies}\quad M_2 \vDash \mathbf{A}_o \text{ for all models } M_2 \text{ of } T_2 \tag{1}$$
$$\text{implies}\quad M_1 \vDash \mathbf{A}_o \text{ for all models } M_1 \text{ of } T_1 \tag{2}$$
$$\text{implies}\quad T_1 \vDash \mathbf{A}_o \tag{3}$$

(1) and (3) are by the definition of being valid in a theory; and (2) is by Lemma 5.14 and $T_1 \trianglelefteq_m T_2$. Therefore, $T_1 \trianglelefteq T_2$.   □

Later in this section, we will define an extension PA^+ of PA that we will prove in Theorem 9.35 is conservative, but not model conservative.

Remark 9.23 (Theory Conservativity) *Lemma 9.22 allows us to show, in many cases, that a theory extension is conservative by showing that it is model conservative — which is often much easier.*

T_2 is a *simple definitional extension* of T_1, written $T_1 \unlhd_{sd} T_2$, if $\mathbf{c}_\alpha \notin C_1$, $L_2 = (\mathcal{B}_1, C_1 \cup \{\mathbf{c}_\alpha\})$, $\mathbf{A}_\alpha$ is a closed expression of L_1, $T_1 \vDash \mathbf{A}_\alpha\!\downarrow$, and $\Gamma_2 = \Gamma_1 \cup \{\mathbf{c}_\alpha = \mathbf{A}_\alpha\}$. T_2 is a *definitional extension* of T_1, written $T_1 \unlhd_d T_2$, if there is a sequence of theories $U_1, \ldots, U_n$ with $n \geq 2$ such that $T_1 = U_1 \unlhd_{sd} \cdots \unlhd_{sd} U_n = T_2$.

Example 9.24 (Simple Definitional Extension) Let us define the following extension PA-with-1 of PA that defines a constant for 1:

Theory Extension 9.25 (Peano Arithmetic with 1)

    Name: PA-with-1.

    Extends PA.

    New base types: (none).

    New constants: 1_N.

    New axioms:

    5. $1 = S\,0$.

PA $\unlhd_{sd}$ PA-with-1 since axiom 1 of PA implies PA $\vDash (S\,0)\!\downarrow$.   □

T_2 is a *simple specificational extension* of T_1, written $T_1 \unlhd_{ss} T_2$, if $\mathbf{c}_\alpha \notin C_1$, $L_2 = (\mathcal{B}_1, C_1 \cup \{\mathbf{c}_\alpha\})$, $\exists \mathbf{x} : \alpha \,.\, \mathbf{B}_o$ is a sentence of L_1, $T_1 \vDash \exists \mathbf{x} : \alpha \,.\, \mathbf{B}_o$, and $\Gamma_2 = \Gamma_1 \cup \{\mathbf{B}_o[(\mathbf{x} : \alpha) \mapsto \mathbf{c}_\alpha]\}$. T_2 is a *specificational extension* of T_1, written $T_1 \unlhd_s T_2$, if there is a sequence of theories $U_1, \ldots, U_n$ with $n \geq 2$ such that $T_1 = U_1 \unlhd_{ss} \cdots \unlhd_{ss} U_n = T_2$.

Example 9.26 (Simple Specificational Extension) Let us define the following extension PA-with-a-successor of PA that defines a constant for an unspecified value that is a successor (i.e., not 0):

Theory Extension 9.27 (Peano Arithmetic with a Successor)

    Name: PA-with-a-successor.

    Extends PA.

    New base types: (none).

New constants: a-successor$_N$.

New axioms:

5. $0 \neq$ a-successor.

PA $\trianglelefteq_{ss}$ PA-with-a-successor since axioms 1–2 imply PA $\vDash \exists x : N . 0 \neq x$. $\square$

Lemma 9.28

1. $T_1 \trianglelefteq_d T_2$ implies that, for all formulas $\mathbf{A}_o$ of L_2, there is formula $\mathbf{B}_o$ of L_1 such that $T_2 \vDash \mathbf{A}_o \Leftrightarrow \mathbf{B}_o$.

2. $T_1 \trianglelefteq_d T_2$ implies $T_1 \trianglelefteq_s T_2$.

3. $T_1 \trianglelefteq_d T_2$ implies that every model of T_1 can be expanded to a unique model of T_2.

4. $T_1 \trianglelefteq_s T_2$ implies that every model of T_1 can be expanded to a model of T_2.

5. $T_1 \trianglelefteq_d T_2$ implies $T_1 \trianglelefteq_m T_2$.

6. $T_1 \trianglelefteq_s T_2$ implies $T_1 \trianglelefteq_m T_2$.

Proof

Part 1   Assume $T_1 \trianglelefteq_d T_2$ and let $\mathbf{A}_o$ be a formula of L_2. Then the formula $\mathbf{B}_o$ of L_1, obtained from $\mathbf{A}_o$ by repeatedly replacing the constants in $\mathcal{C}_2 \setminus \mathcal{C}_1$ by the expressions that define them until there are none to replace, is logically equivalent to $\mathbf{A}_o$ in T_2.

Part 2   Assume $T_1 \trianglelefteq_{sd} T_2$. Then, for some $\mathbf{c}_\alpha$ and $\mathbf{A}_\alpha$, $\mathbf{c}_\alpha \notin \mathcal{C}_1$, $L_2 = (\mathcal{B}_1, \mathcal{C}_1 \cup \{\mathbf{c}_\alpha\})$, $\mathbf{A}_\alpha$ is closed, $T_1 \vDash \mathbf{A}_\alpha \!\downarrow$, and $\Gamma_2 = \Gamma_1 \cup \{\mathbf{c}_\alpha = \mathbf{A}_\alpha\}$. Let $\mathbf{B}_o$ be $(\mathbf{x} : \alpha) = \mathbf{A}_\alpha$. Then $T_1 \vDash \exists \mathbf{x} : \alpha . \mathbf{B}_o$ and $\mathbf{B}_o[(\mathbf{x} : \alpha) \mapsto \mathbf{c}_\alpha]$ is $\mathbf{c}_\alpha = \mathbf{A}_\alpha$. Therefore, $T_1 \trianglelefteq_{ss} T_2$.

   Now assume $T_1 \trianglelefteq_d T_2$. $T_1 \trianglelefteq_s T_2$ follows immediately since $U_1 \trianglelefteq_{sd} U_2$ implies $U_1 \trianglelefteq_{ss} U_2$.

Part 3   Assume $T_1 \trianglelefteq_{sd} T_2$. Then, for some $\mathbf{c}_\alpha$ and $\mathbf{A}_\alpha$, $\mathbf{c}_\alpha \notin \mathcal{C}_1$, $L_2 = (\mathcal{B}_1, \mathcal{C}_1 \cup \{\mathbf{c}_\alpha\})$, $\mathbf{A}_\alpha$ is closed, $T_1 \vDash \mathbf{A}_\alpha \!\downarrow$, and $\Gamma_2 = \Gamma_1 \cup \{\mathbf{c}_\alpha = \mathbf{A}_\alpha\}$. Let $M_1 = (\mathcal{D}_1, I_1)$ be a model of T_1. Then $M_2 = (\mathcal{D}_1, I_2)$, where I_2 is the extension of I_1 such that $I_2(\mathbf{c}_\alpha) = V_\varphi^{M_1}(\mathbf{A}_\alpha)$, is clearly the unique expansion of M_1 to L_2 that is a model of T_2.

Now assume $T_1 \trianglelefteq_d T_2$. Then it follows immediately from the result above that every model of T_1 can be expanded to a unique model of T_2.

Part 4 Similar to Part 3.

Part 5 Follows immediately from Part 3.

Part 6 Follows immediately from Part 4. $\qquad\square$

Example 9.29 (Recursive Definitions) To see how a function can be recursively defined in a theory by creating an extension of the theory, let us consider how the addition function can be recursively defined in PA. There are three ways of creating the extension.

The first is to add $+_{N \to N \to N}$ to the constants of PA and the following two sentences to the axioms of PA:

1. $\forall x : N \,.\, x + 0 = x.$

2. $\forall x, y : N \,.\, x + S\,y = S\,(x + y).$

(These are the same axioms that are used in RA to define addition.) The resulting theory is a model conservative extension, but some effort is required to prove this.

The second is to add $+_{N \to N \to N}$ to the constants of PA and the following single sentence to the axioms of PA:

$$\forall x, y : N \,.\, x + 0 = x \land x + S\,y = S\,(x + y)$$

The resulting theory is a simple specificational extension, but to prove that requires showing

$$\exists f : N \to N \to N \,.\, \forall x, y : N \,.\, f\,x\,0 = x \land f\,x\,(S\,y) = S\,(f\,x\,y)$$

is valid in PA.

The third is to add $+_{N \to N \to N}$ to the constants of PA and the following single sentence to the axioms of PA:

$$+ = \mathrm{I} f : N \to N \to N \,.\, \forall x, y : N \,.\, f\,x\,0 = x \land f\,x\,(S\,y) = S\,(f\,x\,y).$$

The resulting theory is a simple definitional extension, but to prove that requires showing

$$\exists! f : N \to N \to N \,.\, \forall x, y : N \,.\, f\,x\,0 = x \land f\,x\,(S\,y) = S\,(f\,x\,y)$$

is valid in PA, which implies that the definite description on the right-hand side of the definition is defined in PA. $\qquad\square$

Example 9.30 (Peano Arithmetic with + and ∗)   Let us define the following extension PA′ of PA:

Theory Extension 9.31 (Peano Arithmetic with + and ∗)

Name: PA′.

Extends PA.

New base types: (none).

New constants: $P_{N \to N}$, $+_{N \to N \to N}$, $*_{N \to N \to N}$, $\leq_{N \to N \to o}$, $N_{\{N\}}$.

New axioms:

5. $P = \lambda x : N . I y : N . S y = x$ (definition of predecessor function).

6. $+ = I f : N \to N \to N . \forall x, y : N . f x 0 = x \wedge f x (S y) = S (f x y)$
                                    (definition of addition function).

7. $* = I f : N \to N \to N . \forall x, y : N . f x 0 = 0 \wedge f x (S y) = (f x y) + x$
                                    (definition of multiplication function).

8. $\leq = \lambda x : N . \lambda y : N . \exists z : N . x + z = y$
                                    (definition of weak total order).

9. $N_{\{N\}} = U_{\{N\}}$ (definition of quasitype for the type N).

We claim that PA $\unlhd_d$ PA′. In order to prove this we need to show that the right-hand side of each of the five axioms above is defined. The right-hand sides of 5, 8, and 9 are defined by part 6 of Lemma 5.4. The right-hand sides of 6 and 7 are defined since the properties of the definite descriptions in 6 and 7 are uniquely satisfied by the addition and multiplication functions on the natural numbers, respectively. To actually prove that the definite descriptions in 6 and 7 are defined requires showing the functions exist and are unique. Uniqueness is easily shown using axiom 4 of PA, the induction principle. However, existence is more difficult to show. It requires proving that, if a function on the natural numbers is defined by *primitive recursion* (as in 6 and 7), then the function must exist.

It is easy to see that the following sentences are valid in PA′:

1. $\forall x : N . x = 0 \mapsto (P x)\!\uparrow \mid (P x)\!\downarrow$.

2. $\forall x : N . P (S x) = x$.

3. TOTAL2(+).

4. TOTAL2($*$). □

Proposition 9.32 RA $\preceq$ PA' *(i.e., $\overline{\text{RA}} \subseteq \overline{\text{PA'}}$).*

Proof Clearly, the language of RA is a sublanguage of the language of PA'. Each of the axioms of RA is valid in PA', and so $\overline{\text{RA}} \subseteq \overline{\text{PA'}}$. □

Example 9.33 (Peano Arithmetic with an Infinite Element) Let us define the following extension PA$^+$ of PA having infinitely many axioms[3]:

Theory Extension 9.34 (Peano Arithmetic with ∞ Element)

    Name: PA$^+$.

    Extends PA.

    New base types: (none).

    New constants: C_N.

    New axioms:

    5. $C \neq 0$.

    6. $C \neq S\,0$.

    7. $C \neq S\,(S\,0)$.

    $\vdots$

That is, the set of axioms contains the members of $\{C \neq S^m\,0 \mid m \in \mathbb{N}\}$ where $S^m\,0$ is the expression formed by applying $S_{N \to N}$ to 0_N $m \geq 0$ times. The new axioms of PA$^+$ say that every model of PA$^+$ (if there are any) has an infinite element. □

Theorem 9.35 (Properties of PA$^+$)

    1. PA$^+$ is satisfiable.

    2. Every model of PA$^+$ has an infinite element.

    3. Every model of PA$^+$ is a nonstandard model.

    4. PA $\trianglelefteq$ PA$^+$.

---

[3]We are abusing our notation here since an theory extension module normally has just finitely many axioms.

5. PA $\not\leq_m$ PA$^+$.

Proof

Part 1 By Proposition 9.8, there is a standard model M of PA that defines the structure $(\mathbb{N}, 0, S)$. For every finite subset Σ of the axioms of PA$^+$, there is clearly an expansion of M that satisfies Σ. Therefore, PA$^+$ is satisfiable by the Compactness Theorem (Corollary 8.16).

Part 2 The axioms of PA$^+$ force every model of PA$^+$ to have an infinite element.

Part 3 Suppose M is a standard model of PA$^+$. Let M' be the reduct of M to the language of PA. Then M' is a standard model of PA, and so M' has no infinite elements by Theorem 9.9. However, by part 2, M has an infinite element, and so M' has an infinite element, which is a contradiction. Therefore, there must be no standard models of PA$^+$.

Part 4 Let $\mathbf{B}_o$ be a formula of the language of PA, Γ_{PA} be the set of axioms of PA, and $\Gamma_{\mathsf{PA}+}$ be the set of axioms of PA$^+$.

$$\Gamma_{\mathsf{PA}+} \vDash \mathbf{B}_o$$

implies $\Gamma_{\mathsf{PA}+} \cup \{\neg \mathbf{B}_o\}$ is unsatisfiable $\hfill (1)$

implies $\Sigma \cup \Gamma_{\mathsf{PA}} \cup \{\neg \mathbf{B}_o\}$ is unsatisfiable for some finite $\Sigma \subseteq \Gamma_{\mathsf{PA}+}$

$\hfill (2)$

implies $\Sigma \cup \Gamma_{\mathsf{PA}} \vDash \mathbf{B}_o$ for some finite $\Sigma \subseteq \Gamma_{\mathsf{PA}+}$ $\hfill (3)$

implies $\Gamma_{\mathsf{PA}} \vDash \mathbf{B}_o$ $\hfill (4)$

(1) and (3) are by Lemma 6.3; (2) is by the Compactness Theorem (Corollary 8.16); and (4) follows from Lemma 5.14 and the fact that every model of Γ_{PA} expands to a model of $\Sigma \cup \Gamma_{\mathsf{PA}}$ since Σ contains only finitely sentences of the form $C \neq S^m 0$. Hence PA$^+$ is a conservative extension of PA.

Part 5 There is a (standard) model M of PA in which the domain for N is $\mathbb{N}$ by Proposition 9.8, but there is no such model of PA$^+$ by part 2. Hence M does not expand to a model of PA$^+$, and so PA$^+$ is not a model conservative extension of PA. $\hfill \square$

Corollary 9.36 PA *has a nonstandard model that has an infinite element.*

Proof By parts 1–3 of Theorem 9.35, there is a nonstandard model M of PA$^+$ that has an infinite element. The reduct of M to the language of PA is then a nonstandard model of PA that has an infinite element. $\hfill \square$

Corollary 9.37 *There is a conservative extension that is not a model conservative extension.*

Proof Follows immediately from parts 4 and 5 of Theorem 9.35. □

Remark 9.38 (Models of PA) *We have seen that PA has both standard models and nonstandard models. As an axiom of PA, the induction principle is designed to guarantee that the models of PA do not have infinite elements. The standard models do not have infinite elements, but some of the nonstandard models do. How is it possible that a nonstandard model of PA has infinite elements?*

Let M be a model of PA. As in the proof of Theorem 9.9, let $q : D_N^M \to \mathbb{B}$ be the predicate such that $q(x)$ holds iff x is a finite element. If $q \in D_{N\to o}^M$, then M has no infinite elements as shown in the proof of Theorem 9.9. However, it is not necessary that $q \in D_{N\to o}^M$. In particular, there is no closed expression $\mathbf{Q}_{N\to o}$ in the language of PA such that $V_\varphi^M(\mathbf{Q}_{N\to o}) = q$ (for any $\varphi \in \mathsf{assign}(M)$). If M is a standard model, then $D_{N\to o}^M$ is full, $q \in D_{N\to o}$, and M has no infinite elements. If M is a nonstandard model that has infinite elements, then $q \notin D_{N\to o}^M$ and the induction principle is true in M since $D_{N\to o}^M$ is missing all the predicates like q that would prevent M from having infinite elements. Hence the fact that $D_{N\to o}^M$ does not contain q and similar predicates simultaneously enables (1) the induction principle to hold in M and (2) M to have infinite elements.

9.4 Categorical Theories

A theory is *categorical (in the general sense)* if it has a unique model (up to isomorphism). A categorical theory thus specifies a single structure if we consider isomorphic structures to be equal. A theory is *categorical in the standard sense* if it has a unique standard model (up to isomorphism). A categorical theory in the standard sense thus specifies a single structure if we ignore the structures defined by nonstandard models and if we consider isomorphic structures to be equal.

This is the Alonzo analog of Löwenheim-Skolem theorem for first-order logic:

Theorem 9.39 (Löwenheim-Skolem Theorem) *Let T be a theory. If T has an infinite general model, then T has a general model of size and power κ for every cardinal $\kappa \geq \|L\|$.*

Proof Let $T = (L, \Gamma)$ be a theory where $L = (\mathcal{B}, \mathcal{C})$. Assume T has an infinite general model $M = (\mathcal{D}, I)$ where $\mathcal{D} = \{D_\alpha \mid \alpha \in \mathcal{T}(L)\}$. Then $D_{\mathbf{a}}$ is infinite for some $\mathbf{a} \in \mathcal{B}$. Also assume $\kappa \geq \|L\|$ for some cardinal κ. Let $\mathcal{C}' = \{C_{\mathbf{a}}^\xi \mid \xi < \kappa\}$ be a set of distinct constants not in $\mathcal{C}$ and

$$\Gamma' = \{C_{\mathbf{a}}^\xi \neq C_{\mathbf{a}}^\zeta \mid \xi < \zeta < \kappa\}.$$

Define $T' = (L', \Gamma \cup \Gamma')$ where $L' = (\mathcal{B}, \mathcal{C} \cup \mathcal{C}')$. Every finite subset Σ of $\Gamma \cup \Gamma'$ involves at most finitely many of the constants in $\mathcal{C}'$. Hence M can be expanded to a model of every such Σ. By the Compactness Theorem (Theorem 8.16), T' is satisfiable, and so by the Consistency Theorem (Corollary 8.7) and Henkin's Theorem (Theorem 8.11), T has a model M' such that $|M'| \leq \|M'\| \leq \|L'\| = |\mathcal{C} \cup \mathcal{C}'| = \kappa$. However, the sentences in Γ' guarantee that the interpretations of the members of $\mathcal{C}'$ in M' are distinct. Hence $\kappa = |\mathcal{C}'| \leq |\bigcup_{\mathbf{b} \in \mathcal{B}} D_{\mathbf{b}}^{M'}| = |M'|$, and so $|M'| = \|M'\| = \kappa$. Therefore, the reduct of M' to L is a model of T with size and power κ. □

Corollary 9.40 *Every theory that has an infinite general model is noncategorical.*

Proof Two models of different size or power cannot be isomorphic. □

Corollary 9.41 *Let T be a theory. Then the following are equivalent:*

1. *T is categorical.*

2. *T has a unique finite general model (up to isomorphism) and no infinite general models.*

3. *T is categorical in the standard sense and has no infinite general models.*

Proof $(1 \Rightarrow 2)$ by Corollary 9.40. $(2 \Rightarrow 3)$ by Theorem 5.9. And $(3 \Rightarrow 1)$ by Theorem 5.9 again. □

Proposition 9.42

1. *Let $T_1 \trianglelefteq_{\mathbf{d}} T_2$. Then T_1 is categorical iff T_2 is categorical.*

2. *Let $T_1 \trianglelefteq_{\mathbf{d}} T_2$. Then T_1 is categorical in the standard sense iff T_2 is categorical in the standard sense.*

Proof

Part 1 Assume $T_1 \trianglelefteq_d T_2$. By part 3 of Lemma 9.28, $(\star)$ every model of T_1 expands to a unique model of T_2. Let T_1 be categorical. Each model of T_2 must reduce to the unique model of T_1. Then T_2 must be categorical by $(\star)$. Now let T_2 be categorical. Each model of T_1 must expand to the unique model of T_2 by $(\star)$. Then T_1 must be categorical. Therefore, T_1 is categorical iff T_2 is categorical.

Part 2 Similar to Part 1. □

Remark 9.43 (Specification of Infinite Structures) *A finite structure can be uniquely specified by a theory as a specification in the general sense, but an infinite structure cannot. However, there are infinite structures that can be uniquely specified by a theory as a specification in the standard sense. That is, a theory categorical in the standard sense uniquely specifies the structure defined by its unique standard model. So, if we consider theories as specifications in the standard sense, we can use theories to uniquely specify some infinite structures — something that cannot be done in first-order logic.*

Theorem 9.44 (Categoricity of PA) PA *is categorical in the standard sense.*

Proof Richard Dekekind proved in 1888 [35] that all structures satisfying Peano's axioms are isomorphic to the structure $(\mathbb{N}, 0, S)$ — which implies the theorem. However, the theorem follows easily from Theorem 9.9 since any two standard models of PA that have no infinite elements must be isomorphic. □

Corollary 9.45 PA′ *is categorical in the standard sense.*

Proof By the claim that PA $\trianglelefteq_d$ PA′ in Example 9.30 and part 2 of Proposition 9.42. □

Corollary 9.46 *There is a theory categorical in the standard sense that uniquely specifies* $(\mathbb{N}, 0, 1, +, *)$, *the structure of natural number arithmetic.*

Proof For the theory, take a definitional extension of PA that defines $(\mathbb{N}, 0, 1, +, *)$, and then use part 2 of Proposition 9.42. □

There are also theories categorical in the standard sense that uniquely specify $(\mathbb{Q}, 0, 1, +, *, -, \cdot^{-1})$, the structure of rational number arithmetic, and $(\mathbb{R}, 0, 1, +, *, -, \cdot^{-1})$, the structure of real number arithmetic. A theory that uniquely specifies $(\mathbb{R}, 0, 1, +, *, -, \cdot^{-1})$ is presented in Section 12.1.

Example 9.47 (Rational Numbers Order) In Example 9.18, we defined DWTOWE, a theory of dense weak total orders without endpoints. Let us now define an extension RAT of DWTOWE called *Rational Numbers Order*.

Theory Extension 9.48 (Rational Numbers Order)

> **Name:** RAT.
>
> **Extends** DWTOWE.
>
> **New base types:** (none).
>
> **New constants:** (none).
>
> **New axioms:**
>
> 9. $\forall s : \{S\} . (\mathbf{A}_o^1 \wedge \mathbf{A}_o^2 \wedge \mathbf{A}_o^3 \wedge \mathbf{A}_o^4 \wedge \mathbf{A}_o^5 \wedge \mathbf{A}_o^6 \wedge \mathbf{A}_o^7 \wedge \mathbf{A}_o^8) \Rightarrow$
> $(\exists f : S \to S . \text{BIJ-ON}(f, s, U_{\{S\}}) \wedge \forall x, y : s . x \leq y \Rightarrow f x \leq f y)$
>   (subdomain isomorphism property)

where:

$$\mathbf{A}_o^1 \text{ is } \forall x : s . x \leq x \qquad \text{(reflexivity).}$$

$$\mathbf{A}_o^2 \text{ is } \forall x, y : s . (x \leq y \wedge y \leq x) \Rightarrow x = y \qquad \text{(antisymmetry).}$$

$$\mathbf{A}_o^3 \text{ is } \forall x, y, z : s . (x \leq y \wedge y \leq z) \Rightarrow x \leq z \qquad \text{(transitivity).}$$

$$\mathbf{A}_o^4 \text{ is } \forall x, y : s . x \leq y \vee y \leq x \qquad \text{(totality).}$$

$$\mathbf{A}_o^5 \text{ is } \exists x, y : s . x \neq y \qquad \text{(more than one element).}$$

$$\mathbf{A}_o^6 \text{ is } \forall x, y : s . x < y \Rightarrow (\exists z : s . x < z < y) \qquad \text{(density).}$$

$$\mathbf{A}_o^7 \text{ is } \forall x : s . \exists y : s . y < x \qquad \text{(no minimum).}$$

$$\mathbf{A}_o^8 \text{ is } \forall x : s . \exists y : s . x < y \qquad \text{(no maximum).}$$

Let M be a standard model of RAT. Axiom 9 of RAT says that, if $D \subseteq D_S^M$ and $\leq_D$ is $I^M(\leq)$ restricted to D such that $(D, \leq_D)$ is a dense weak total order without endpoints, then $(D, \leq_D)$ and $(D_S^M, I^M(\leq))$ are order isomorphic. This implies that $(D_S^M, I^M(\leq))$ is order isomorphic to $(\mathbb{Q}, \leq)$ since $(\mathbb{Q}, \leq)$ can be embedded in every dense weak total order. □

Theorem 9.49 (Categoricity of RAT) RAT *is categorical in the standard sense.*

Proof As shown in Example 9.47, every standard model of RAT is isomorphic to the standard model of RAT that defines $(\mathbb{Q}, \leq)$. □

Corollary 9.50 RAT *is a specification in the standard sense of all countable dense weak total orders without endpoints.*

Proof Georg Cantor proved in 1895 that every two countable dense weak total orders without endpoints are order isomorphic [20]. Since RAT is categorical in the standard sense and it specifies in the standard sense all dense weak total orders without endpoints order isomorphic to $(\mathbb{Q}, \leq)$, which is countable, it specifies in the standard sense, without restriction, all countable dense weak total orders without endpoints. □

9.5 Complete Theories

A theory $T = (L, \Gamma)$ is *(semantically) complete* if either $T \vDash \mathbf{A}_o$ or $T \vDash \neg\mathbf{A}_o$ holds for all sentences $\mathbf{A}_o$ of L. By Theorem 8.12, T is *semantically complete* iff T is *syntactically complete in* $\mathfrak{A}$, i.e., either $T \vdash_\mathfrak{A} \mathbf{A}_o$ or $T \vdash_\mathfrak{A} \neg\mathbf{A}_o$ holds for all sentences $\mathbf{A}_o$ of L.

Example 9.51 (Theory of False) Let $T = (L, \{F_o\})$ where L is any language. T is clearly unsatisfiable. Then $T \vDash \mathbf{A}_o$ holds for all formulas $\mathbf{A}_o$ of L, and so T is complete. □

This example suggests the following proposition:

Proposition 9.52 *Every unsatisfiable theory is complete.*

Example 9.53 (Complete Theory of a General Model) Let M be a general model of a language L and Γ be the set of all sentences $\mathbf{A}_o$ of L such that $M \vDash \mathbf{A}_o$. Then $T = (L, \Gamma)$ is a complete theory since every sentence of L is either true or false in M. T is called the *complete theory of* M. □

This example proves:

Proposition 9.54 *The complete theory of any general model is complete.*

This proposition shows that there are many complete theories, one for each general model. However, a complete theory of an infinite general model is usually not recursively axiomatizable and thus not suitable as a theory to be used in practice. A practical theory is one that is both satisfiable and recursively axiomatizable. So what practical theories are complete?

Example 9.55 (Theory of a Singleton Domain) Let us define the following theory:

Theory Definition 9.56 (Singleton Domain)

    Name: SD.

    Base types: A.

    Constants: (none).

    Axioms:

    1. $\forall x, y : A . x = y$ (all values are equal).

In every model of SD, the domain of the base type A is a singleton set. Thus (a) all the models of SD are finite and (b) all the standard models of SD are isomorphic. (a) implies (c) all the models of SD are standard by Theorem 5.9. (b) and (c) then imply SD is categorical. Moreover, every sentence of L is either true or false in the unique model of SD (up to isomorphism), and so SD is complete.   □

This example suggests the following proposition:

Proposition 9.57 *Every categorical theory is complete.*

Proof A categorical theory T is equivalent to the complete theory of any of the isomorphic models of T, and so T is complete itself.   □

Proposition 9.58 *Let T be a complete theory that has a finite general model. Then T is categorical.*

Proof The proof is left to the reader as an exercise.   □

Proposition 9.59 *Let $T_1 \trianglelefteq_d T_2$. Then T_1 is complete iff T_2 is complete.*

Proof Let $T_1 \trianglelefteq_d T_2$. Then (a) for all formulas $\mathbf{A}_o$ of L_2, there is a formula of $\mathbf{B}_o$ of L_1 such that $T_2 \vDash \mathbf{A}_o \Leftrightarrow \mathbf{B}_o$ by part 1 of Lemma 9.28 and (b) $T_1 \trianglelefteq T_2$ by Lemma 9.22 and part 5 of Lemma 9.28.

T_1 is complete

    implies $T_1 \vDash \mathbf{B}_o$ or $T_1 \vDash \neg\mathbf{B}_o$ for all sentences $\mathbf{B}_o$ in L_1 (1)

    implies $T_2 \vDash \mathbf{B}_o$ or $T_2 \vDash \neg\mathbf{B}_o$ for all sentences $\mathbf{B}_o$ in L_1 (2)

    implies $T_2 \vDash \mathbf{A}_o$ or $T_2 \vDash \neg\mathbf{A}_o$ for all sentences $\mathbf{A}_o$ in L_2 (3)

    implies T_2 is complete   (4)

(1) and (4) are by the definition of a complete theory; (2) is by $T_1 \leq T_2$; and (3) is by (a). This proves the $\Rightarrow$ direction.

$$T_2 \text{ is complete}$$

$$\text{implies } T_2 \vDash \mathbf{A}_o \text{ or } T_2 \vDash \neg\mathbf{A}_o \text{ for all sentences } \mathbf{A}_o \text{ in } L_2 \tag{1}$$

$$\text{implies } T_2 \vDash \mathbf{B}_o \text{ or } T_2 \vDash \neg\mathbf{B}_o \text{ for all sentences } \mathbf{B}_o \text{ in } L_1 \tag{2}$$

$$\text{implies } T_1 \vDash \mathbf{B}_o \text{ or } T_1 \vDash \neg\mathbf{B}_o \text{ for all sentences } \mathbf{B}_o \text{ in } L_1 \tag{3}$$

$$\text{implies } T_1 \text{ is complete} \tag{4}$$

(1) and (4) are by the definition of a complete theory; (2) is by $L_1 \leq L_2$; and (3) is by (b). This proves the $\Leftarrow$ direction. $\qquad\square$

By Corollary 9.41, every categorical theory is a theory of a finite general model, and so by Proposition 9.57, there are complete, satisfiable, recursively axiomatizable theories like SD that have only finite general models. Are there complete, satisfiable, recursively axiomatizable theories that have infinite general models? The next example is a good candidate for such a theory if we assume the *generalized continuum hypothesis (GCH)*.

Example 9.60 (Theory of a Countably Infinite Domain)
Assume that GCH holds. Let $L = (\{A\}, \emptyset)$ and CID be the theory

$$(L, \{\mathsf{INF}(U_{\{A\}}), \mathsf{COUNT}(U_{\{A\}})\} \cup \{\mathbf{X}_o^{\alpha,\beta} \wedge \mathbf{Y}_o^{\alpha,\beta} \wedge \mathbf{Z}_o^{\alpha,\beta} \mid \alpha, \beta \in \mathcal{T}(L)\})$$

where $\mathbf{X}_o^{\alpha,\beta}$ is the sentence

$$\exists f : \alpha \to (\alpha \to \beta) \,.\, \mathsf{TOTAL}(f) \wedge \mathsf{INJ}(f),$$

which says the cardinality of the denotation of α is less than or equal to the cardinality of the denotation of $\alpha \to \beta$; $\mathbf{Y}_o^{\alpha,\beta}$ is the sentence

$$\neg(\exists f : \alpha \to (\alpha \to \beta) \,.\, \mathsf{BIJ}(f)),$$

which says the cardinality of the denotation of α is not equal to the cardinality of the denotation of $\alpha \to \beta$; and $\mathbf{Z}_o^{\alpha,\beta}$ is the sentence

$$\forall s : \{\alpha \to \beta\} \,.$$
$$\mathsf{INF}(U_{\{\alpha\}}) \Rightarrow$$
$$(\exists f : \alpha \to (\alpha \to \beta) \,.\, \mathsf{BIJ\text{-}ON}(f, U_{\{\alpha\}}, s) \vee$$
$$\exists g : (\alpha \to \beta) \to (\alpha \to \beta) \,.\, \mathsf{BIJ\text{-}ON}(g, s, U_{\{\alpha \to \beta\}})),$$

which says, if the denotation of α is infinite, then there is no subset of the denotation of $\alpha \to \beta$ that has cardinality strictly greater than the cardinality of the denotation of α and strictly less than the cardinality of the denotation of $\alpha \to \beta$. Together, $\mathbf{X}_o^{\alpha,\beta}$, $\mathbf{Y}_o^{\alpha,\beta}$, and $\mathbf{Z}_o^{\alpha,\beta}$ say the cardinality of the denotation of $\alpha \to \beta$ is the successor of the cardinality of the denotation of α. CID is clearly recursively axiomatizable. A standard model M of L where D_A^M is a countably infinite set is a model of CID by GCH, and so CID is satisfiable. Every standard model of CID is isomorphic to M, and so CID is categorical in the standard sense. Therefore, CID is a satisfiable, recursively axiomatizable theory that has infinite general models and is categorical in the standard sense (provided we assume GCH). We claim, without proof, that the axioms of CID guarantee that CID is complete.   □

Gödel's first incompleteness theorem tells us that any satisfiable, recursively axiomatizable theory that is "sufficiently strong" must be "essentially incomplete". Thus satisfiable, recursively axiomatizable theories with infinite general models can be complete only if their expressivity is severely limited like the theory CID.

A theory T is *sufficiently strong* if $\mathsf{RA} \preceq T$ (i.e., $\overline{\mathsf{RA}} \subseteq \overline{T}$). A theory T is *essentially incomplete* if T is incomplete and every satisfiable, recursively axiomatizable extension of T is incomplete.

Theorem 9.61 (Gödel's First Incompleteness Thm. for Alonzo)
Let T be a theory of Alonzo that is (1) satisfiable, (2) recursively axiomatizable, and (3) sufficiently strong. Then T is essentially incomplete.

Proof This theorem follows from Gödel's First Incompleteness Theorem [56].   □

Corollary 9.62 (Robinson Arithmetic Incompleteness) RA *is essentially incomplete.*

Proof RA is clearly satisfiable, recursively axiomatizable, and sufficiently strong, and so RA is essentially incomplete by Gödel's First Incompleteness Theorem for Alonzo.   □

Corollary 9.63 (Peano Arithmetic Incompleteness) PA *is essentially incomplete.*

Proof PA is satisfiable by Proposition 9.8 and clearly recursively axiomatizable. Let T be a satisfiable, recursively axiomatizable extension of PA, and let T' extend T in the same way that PA' extends PA. Then (a) $T \trianglelefteq_d T'$ and (b) PA' $\leq T'$. T' is satisfiable by part 3 of Lemma 9.28. T' is recursively axiomatizable by (a). T' is sufficiently strong by Propositions 9.12 and 9.32 and (b). Thus T' is incomplete by Gödel's First Incompleteness Theorem for Alonzo, and so T is incomplete by (a) and Proposition 9.59. Therefore, PA is incomplete since T can be PA, and PA is essentially incomplete since T can be any satisfiable, recursively axiomatizable extension of PA. □

We know now that PA is both categorical in the standard sense and incomplete. From these facts we can prove the following theorem stated in Chapter 8 but not proved there:

Proof of Theorem 8.2 *(There is no sound proof system for Alonzo that is complete in the standard sense.)* Suppose $\mathfrak{P}$ were a proof system for Alonzo that is sound and complete in the standard sense. Recall the theory PA $= (L, \Gamma)$ presented in Example 9.6. By Theorem 9.44, PA is categorical in the standard sense, so it has a standard model M that is unique up to isomorphism. By Corollary 9.63, PA is incomplete. Then there is a sentence $\mathbf{A}_o$ of L such that (a) $M \vDash \mathbf{A}_o$ and (b) not $\Gamma \vDash \mathbf{A}_o$. Since $\mathfrak{P}$ is sound, (b) implies (c) not $\Gamma \vdash_{\mathfrak{P}} \mathbf{A}_o$, and since $\mathfrak{P}$ is complete in the standard sense, (c) implies (d) not $\Gamma \vDash^s \mathbf{A}_o$. (d) implies (e) not $M \vDash \mathbf{A}_o$ since M is the only standard model of PA. (e) contradicts (a). Therefore, there is no such proof system $\mathfrak{P}$. □

There are several well-known first-order theories — such as the theories of Presburger arithmetic, real closed fields, and algebraically closed fields — that can be shown to be complete in first-order logic using the method of *quantifier elimination*. The analogs of these theories expressed in Alonzo are usually incomplete. For example, the theory of dense weak total orders without endpoints is complete in first-order logic but incomplete in Alonzo as shown by the following proposition:

Theorem 9.64 (DWTOWE **Incompleteness**) DWTOWE, *the Alonzo theory of dense weak total orders without endpoints, is incomplete.*

Proof Let $\mathbf{C}_o$ be axiom 9 of RAT from Example 9.47. $\mathbf{C}_o$ is satisfied by the standard models of DWTOWE that are countable but not by those that are uncountable. Thus neither DWTOWE $\vDash \mathbf{C}_o$ nor DWTOWE $\vDash \neg\mathbf{C}_o$ holds, and so DWTOWE is incomplete. □

9.6   Fundamental Form of a Mathematical Problem

Most mathematical problems can be expressed in Alonzo as statements of the form

$$T \vDash^s \mathbf{A}_o$$

where $T = (L, \Gamma)$ is a theory of Alonzo and $\mathbf{A}_o$ is a sentence of L. For example, the problem of whether or not there is an odd perfect number is expressed by the statement

$$\mathsf{PA}' \vDash^s \exists x : N . \, \mathsf{odd}\, x \wedge \mathsf{perfect}\, x,$$

which asserts that $\exists x : N . \, \mathsf{odd}\, x \wedge \mathsf{perfect}\, x$ is true in the unique standard model of PA', i.e., the model that defines natural number arithmetic. We call $T \vDash^s \mathbf{A}_o$ the *fundamental form of a mathematical problem* in Alonzo. The problem is *positively* solved by proving that this statement holds and is *negatively* solved by proving that this statement does not hold.

There are three principal approaches for solving a problem of the form $T \vDash^s \mathbf{A}_o$:

1. **Model-theoretic approach.**

   a. *Positive solution.* Prove that $M \vDash \mathbf{A}_o$ for every standard model M of T. Then $T \vDash^s \mathbf{A}_o$ holds by part 2 of Proposition 5.10. Constructing the proof may not be feasible if there are a large finite number or an infinite number of nonisomorphic standard models of T.

   b. *Negative solution.* Prove that $M \vDash \neg \mathbf{A}_o$ for some standard model M of T. Then M is a counterexample to the problem, and so $T \vDash^s \mathbf{A}_o$ obviously does not hold.

2. **Proof-theoretic approach.**

   a. *Positive solution.* Prove that $T \vdash_{\mathfrak{P}} \mathbf{A}_o$ for some sound proof system $\mathfrak{P}$ for Alonzo. Then $T \vDash \mathbf{A}_o$ holds since $\mathfrak{P}$ is sound, and this trivially implies $T \vDash^s \mathbf{A}_o$. Note that $T \vdash_{\mathfrak{P}} \mathbf{A}_o$ will not hold, even if $\mathfrak{P}$ is complete, if there is some model M (standard or nonstandard) such that $M \vDash \neg \mathbf{A}_o$.

   b. *Negative solution.* Prove that $T \vdash_{\mathfrak{P}} \neg \mathbf{A}_o$ for some sound proof system $\mathfrak{P}$ for Alonzo. Then $T \vDash \neg \mathbf{A}_o$ holds since $\mathfrak{P}$ is sound. This implies $M \vDash \neg \mathbf{A}_o$ for every general model M of T by part 1 of Proposition 5.10, and so obviously $T \vDash^s \mathbf{A}_o$ does not hold.

Thus we see that, in Alonzo (and other versions of simple type theory with the general models semantics), mathematical problems are about standard models, but some mathematical problems can be solved by considering all general models via a sound proof system.

3. **Decision procedure approach.** A *decision procedure* for a theory $T = (L, \Gamma)$ and a set $\Sigma \subseteq \Gamma$ of sentences is an algorithm that, given $\mathbf{B}_o \in \Sigma$ as input, returns $b \in \mathbb{B}$ as output such that $T \vDash^s \mathbf{B}_o$ holds iff $b = \text{T}$. Then, if $\mathbf{A}_o \in \Sigma$, $T \vDash^s \mathbf{A}_o$ is solved positively if the decision procedure applied to $\mathbf{A}_o$ returns T and negatively if it returns F. Decision procedures only exist for certain sets of sentences in certain theories. For example, if $T = (L, \Gamma)$ is any theory of Alonzo and $\Sigma \subseteq \Gamma$ is the set of propositional formulas in L, then there is a decision procedure for T and Σ since propositional logic is decidable.

Remark 9.65 (Fundamental Form of a Mathematical Problem)
The fundamental form of a mathematical problem, as shown above, can generally not be expressed in first-order logic since, as we stated in section 5.8, there is no precise way to distinguish between the standard and nonstandard models of a first-order theory without a higher-order perspective [79]. *So, for example, $T \vDash A$, where T is first-order Peano arithmetic and A is a sentence that says there is an odd perfect number, does not adequately express the problem of whether or not there is an odd perfect number in the standard model of natural number arithmetic because T has nonstandard models as well as the standard model.*

9.7 Model Theory

Model theory [66] is the study of the relationship between theories and their models. The model theory of first-order logic is highly developed. Elementary first-order model theory is covered in several introductory textbooks on first-order logic. There are also good survey articles on first-order model theory, such as [67, 74, 82], as well as good textbooks, such as [28, 65, 75, 83, 100, 123].

In contrast, the model theory of simple type theory with the general models semantics — which we will call *higher-order model theory* — is not well developed. The literature is sparse; there are a few papers on higher-order model theory including [3, 4, 14, 61, 63, 69, 94], but no survey articles or textbooks.

Model theorists have put all of their attention on first-order model theory and have almost entirely ignored higher-order model theory. This is bit surprising since the nonstandard models of a higher-order theory (e.g., higher-order Peano arithmetic with the *induction principle*) are useful for understanding the nonstandard models of the corresponding first-order theory (first-order Peano arithmetic with the *induction schema*). As a result of the almost exclusive focus on first-order model theory, higher-order model theory is largely an unexplored research area. Little is known about the structure of nonstandard models, particularly nonfrugal nonstandard models, and very few of the ideas of first-order model theory have been applied to simple type theory.

We have begun in this book the study of the model theory of Alonzo. We have examined different kinds of general models: finite, infinite, standard, nonstandard, and frugal. We have examined different kinds of theories: categorical in the general sense, categorical in the standard sense, and complete. And we have investigated the models of an important theory of Alonzo: PA (Peano Arithmetic). Further study will require venturing deeper into the uncharted territory of higher-order model theory.

9.8   Conclusions

A theory of Alonzo is a fundamental unit of mathematical knowledge. A theory T specifies a collection of intended structures that have similar components and satisfy similar properties. These are the structures defined by the standard models of T. A theory also specifies a collection of sentences that are valid in the structures. These are the theorems of T. The standard semantics is best suited for studying the structures, while the general semantics is best suited for studying the theorems. Most mathematical problems can be expressed in Alonzo in the fundamental form $T \vDash^s \mathbf{A}_o$. And much of the model theory of Alonzo remains unexplored.

Table 9.3 lists the theories presented in this section. Figure 9.1 shows pictorially how the different kinds of theories are related according to Corollary 9.41, Proposition 9.57, and Proposition 9.58.

The theories of Alonzo are the basis of a structure

$$(\mathbb{T}, T_{\mathsf{min}}, \equiv, \preceq, \leq, \trianglelefteq, \trianglelefteq_{\mathsf{m}}, \trianglelefteq_{\mathsf{d}}, \trianglelefteq_{\mathsf{s}}, \mathsf{sat}, \mathsf{cat\text{-}g}, \mathsf{cat\text{-}s}, \mathsf{complete})$$

where:

- $\mathbb{T}$ is the set of theories of Alonzo.

Theory	Name
Theory of monoids	MON
Theory of weak partial orders	WPO
Peano Arithmetic	PA
Robinson Arithmetic	RA
Theory of groups	GRP
Theory of weak total orders	WTO
Dense weak total orders without endpoints	DWTOWE
Peano Arithmetic with 1	PA-with-1
Peano Arithmetic with a successor	PA-with-a-successor
Peano Arithmetic with + and ∗	PA′
Peano Arithmetic with an infinite element	PA⁺
Rational numbers order	RAT
Theory of a singleton domain	SD
Theory of countably infinite domains	CID

Table 9.3: Theories Presented in Chapter 9

- $T_{\min} \in \mathbb{T}$ is the empty theory.

- $\equiv : (\mathbb{T} \times \mathbb{T}) \to \mathbb{B}$ such that $T_1 \equiv T_2$ holds iff T_1 and T_2 are equivalent theories..

- $\preceq : (\mathbb{T} \times \mathbb{T}) \to \mathbb{B}$ such that $T_1 \preceq T_2$ holds iff T_1 is included in T_2.

- $\leq : (\mathbb{T} \times \mathbb{T}) \to \mathbb{B}$ such that $T_1 \leq T_2$ holds iff T_2 is an extension of T_1.

- $\unlhd : (\mathbb{T} \times \mathbb{T}) \to \mathbb{B}$ such that $T_1 \unlhd T_2$ holds iff T_2 is a conservative extension of T_1.

- $\unlhd_{\mathsf{m}} : (\mathbb{T} \times \mathbb{T}) \to \mathbb{B}$ such that $T_1 \unlhd_{\mathsf{m}} T_2$ holds iff T_2 is a model conservative extension of T_1.

- $\unlhd_{\mathsf{d}} : (\mathbb{T} \times \mathbb{T}) \to \mathbb{B}$ such that $T_1 \unlhd_{\mathsf{d}} T_2$ holds iff T_2 is a definitional extension of T_1.

- $\unlhd_{\mathsf{s}} : (\mathbb{T} \times \mathbb{T}) \to \mathbb{B}$ such that $T_1 \unlhd_{\mathsf{s}} T_2$ holds iff T_2 is a specificational extension of T_1.

- $\mathsf{sat} : \mathbb{T} \to \mathbb{B}$ such that $\mathsf{sat}(T)$ holds iff T is satisfiable.

- $\mathsf{cat\text{-}g} : \mathbb{T} \to \mathbb{B}$ such that $\mathsf{cat\text{-}g}(t)$ holds iff T is categorical in the general sense.

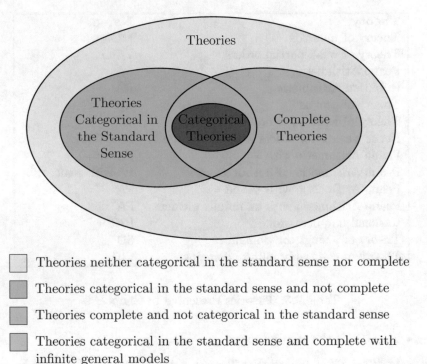

Theories neither categorical in the standard sense nor complete

Theories categorical in the standard sense and not complete

Theories complete and not categorical in the standard sense

Theories categorical in the standard sense and complete with
infinite general models

Figure 9.1: Kinds of Theories

- cat-s : $\mathbb{T} \to \mathbb{B}$ such that cat-s(T) holds iff T is categorical in the
  standard sense.

- complete : $\mathbb{T} \to \mathbb{B}$ such that complete(T) holds iff T is complete.

9.9   Exercises

1. Prove that two theories having the same language are equivalent iff
   they have exactly the same models.

2. Prove that, if T_1 and T_2 are theories with the same language and
   $T_1 \trianglelefteq T_2$, then $T_1 \equiv T_2$.

3. Prove that each of the following binary relations on theories is a weak
   partial order:

   a. $\leq$.

    b. $\unlhd$.

    c. $\unlhd_m$.

    d. Reflexive closure of $\unlhd_d$.

    e. Reflexive closure of $\unlhd_s$.

4. Let $\mathbb{T}$ be the set of theories of Alonzo and $\leq$ be the subtheory relation over $\mathbb{T}$. Prove that $(\mathbb{T}, \leq)$ is a meet-semilattice with a bottom element.

5. Prove that, if $L_1 = (\mathcal{B}_1, \mathcal{C}_1)$, $T_1 = (L_1, \Gamma_1)$, $L_2 = (\mathcal{B}_1 \cup \{a\}, \mathcal{C}_1)$, and $T_2 = (L_2, \Gamma_1)$, then $T_1 \unlhd_m T_2$.

6. Prove that, if $T_1 \leq T_2 \leq T_3$ and $T_1 \unlhd_m T_3$, then $T_1 \unlhd_m T_2$, but it is not necessary that $T_2 \unlhd_m T_3$.

7. Prove the claim $\mathsf{PA} \unlhd_d \mathsf{PA}'$ stated in Example 9.30.

8. Express in Alonzo a theory of well-founded relations.

9. Express in Alonzo a theory of stacks of abstract elements. Use two base types, E and S, to represent the domain of abstract elements and the domain of stacks of abstract elements, respectively. Model your theory on the Alonzo theory PA.

10. Express in Alonzo a theory of two abstract monoids. Use two base types of individuals, B_1 and B_2, to represent the domains of the two monoids. Define in the theory the notion of a homomorphism from the first monoid to the second and the notion of the kernel of a homomorphism.

11. Express in Alonzo a theory of categories.

12. Using the Compactness Theorem (Corollary 8.16), prove that a theory of Alonzo that has general models of arbitrarily large finite size has an infinite general model.

13. Express in Alonzo a theory that has general models of every finite size but no standard models of infinite size.

14. Let $T = (L, \Gamma)$ be a theory such that $\|L\| = \omega$ and Γ is the set of all sentences that are true in all standard models of L. Prove that T has a nonstandard model. (Exercise X8028 in [5].)

15. Let T be a theory that has an infinite general model. Prove that T has a nonstandard model.

16. Let T be a theory that has an infinite general model. Prove that T has infinitely many nonisomorphic nonstandard models.

17. Prove Proposition 9.58. Hint: Prove that T is categorical assuming that the language of T contains only finitely many constants, and then use the Compactness Theorem (Corollary 8.16).

Chapter 10

Sequences

This is a very short chapter about a very important topic: sequences. An infinite sequence of values from a set A can be represented as a total function $s : \mathbb{N} \to A$. Likewise, a finite sequence can be represented as a partial function $s : \mathbb{N} \to A$ such that, for some $n \in \mathbb{N}$, $s(m)$ is defined iff $m \leq n$. This chapter introduces compact notation for infinite and finite sequences represented in this manner in theories that include the natural numbers.

10.1 Systems of Natural Numbers

Let $T = (L, \Gamma)$ be a theory. A set

$$\mathcal{E}_{\mathsf{nat}} = \{\mathbf{C}^N_{\{\alpha\}}, \mathbf{C}^0_\alpha, \mathbf{C}^S_{\alpha \to \alpha}, \mathbf{C}^P_{\alpha \to \alpha}, \mathbf{C}^+_{\alpha \to \alpha \to \alpha}, \mathbf{C}^*_{\alpha \to \alpha \to \alpha}, \mathbf{C}^{\leq}_{\alpha \to \alpha \to o}\}$$

of closed expressions is a *system of natural numbers in* T if $\mathcal{E}_{\mathsf{nat}} \subseteq \mathcal{E}(L)$ and the following sentences are theorems of T:

1. $\mathbf{C}^0_\alpha \in \mathbf{C}^N_{\{\alpha\}}$.

2. $\mathbf{C}^S_{\alpha \to \alpha} \in \mathbf{C}^N_{\{\alpha\}} \to \mathbf{C}^N_{\{\alpha\}}$.

3. $\mathsf{TOTAL\text{-}ON}(\mathbf{C}^S_{\alpha \to \alpha}, \mathbf{C}^N_{\{\alpha\}}, \mathbf{C}^N_{\{\alpha\}})$.

4. $\forall x : \mathbf{C}^N_{\{\alpha\}} \cdot \mathbf{C}^0_\alpha \neq \mathbf{C}^S_{\alpha \to \alpha} x$.

5. $\forall x, y : \mathbf{C}^N_{\{\alpha\}} \cdot \mathbf{C}^S_{\alpha \to \alpha} x = \mathbf{C}^S_{\alpha \to \alpha} y \Rightarrow x = y$.

6. $\forall p : \mathbf{C}^N_{\{\alpha\}} \to o \,.$
   $(p\,\mathbf{C}^0_\alpha \wedge \forall x : \mathbf{C}^N_{\{\alpha\}} \cdot (p\,x \Rightarrow p\,(\mathbf{C}^S_{\alpha \to \alpha} x))) \Rightarrow \forall x : \mathbf{C}^N_{\{\alpha\}} \cdot p\,x.$

© The Author(s), under exclusive license to Springer Nature Switzerland AG 2023
W. M. Farmer, *Simple Type Theory*, Computer Science Foundations
and Applied Logic, https://doi.org/10.1007/978-3-031-21112-6_10

7. $\mathbf{C}^P_{\alpha\to\alpha} = \lambda\,x : \mathbf{C}^N_{\{\alpha\}}\,.\,\mathrm{I}\,y : \mathbf{C}^N_{\{\alpha\}}\,.\,\mathbf{C}^S_{\alpha\to\alpha}\,y = x.$

8. $\mathbf{C}^+_{\alpha\to\alpha\to\alpha} = \mathrm{I}\,f : \mathbf{C}^N_{\{\alpha\}} \to \mathbf{C}^N_{\{\alpha\}} \to \mathbf{C}^N_{\{\alpha\}}\,.\,\forall\,x,y : \mathbf{C}^N_{\{\alpha\}}\,.$
$$f\,x\,0 = x \wedge f\,x\,(\mathbf{C}^S_{\alpha\to\alpha}\,y) = \mathbf{C}^S_{\alpha\to\alpha}\,(f\,x\,y).$$

9. $\mathbf{C}^*_{\alpha\to\alpha\to\alpha} = \mathrm{I}\,f : \mathbf{C}^N_{\{\alpha\}} \to \mathbf{C}^N_{\{\alpha\}} \to \mathbf{C}^N_{\{\alpha\}}\,.\,\forall\,x,y : \mathbf{C}^N_{\{\alpha\}}\,.$
$$f\,x\,0 = 0 \wedge f\,x\,(\mathbf{C}^S_{\alpha\to\alpha}\,y) = \mathbf{C}^+_{\alpha\to\alpha\to\alpha}\,(f\,x\,y)\,x.$$

10. $\mathbf{C}^{\leq}_{\alpha\to\alpha\to o} = \lambda\,x : \mathbf{C}^N_{\{\alpha\}}\,.\,\lambda\,y : \mathbf{C}^N_{\{\alpha\}}\,.\,\exists\,z : \mathbf{C}^N_{\{\alpha\}}\,.\,\mathbf{C}^+_{\alpha\to\alpha\to\alpha}\,x\,z = y.$

Example 10.1 (System of Natural Numbers in PA′) Let PA′ be the definitional extension of PA defined in Example 9.30. Then

$$\{N_{\{N\}},\ 0_N,\ S_{N\to N},\ P_{N\to N},\ +_{N\to N\to N},\ *_{N\to N\to N},\ \leq_{N\to N\to o}\}$$

is a system of natural numbers in PA′. Thus any extension of PA′ contains a system of natural numbers. Moreover, if T contains a system of natural numbers, then a copy of PA′ is effectively embedded in T.   □

10.2   Notation for Sequences

In this section, let T_{nat} be a theory in which

$$\{\mathbf{C}^N_{\{\alpha\}}, \mathbf{C}^0_\alpha, \mathbf{C}^S_{\alpha\to\alpha}, \mathbf{C}^P_{\alpha\to\alpha}, \mathbf{C}^+_{\alpha\to\alpha\to\alpha}, \mathbf{C}^*_{\alpha\to\alpha\to\alpha}, \mathbf{C}^{\leq}_{\alpha\to\alpha\to o}\}$$

is a system of natural numbers.

In Table 10.1, we introduce notation for sequences in T_{nat}. $\langle\!\langle\beta\rangle\!\rangle$, $\langle\beta\rangle$, $[\beta]$, and $[\beta]^n$ are quasitypes within $\alpha \to \beta$. Let M be a model of T_{nat}. Then $\langle\!\langle\beta\rangle\!\rangle$ denotes the set $D \subseteq D^M_{\alpha\to\beta}$ of functions whose domains are subsets of the denotation of $\mathbf{C}^N_{\{\alpha\}}$. $\langle\beta\rangle$ denotes the set of functions in D that represent *infinite sequences* (or *streams*) over D^M_β. $[\beta]$ denotes the set of functions in D that represent *finite sequences* (or *lists*) over D^M_β. And $[\beta]^n$ denotes the set of functions in D that represent finite sequences (or lists) over D^M_β of length n.

Proposition 10.2 *The following sentences are valid in T_{nat} for every type β of T_{nat}:*

1. $\forall\,x : \beta,\,s : \langle\beta\rangle\,.\,(x :: s) \in \langle\beta\rangle.$

2. $\forall\,x : \beta,\,s : [\beta]\,.\,(x :: s) \in [\beta].$

sequences$_{\{\alpha\rightarrow\beta\}}$	stands for	$\mathbf{C}^N_{\{\alpha\}} \rightarrow \beta$.
$\langle\!\langle\beta\rangle\!\rangle$	stands for	sequences$_{\{\alpha\rightarrow\beta\}}$.
streams$_{\{\alpha\rightarrow\beta\}}$	stands for	$\{s : \langle\!\langle\beta\rangle\!\rangle \mid \mathsf{TOTAL}(s)\}$.
$\langle\beta\rangle$	stands for	streams$_{\{\alpha\rightarrow\beta\}}$.
lists$_{\{\alpha\rightarrow\beta\}}$	stands for	$\{s : \langle\!\langle\beta\rangle\!\rangle \mid \exists n : \mathbf{C}^N_{\{\alpha\}} . \forall m : \mathbf{C}^N_{\{\alpha\}} .$ $(s\,m){\downarrow} \Leftrightarrow \mathbf{C}^{\leq}_{\alpha\rightarrow\alpha\rightarrow o}\, m\, n\}$.
$[\beta]$	stands for	lists$_{\{\alpha\rightarrow\beta\}}$.
cons$_{\beta\rightarrow(\alpha\rightarrow\beta)\rightarrow(\alpha\rightarrow\beta)}$	stands for	$\lambda x : \beta . \lambda s : \langle\!\langle\beta\rangle\!\rangle . \lambda n : \mathbf{C}^N_{\{\alpha\}} .$ $n = \mathbf{C}^0_\alpha \mapsto x \mid s\,(\mathbf{C}^P_{\alpha\rightarrow\alpha}\, n)$.
$(\mathbf{A}_\beta :: \mathbf{B}_{\alpha\rightarrow\beta})$	stands for	cons$_{\beta\rightarrow(\alpha\rightarrow\beta)\rightarrow(\alpha\rightarrow\beta)}\, \mathbf{A}_\beta\, \mathbf{B}_{\alpha\rightarrow\beta}$.
hd$_{(\alpha\rightarrow\beta)\rightarrow\beta}$	stands for	$\lambda s : \langle\!\langle\beta\rangle\!\rangle . \mathrm{I}\, x : \beta . \exists t : \langle\!\langle\beta\rangle\!\rangle .$ $s = (x :: t)$.
tl$_{(\alpha\rightarrow\beta)\rightarrow(\alpha\rightarrow\beta)}$	stands for	$\lambda s : \langle\!\langle\beta\rangle\!\rangle . \mathrm{I}\, t : \langle\!\langle\beta\rangle\!\rangle . \exists x : \beta .$ $s = (x :: t)$.
nil$_{\alpha\rightarrow\beta}$	stands for	$\Delta_{\alpha\rightarrow\beta}$.
$[\,]_{\alpha\rightarrow\beta}$	stands for	nil$_{\alpha\rightarrow\beta}$.
$[\mathbf{A}_\beta]$	stands for	$(\mathbf{A}_\beta :: [\,]_{\alpha\rightarrow\beta})$.
$[\mathbf{A}^1_\beta, \ldots, \mathbf{A}^n_\beta]$	stands for	$(\mathbf{A}^1_\beta :: [\mathbf{A}^2_\beta, \ldots, \mathbf{A}^n_\beta])$ where $n \geq 2$.
len$_{(\alpha\rightarrow\beta)\rightarrow\alpha}$	stands for	$\mathrm{I}\, f : [\beta] \rightarrow \mathbf{C}^N_{\{\alpha\}} . f\,[\,]_{\alpha\rightarrow\beta} = \mathbf{C}^0_\alpha \wedge$ $\forall x : \beta,\, s : [\beta] . f\,(x :: s) =$ $\mathbf{C}^+_{\alpha\rightarrow\alpha\rightarrow\alpha}\,(f\, s)\,(\mathbf{C}^S_{\alpha\rightarrow\alpha}\, \mathbf{C}^0_\alpha)$.
$\lvert\mathbf{A}_{\alpha\rightarrow\beta}\rvert$	stands for	len$_{(\alpha\rightarrow\beta)\rightarrow\alpha}\, \mathbf{A}_{\alpha\rightarrow\beta}$.
nlists$_{\alpha\rightarrow\{\alpha\rightarrow\beta\}}$	stands for	$\lambda n : \mathbf{C}^N_{\{\alpha\}} . \{s : [\beta] \mid \lvert s \rvert = n\}$.
$[\beta]^{\mathbf{N}_\alpha}$	stands for	nlists$_{\alpha\rightarrow\{\alpha\rightarrow\beta\}}\, \mathbf{N}_\alpha$.

Table 10.1: Notational Definitions for Sequences

3. $\forall s : \langle \beta \rangle \, . \, (\mathsf{tl}_{(\alpha \to \beta) \to (\alpha \to \beta)} \, s) \in \langle \beta \rangle.$

4. $\forall s : [\beta] \, . \, s \neq [\,]_{\alpha \to \beta} \Rightarrow (\mathsf{tl}_{(\alpha \to \beta) \to (\alpha \to \beta)} \, s) \in [\beta].$

5. $(\mathsf{hd}_{(\alpha \to \beta) \to \beta} \, [\,]_{\alpha \to \beta}) \mathord{\uparrow}.$

6. $(\mathsf{tl}_{(\alpha \to \beta) \to (\alpha \to \beta)} \, [\,]_{\alpha \to \beta}) \mathord{\uparrow}.$

7. $\forall s : \langle \beta \rangle \, . \, (\mathsf{len}_{(\alpha \to \beta) \to \alpha} \, s) \mathord{\uparrow}.$

8. $\forall s : [\beta] \, . \, (\mathsf{len}_{(\alpha \to \beta) \to \alpha} \, s) \mathord{\downarrow}.$

Proof The proof is left to the reader as an exercise. □

Example 10.3 (Dependent Quasitypes of Lists) Let β be any type of T_{nat} and $n \geq 0$. If $[\beta]$ denotes D_β^*, the set of lists over D_β, then $[\beta]^n$ denotes D_β^n, the set of lists over D_β of length n. Thus, for all types β of T_{nat} and all $n \geq 0$, $[\beta]^n$ is a dependent quasitype of lists of length n that is a subquasitype of $[\beta]$ and $\langle\!\langle \beta \rangle\!\rangle$ and a subtype of $\alpha \to \beta$. □

10.3   Conclusions

A theory in which there is a system of natural numbers effectively contains a copy of the theory PA$'$, which is a definitional extension of Peano arithmetic. Notation for infinite and finite sequences is available for any theory that has a system of natural numbers.

10.4   Exercises

1. Prove Proposition 10.2.

2. Express in Alonzo a theory of labeled finitely branching trees as an extension of PA$'$ where the labels are members of the type N. Model your theory on the Alonzo theory PA.

Chapter 11

Developments

This chapter introduces the notion of a development of a theory.

11.1 Theory Developments

A key activity of mathematical practice is the creation of narratives based on series of definitions, theorems, and proofs. In Alonzo, we represent a narrative of this kind as a sequential "development" of a theory in which definitions are made and theorems are stated and proved. We will make the notion of a "theory development" precise in just a moment, but first we have to introduce some definitions.

A *definition package* is a tuple $P = (n, \mathbf{c}_\alpha, \mathbf{A}_\alpha, \pi)$ where n is the name of the package, $\mathbf{c}_\alpha$ is a constant, $\mathbf{A}_\alpha$ is a closed expression, and π is a proof (either a traditional proof or a formal proof). P is *valid in a theory* $T = ((\mathcal{B}, \mathcal{C}), \Gamma)$ if

$$T[P] = ((\mathcal{B}, \mathcal{C} \cup \{\mathbf{c}_\alpha\}), \Gamma \cup \{\mathbf{c}_\alpha = \mathbf{A}_\alpha\})$$

is a simple definitional extension of T and π is a proof of $\mathbf{A}_\alpha{\downarrow}$ from Γ.

A *theorem package* is a tuple $P = (n, \mathbf{A}_o, \pi)$ where n is the name of the package, $\mathbf{A}_o$ is a sentence, and π is a proof. P is *valid in a theory* $T = (L, \Gamma)$ if π is a proof of $\mathbf{A}_o$ from Γ.

A *theory development* (or *development* for short) of Alonzo is a pair $D = (T, \Xi)$ such that:

1. T is a theory called the *bottom theory* of D.

2. Ξ is a (possibly empty) list $[P_1, \ldots, P_n]$ of definition and theorem packages where $n \geq 0$.

3. For each i with $1 \le i \le n$, $\mathbf{B}_o^i$ is $\mathbf{c}_\alpha = \mathbf{A}_\alpha$ if $P_i = (n, \mathbf{c}_\alpha, \mathbf{A}_\alpha, \pi)$ and is $\mathbf{A}_o$ if $P_i = (n, \mathbf{A}_o, \pi)$. We say in the former case that $\mathbf{B}_o^i$ is a *definition of D* and in the latter case that $\mathbf{B}_o^i$ is a *theorem of D*.

4. There is a list $[T_0, T_1, \ldots, T_n]$ of theories such that:

    a. $T_0 = T$.

    b. For all i with $0 \le i \le n-1$, P_{i+1} is valid in T_i and $T_{i+1} = T[P_{i+1}]$ if P_{i+1} is a definition package and $T_{i+1} = T_i$ if P_{i+1} is a theorem package.

5. T_n is called the *top theory* of D.

We say that D is a *development of T*. D is a *trivial* if $\Xi = [\,]$ and so $T_n = T$. We identify a theory T with its trivial development $(T, [\,])$.

Proposition 11.1 *Let $D = (T, \Xi)$ be a nontrivial development that defines the list $[T_0, T_1, \ldots, T_n]$ of theories. Then $T_i \trianglelefteq_{\mathsf{sd}} T_{i+1}$ or $T_i = T_{i+1}$ for all i with $0 \le i \le n-1$, and hence the top theory of D is a definitional extension of the bottom theory of D.*

Proof Follows immediately from the definition of D.   □

Thus we see that a nontrivial development of a theory T is an enrichment of T with new defined constants and new proved theorems. A trivial development is an unenriched theory.

A *development definition module* is used to present a development. It has the following form:

Development Definition X.Y

    Name: Name.

    Bottom theory: Name-of-bottom-theory.

    Definitions and theorems:

        P_1 (description of P_1).

        ⋮

        P_n (description of P_n).

The development definition module is well-formed if Name-of-bottom-theory is the name of a theory T; $P_1, \ldots, P_n$ are definition or theorem packages with $n \geq 0$; and $D = (T, [P_1, \ldots, P_n])$ is a development. Name is a name for the development D.

Let $D_i = (T, \Xi_i)$ be a development of T for $i \in \{1, 2\}$. D_2 is an *extension* of D_1 (or D_1 is a *subdevelopment* of D_2), written $D_1 \leq D_2$, if there is a list Ξ of definition and theorem packages such that $\Xi_2 = \Xi_1 \mathbin{++} \Xi$.

A *development extension module* is used to present a development extension. It has the following form:

Development Extension X.Y

　Name: Name.

　Extends　Name-of-subdevelopment.

　New definitions and theorems:

　　P_1 (description of P_1).

　　$\vdots$

　　P_n (description of P_n).

The development extension module is well-formed if Name-of-subdevelopment is the name of a development $D_1 = (T, \Xi_1)$; $P_1, \ldots, P_n$ are definition or theorem packages with $n \geq 0$; and $D_2 = (T, \Xi_1 \mathbin{++} [P_1, \ldots, P_n])$ is a development. Thus $D_1 \leq D_2$. Name is a name for the development D_2.

In this and subsequent chapters, we will present several examples of developments and development extensions using development definition and development extension modules, respectively. For the sake of readability, we will not include the proofs of the definition and theorem packages in these examples since the proofs are either straightforward or easily found in the literature and, for that reason, are left to the reader to fill in. In the modules, a definition package $(\mathsf{Def}n, \mathbf{c}_\alpha, \mathbf{A}_\alpha, \pi)$ is presented as

　　$\mathsf{Def}n$: $\mathbf{c}_\alpha = \mathbf{A}_\alpha$ (description of $\mathsf{Def}n$).

with the proof π omitted, and a theorem package $(\mathsf{Thm}n, \mathbf{A}_o, \pi)$ is presented as

　　$\mathsf{Thm}n$: $\mathbf{A}_o$ (description of $\mathsf{Thm}n$).

again with the proof π omitted.

11.2   Development of Natural Number Arithmetic

The purpose of this section is to produce, as an example, an Alonzo theory of natural number arithmetic named NAT by developing the theory PA. We will define a development of PA as a series of extensions of an initial development and then NAT will be the top theory of the development. NAT will be a relatively modest theory of natural number arithmetic; it will include only a small portion of the basic concepts and facts of natural number arithmetic.

11.2.1   Basic Definitions and Theorems

We will begin the development of PA by adding some basic definitions and theorems to PA. The definitions define constants that denote the number one, the predecessor function, and the addition and multiplication functions in Curryed form. The theorems present facts about the definedness of the new constants.

Development Definition 11.2 (PA Development 1)

  Name: PA-dev-1.

  Bottom theory: PA.

  Definitions and theorems:

  Def1: $1_N = S\,0$ (number one).

  Def2: $P_{N \to N} = \lambda x : N \,.\, \mathrm{I}\,y : N \,.\, S\,y = x$ (predecessor function).

  Def3: $+_{N \to N \to N} =$
   $\mathrm{I}\,f : N \to N \to N \,.\, \forall x, y : N \,.\, f\,x\,0 = x \wedge f\,x\,(S\,y) = S\,(f\,x\,y)$
                                                                              (addition function).

  Def4: $*_{N \to N \to N} =$
   $\mathrm{I}\,f : N \to N \to N \,.\, \forall x, y : N \,.\, f\,x\,0 = 0 \wedge f\,x\,(S\,y) = (f\,x\,y) + x$
                                                                         (multiplication function).

  Thm1: $\forall x : N \,.\, x = 0 \mapsto (P\,x)\!\uparrow \mid (P\,x)\!\downarrow$ (P is almost total).

  Thm2: $\forall x : N \,.\, P\,(S\,x) = x$ (P is a left inverse of S).

  Thm3: $\mathrm{TOTAL2}(+)$.   ($+$ is total).

  Thm4: $\mathrm{TOTAL2}(*)$.   ($*$ is total).

11.2.2 Commutative Semiring

Example 3.8 states that the structure $(\mathbb{N}, 0, 1, +, *)$ of natural number arithmetic is a commutative semiring. We will now extend the development of PA to include as theorems the axioms of a commutative semiring.

Development Extension 11.3 (PA Development 2)

Name: PA-dev-2.

Extends PA-dev-1.

New definitions and theorems:

Thm5: $\forall x, y, z : N \,.\, (x + y) + z = x + (y + z)$ (associativity of $+$).

Thm6: $\forall x : N \,.\, 0 + x = x = x + 0$ (0 is a $+$-identity).

Thm7: $\forall x, y : N \,.\, x + y = y + x$ (commutativity of $+$).

Thm8: $\forall x, y, z : N \,.\, (x * y) * z = x * (y * z)$ (associativity of $*$).

Thm9: $\forall x : N \,.\, 1 * x = x = x * 1$ (1 is a $*$-identity).

Thm10: $\forall x, y : N \,.\, x * y = y * x$ (commutativity of $*$).

Thm11: $\forall x, y, z : N \,.\, x * (y + z) = (x * y) + (x * z)$
                                     (left distributivity).

Thm12: $\forall x, y, z : N \,.\, (x + y) * z = (x * z) + (y * z)$
                                     (right distributivity).

Thm13: $\forall x : N \,.\, 0 * x = 0 = x * 0$ (0 is an annihilator).

Def1, Def3–Def4, and Thm3–Thm13 guarantee that the reduct of any model of the top theory of PA-dev-2 to the language

$$(N, 0_N, 1_N, +_{N \to N \to N}, *_{N \to N \to N})$$

is a commutative semiring.

11.2.3   Weak Total Order

Example 3.7 states that $(\mathbb{N}, \leq)$ is a weak total order. We will continue the development of PA with the definition of a constant that denotes $\leq$ in Curryed form and theorems that state the axioms of a weak total order.

Development Extension 11.4 (PA Development 3)

Name: PA-dev-3.

Extends PA-dev-2.

New definitions and theorems:

Def5: $\leq_{N \to N \to o} = \lambda x : N \,.\, \lambda y : N \,.\, \exists z : N \,.\, x + z = y$ (weak order).

Thm14: $\forall x : N \,.\, x \leq x$ (reflexivity).

Thm15: $\forall x, y : N \,.\, (x \leq y \wedge y \leq x) \Rightarrow x = y$ (antisymmetry).

Thm16: $\forall x, y, z : N \,.\, (x \leq y \wedge y \leq z) \Rightarrow x \leq z$ (transitivity).

Thm17: $\forall x, y : N \,.\, x \leq y \vee y \leq x$ (totality).

Def5 and Thm14–Thm17 guarantee that the reduct of any model of the top theory of PA-dev-3 to the language $(N, \leq_{N \to N \to o})$ is a weak total order.

11.2.4   Divides Lattice

Example 3.9 states that $(\mathbb{N}, |, \mathsf{lcm}, \mathsf{gcd}, 1, 0)$ is a complete bounded lattice. We will complete the development of PA with definitions of constants that denote the $|$, lcm, and gcd functions in Curryed form as well as a several other functions and theorems that state the axioms of a complete bounded lattice.

Development Extension 11.5 (PA Development 4)

Name: PA-dev-4.

Extends PA-dev-3.

New definitions and theorems:

Def6: $|_{N \to N \to o} = \lambda x : N \,.\, \lambda y : N \,.\, \exists z : N \,.\, x * z = y$ (divides).

Thm18: $\forall x : N \,.\, x \mid x$ (reflexivity).

Thm19: $\forall x, y : N . (x \mid y \wedge y \mid x) \Rightarrow x = y$ (antisymmetry).

Thm20: $\forall x, y, z : N . (x \mid y \wedge y \mid z) \Rightarrow x \mid z$ (transitivity).

Thm21: $\forall x : N . 1 \mid x$ (1 is bottom).

Thm22: $\forall x : N . x \mid 0$ (0 is top).

Def7: $\mathsf{cm}_{N \to N \to N \to o} = \lambda x : N . \lambda y : N . \lambda z : N . y \mid x \wedge z \mid x$
         (common multiple).

Def8: $\mathsf{cd}_{N \to N \to N \to o} = \lambda x : N . \lambda y : N . \lambda z : N . x \mid y \wedge x \mid z$
         (common divisor).

Def9: $\mathsf{lcm}_{N \to N \to N} =$
   $\lambda x : N . \lambda y : N . I z : N . \mathsf{cm}\, z\, x\, y \wedge (\forall z' : N . \mathsf{cm}\, z'\, x\, y \Rightarrow z \mid z')$
         (least common multiple).

Def10: $\mathsf{gcd}_{N \to N \to N} =$
   $\lambda x : N . \lambda y : N . I z : N . \mathsf{cd}\, z\, x\, y \wedge (\forall z' : N . \mathsf{cd}\, z'\, x\, y \Rightarrow z' \mid z)$
         (greatest common divisor).

Thm23: TOTAL2(lcm)   (least common multiples always exist).

Thm24: TOTAL2(gcd)   (greatest common divisors always exist).

Def11: $\mathsf{set\text{-}cm}_{N \to \{N\} \to o} = \lambda x : N . \lambda s : \{N\} . \forall y : s . y \mid x$
         (common multiple).

Def12: $\mathsf{set\text{-}cd}_{N \to \{N\} \to o} = \lambda x : N . \lambda s : \{N\} . \forall y : s . x \mid y$
         (common divisor).

Def13: $\mathsf{set\text{-}lcm}_{\{N\} \to N} = .$
   $\lambda s : \{N\} . I z : N . \mathsf{set\text{-}cm}\, z\, s \wedge (\forall z' : N . \mathsf{set\text{-}cm}\, z'\, s \Rightarrow z \mid z')$
         (set least common multiple).

Def14: $\mathsf{set\text{-}gcd}_{\{N\} \to N} =$
   $\lambda s : \{N\} . I z : N . \mathsf{set\text{-}cd}\, z\, s \wedge (\forall z' : N . \mathsf{set\text{-}cd}\, z'\, s \Rightarrow z' \mid z)$
         (set greatest common divisor).

Thm25: TOTAL(set-lcm)
         (least common multiples of sets always exist).

Thm26: TOTAL(set-gcd)
         (greatest common divisors of sets always exist).

Define NAT to be the top theory of PA-dev-4. Def6 and Thm18–Thm22 guarantee that the reduct of any model of NAT to the language

$$(N, |_{N \to N \to o}, 1_N, 0_N)$$

is a weak partial order with bottom and top elements. The addition of Def7–Def14 and Thm23–Thm26 guarantees that the reduct to the language

$$(N, |_{N \to N \to o}, \text{lcm}_{N \to N \to N}, \text{gcd}_{N \to N \to N}, 1_N, 0_N)$$

is a complete bounded lattice.

11.3 Conclusions

A development is a generalization of a theory that consists of, in addition to a language and a set of axioms, a series of definitions and theorems. It represents a definition-theorem-proof narrative.

11.4 Exercises

1. Develop a theory of binary natural number arithmetic that includes:

    a. PA$'$ as a subtheory.

    b. A theory of binary natural number numerals as a subtheory modeled on PA.

    c. Definitions of $+$ and $*$ defined on binary numerals.

    d. A definition of a function mapping binary numerals to natural numbers.

    e. A definition of a function mapping binary numerals to strings of 0s and 1s.

    f. A theorem stating $+$ on binary numerals correctly implements $+$ on natural numbers.

    g. A theorem stating $*$ on binary numerals correctly implements $*$ on natural numbers.

Chapter 12

Real Number Mathematics

$(\mathbb{R}, 0, 1, +, *, -, \cdot^{-1})$, the structure of real number arithmetic, is certainly one of the most important and useful structures in mathematics. It contains several important substructures:

- $(\mathbb{N}, 0, 1, +, *)$, the structure of natural number arithmetic.

- $(\mathbb{Z}, 0, 1, +, *, -)$, the structure of integer arithmetic.

- $(\mathbb{Q}, 0, 1, +, *, -, \cdot^{-1})$, the structure of rational number arithmetic.

It is also contained in several important superstructures:

- $(\mathbb{C}, 0, 1, +, *, -, \cdot^{-1})$, the structure of complex number arithmetic.

- $(\mathbb{H}, 0, 1, +, *, -, \cdot^{-1})$, the structure of quaternion arithmetic.

- $(\mathbb{O}, 0, 1, +, *, -, \cdot^{-1})$, the structure of octonion arithmetic.

- $(^*\mathbb{R}, 0, 1, +, *, -, \cdot^{-1})$, the structure of hyperreal number arithmetic for any set $^*\mathbb{R}$ of hyperreal numbers.

- $(\mathbb{S}, 0, 1, +, *, -, \cdot^{-1})$, the structure of surreal number arithmetic.

And it is a component of many mathematical models.

The purpose of this chapter is to (1) produce an Alonzo theory of real number mathematics named REAL by developing a theory COF defined below and (2) show that ideas involving the real numbers can be readily expressed in it. We will begin the construction with COF, a theory of real number arithmetic that is categorical in the standard sense whose unique standard model (up to isomorphism) defines a noncreative expansion of

W. M. Farmer, *Simple Type Theory*, Computer Science Foundations
and Applied Logic, https://doi.org/10.1007/978-3-031-21112-6_12

$\mathbf{A}_R{}^{-1}$	stands for	$\cdot^{-1}{}_{R\to R}\,\mathbf{A}_R.$
$\dfrac{\mathbf{A}_R}{\mathbf{B}_R}$	stands for	$/_{R\to R\to R}\,\mathbf{A}_R\,\mathbf{B}_R.$
$\lvert\mathbf{A}_R\rvert$	stands for	$\lvert\cdot\rvert_{R\to R}\,\mathbf{A}_R.$
$\sqrt{\mathbf{A}_R}$	stands for	$\sqrt{\cdot}_{R\to R}\,\mathbf{A}_R.$
$\lVert\mathbf{A}_{R\to R}\rVert$	stands for	$\lVert\cdot\rVert_{(R\to R)\to R}\mathbf{A}_{R\to R}.$
$\displaystyle\sum_{\mathbf{i}=\mathbf{M}_R}^{\mathbf{N}_R}\mathbf{A}_R$	stands for	$\mathsf{sum}_{R\to R\to(R\to R)\to R}\,\mathbf{M}_R\,\mathbf{N}_R\,(\lambda\,\mathbf{i}:R\,.\,\mathbf{A}_R).$
$\displaystyle\prod_{\mathbf{i}=\mathbf{M}_R}^{\mathbf{N}_R}\mathbf{A}_R$	stands for	$\mathsf{prod}_{R\to R\to(R\to R)\to R}\,\mathbf{M}_R\,\mathbf{N}_R\,(\lambda\,\mathbf{i}:R\,.\,\mathbf{A}_R).$
$\displaystyle\lim_{\mathbf{x}\to\mathbf{A}_R}\mathbf{B}_R$	stands for	$\mathsf{lim}_{(R\to R)\to R\to R}\,(\lambda\,\mathbf{x}:R\,.\,\mathbf{B}_R)\,\mathbf{A}_R.$
$\displaystyle\lim_{\mathbf{x}\to\mathbf{A}_R{}^{+}}\mathbf{B}_R$	stands for	$\mathsf{right\text{-}lim}_{(R\to R)\to R\to R}\,(\lambda\,\mathbf{x}:R\,.\,\mathbf{B}_R)\,\mathbf{A}_R.$
$\displaystyle\lim_{\mathbf{x}\to\mathbf{A}_R{}^{-}}\mathbf{B}_R$	stands for	$\mathsf{left\text{-}lim}_{(R\to R)\to R\to R}\,(\lambda\,\mathbf{x}:R\,.\,\mathbf{B}_R)\,\mathbf{A}_R.$
$\displaystyle\lim_{\mathbf{n}\to\infty}\mathbf{B}_R$	stands for	$\mathsf{lim\text{-}seq}_{(R\to R)\to R}\,(\lambda\,\mathbf{n}:N_{\{R\}}\,.\,\mathbf{B}_R).$
$\displaystyle\int_{\mathbf{A}_R}^{\mathbf{B}_R}\mathbf{C}_R\,d\mathbf{x}$	stands for	$\mathsf{integral}_{(R\to R)\to R\to R\to R}\,(\lambda\,\mathbf{x}:R\,.\,\mathbf{C}_R)\,\mathbf{A}_R\,\mathbf{B}_R.$

Table 12.1: Notational Definitions for Real Numbers

$(\mathbb{R},0,1,+,*,-,\cdot^{-1})$. Then we will develop the theory by defining a series of constants and proving a series of theorems in it. Table 12.1 contains various notational definitions that we will employ in the development of COF.

12.1 Complete Ordered Fields

A structure

$$S = (F,0,1,+,*,-,\cdot^{-1},\mathsf{pos},\le,\mathsf{ub},\mathsf{lub})$$

is a *complete ordered field* if:

1. $(F,0,1,+,*,-,\cdot^{-1})$ is a field.

2. $(F,\le)$ is a weak total order.

3. For all $x \in F$, $\mathsf{pos}(x)$ holds iff $0 < x$.

4. $(F,0,1,+,*,-,\cdot^{-1},\mathsf{pos},\le)$ is an ordered field.

5. For all $x \in F$ and $s \in \mathcal{P}(F)$, $\mathsf{ub}(x,s)$ holds iff x is an upper bound of s.

6. For all $x \in F$ and $s \in \mathcal{P}(F)$, $\mathsf{lub}(x, s)$ holds iff x is the least upper bound of s.

7. S is complete (with respect to least upper bounds), i.e., every nonempty subset of F that has an upper bound in F has a least upper bound in F.

There are many examples of ordered fields. Here are some:

- The field of the rational numbers.

- The field of the real numbers.

- Any real closed field.

- Any field of hyperreal numbers.

- The field of surreal numbers.

- Any subfield of an ordered field.

Of these examples, only the field of the real numbers is complete:

Proposition 12.1 *The structure*

$$(\mathbb{R}, 0, 1, +, *, -, \cdot^{-1}, \mathsf{pos}, \leq, \mathsf{ub}, \mathsf{lub})$$

is a complete ordered field.

Proof It is easy to verify, from the properties of the real numbers, that this structure is a complete ordered field. □

Corollary 12.2 $(\mathbb{R}, 0, 1, +, *, -, \cdot^{-1})$ *expands to a complete ordered field.*

Theorem 12.3 (Uniqueness of a Complete Ordered Field) *All complete ordered fields are isomorphic.*

Proof An early proof is given in [71]. □

Let us define the following *theory of complete ordered fields*:

Theory Definition 12.4 (Complete Order Fields)

Name: COF.

Base types: R

Constants: 0_R, 1_R, $+_{R \to R \to R}$, $*_{R \to R \to R}$, $-_{R \to R}$, $\cdot^{-1}{}_{R \to R}$, $\mathsf{pos}_{R \to o}$, $\leq_{R \to R \to o}$, $\mathsf{ub}_{R \to \{R\} \to o}$, $\mathsf{lub}_{R \to \{R\} \to o}$.

Axioms:

1. $\forall x, y, z : R \,.\, (x + y) + z = x + (y + z)$.

2. $\forall x, y : R \,.\, x + y = y + x$.

3. $\forall x : R \,.\, x + 0 = x$.

4. $\forall x : R \,.\, x + (-x) = 0$

5. $\forall x, y, z : R \,.\, (x * y) * z = x * (y * z)$.

6. $\forall x, y : R \,.\, x * y = y * x$.

7. $\forall x : R \,.\, x * 1 = x$.

8. $\forall x : R \,.\, x \neq 0 \Rightarrow x * x^{-1} = 1$.

9. $0^{-1}\uparrow$.

10. $0 \neq 1$.

11. $\forall x, y, z : R \,.\, x * (y + z) = (x * y) + (x * z)$.

12. $\forall x : R \,.\, (x = 0 \wedge \neg(\mathsf{pos}\, x) \wedge \neg(\mathsf{pos}\,(-x))) \,\vee$
    $\quad\quad (x \neq 0 \wedge \mathsf{pos}\, x \wedge \neg(\mathsf{pos}\,(-x))) \,\vee$
    $\quad\quad (x \neq 0 \wedge \neg(\mathsf{pos}\, x) \wedge \mathsf{pos}\,(-x))$.

13. $\forall x, y : R \,.\, (\mathsf{pos}\, x \wedge \mathsf{pos}\, y) \Rightarrow \mathsf{pos}\,(x + y)$.

14. $\forall x, y : R \,.\, (\mathsf{pos}\, x \wedge \mathsf{pos}\, y) \Rightarrow \mathsf{pos}\,(x * y)$.

15. $\forall x, y : R \,.\, x \leq y \Leftrightarrow (x = y \vee \mathsf{pos}\,(y + (-x)))$.

16. $\forall x : R, s : \{R\} \,.\, \mathsf{ub}\, x\, s \Leftrightarrow \forall y : s \,.\, y \leq x$.

17. $\forall x : R, s : \{R\} \,.\, \mathsf{lub}\, x\, s \Leftrightarrow (\mathsf{ub}\, x\, s \wedge (\forall y : R \,.\, \mathsf{ub}\, y\, s \Rightarrow x \leq y))$.

18. $\forall s : \{R\} \,.\, (s \neq \emptyset_{\{R\}} \wedge \exists x : R \,.\, \mathsf{ub}\, x\, s) \Rightarrow \exists x : R \,.\, \mathsf{lub}\, x\, s$.

Proposition 12.5 COF *specifies in the standard sense the set of complete ordered fields.*

Proof Let M be a standard model of COF. Axioms 1–11 say that M is a field; axioms 12-15 say in addition that M is an ordered field; and axioms 16–18 say that M is complete (with respect to least upper bounds). Thus every standard model of COF defines a complete ordered field. Now let S be a complete ordered field. Then S satisfies the axioms of COF, and so there is a standard model of COF that defines S. Therefore, COF specifies in the standard sense the set of complete ordered fields. □

Theorem 12.6 (Categoricity of COF) COF *is categorical in the standard sense.*

Proof Every standard model of COF defines a complete ordered field by Proposition 12.5 and all complete ordered fields are isomorphic by Theorem 12.3. □

Without loss of generality, we may assume that the unique standard model of COF (up to isomorphism) defines

$$(\mathbb{R}, 0, 1, +, *, -, \cdot^{-1}, \mathsf{pos}, \leq, \mathsf{ub}, \mathsf{lub}),$$

the complete ordered field of the real numbers, which is an expansion of

$$(\mathbb{R}, 0, 1, +, *, -, \cdot^{-1}),$$

the structure of real number arithmetic.

12.2 Alternatives to the Construction of COF

We constructed COF in a single step using a theory definition module. There are alternative ways to construct a theory of real number arithmetic categorical in the standard sense.

First, we can construct a different theory of complete ordered fields (with different constants and axioms) in a single step.

Second, we can construct COF (or some other theory of complete ordered fields) as the last theory in a sequence of extensions. For example, the sequence could be something like:

$$((R, 0, +), \Gamma_1) \qquad\qquad \text{(theory of monoids)}$$
$$\leq ((R, 0, +, -), \Gamma_2) \qquad\qquad \text{(theory of groups)}$$
$$\leq ((R, 0, +, -), \Gamma_3) \qquad\qquad \text{(theory of commutative groups)}$$
$$\leq ((R, 0, 1, +, *, -), \Gamma_4) \qquad\qquad \text{(theory of rings)}$$
$$\leq ((R, 0, 1, +, *, -, \cdot^{-1}), \Gamma_5) \qquad\qquad \text{(theory of fields)}$$
$$\leq ((R, 0, 1, +, *, -, \cdot^{-1}, \mathsf{pos}, \leq), \Gamma_6) \qquad\qquad \text{(theory of ordered fields)}$$
$$\leq ((R, 0, 1, +, *, -, \cdot^{-1}, \mathsf{pos}, \leq, \mathsf{ub}, \mathsf{lub}), \Gamma_7) \qquad\qquad \text{(COF)}$$

(We have dropped the types of the constants.)

And third, we can construct a theory of real number arithmetic categorical in the standard sense as the last theory in a sequence of definitional extensions that starts with PA. For example, the sequence could include definitions of natural, integer, rational, and real number arithmetic in succession as follows:

$$((N, 0, S), \Gamma_1) \qquad\qquad \text{(PA)}$$
$$\leq_{\mathsf{d}} ((N, 0, 1, S, +, *), \Gamma_2) \qquad\qquad \text{(natural number arithmetic)}$$
$$\leq_{\mathsf{d}} ((N, 0, 1, S, +, *, Z_{\{\alpha\}}, \ldots), \Gamma_3) \qquad\qquad \text{(integer arithmetic)}$$
$$\leq_{\mathsf{d}} ((N, 0, 1, S, +, *, Z_{\{\alpha\}}, \ldots, Q_{\{\beta\}}, \ldots), \Gamma_4)$$
$$\text{(rational number arithmetic)}$$
$$\leq_{\mathsf{d}} ((N, 0, 1, S, +, *, Z_{\{\alpha\}}, \ldots, Q_{\{\beta\}}, \ldots, R_{\{\gamma\}}, \ldots), \Gamma_5)$$
$$\text{(real number arithmetic)}$$

(We have dropped most of the types of the constants.) The integers, rational numbers, and real numbers are defined, respectively, as quasitypes $Z_{\{\alpha\}}$, $Q_{\{\beta\}}$, and $R_{\{\gamma\}}$ within types α, β, and γ, and the arithmetic operations on these three kinds of numbers are defined, respectively, as constants with types constructed from α, β, and γ. See Edmund Landau's classic *Foundations of Analysis* [80] for the details of this construction. Since PA is categorical in the standard sense and all the extensions are definitional, each theory in the sequence is categorical in the standard sense by part 2 of Proposition 9.42.

12.3   Development of Real Number Mathematics

In this section, we will define a development of COF as a series of development extensions and then define REAL as the top theory of the development.

12.3.1   Some Basic Definitions and Theorems

We will begin by defining a development that introduces some basic definitions and theorems. The definitions introduce constants that denote the subtraction and division functions in Curryed form and the absolute value and square root functions. The theorems present facts concerning the definedness of the original and the newly defined arithmetic operators.

Development Definition 12.7 (COF Development 1)

   Name: COF-dev-1.

   Bottom theory: COF.

   Definitions and theorems:

   Def1: $-_{R \to R \to R} = \lambda x : R . \lambda y : R . x + (-y)$ \qquad (subtraction).

   Def2: $/_{R \to R \to R} = \lambda x : R . \lambda y : R . x * y^{-1}$ \qquad (division).

   Def3: $| \cdot |_{R \to R} = \lambda x : R . 0 \leq x \mapsto x \mid -x$ \qquad (absolute value).

   Def4: $\sqrt{\cdot}_{R \to R} = \lambda x : R . I y : R . 0 \leq y \wedge y * y = x$ \qquad (square root).

   Thm1: $\mathsf{TOTAL2}(+)$ \qquad ($+$ is total).

   Thm2: $\mathsf{TOTAL2}(*)$. \qquad ($*$ is total):

   Thm3: $\mathsf{TOTAL}(-_{R \to R})$ \qquad ($-_{R \to R}$ is total).

   Thm4: $\forall x : R . x^{-1}{\downarrow} \Leftrightarrow x \neq 0$ \qquad (domain of $\cdot^{-1}$).

   Thm5: $\mathsf{TOTAL2}(-_{R \to R \to R})$ \qquad ($-_{R \to R \to R}$ is total).

   Thm6: $\forall x, y : R . (x \ / \ y){\downarrow} \Leftrightarrow y \neq 0$ \qquad (domain of $/$).

   Thm7: $\mathsf{TOTAL}(| \cdot |_{R \to R})$ \qquad ($| \cdot |_{R \to R}$ is total).

   Thm8: $\forall x : R . (\sqrt{x}){\downarrow} \Leftrightarrow x \geq 0$ \qquad (domain of $\sqrt{\cdot}$).

12.3.2   Naturals, Integers, and Rationals

We will next extend the development of COF defined in the previous subsection by defining the natural numbers, integers, and rational numbers as quasitypes within the type R and then listing a series of basic theorems about these three quasitypes.

Development Extension 12.8 (COF Development 2)

Name: COF-dev-2.

Extends COF-dev-1.

New definitions and theorems:

Def5: $N_{\{R\}} =$
$\{x : R \mid \forall p : R \to o \,.\, (p\,0 \wedge \forall y : R \,.\, (p\,y \Rightarrow p\,(y + 1))) \Rightarrow p\,x\}$
(natural numbers quasitype).

Thm9: $\forall p : N_{\{R\}} \to o \,.$
$(p\,0 \wedge \forall x : N_{\{R\}} \,.\, (p\,x \Rightarrow p\,(x + 1))) \Rightarrow \forall x : N_{\{R\}} \,.\, p\,x$
(induction principle for $N_{\{R\}}$).

Thm10: TOTAL-ON2$(+, N_{\{R\}}, N_{\{R\}}, N_{\{R\}})$ ($N_{\{R\}}$ is closed under $+$).

Thm11: TOTAL-ON2$(*, N_{\{R\}}, N_{\{R\}}, N_{\{R\}})$ ($N_{\{R\}}$ is closed under $*$).

Def6: $Z_{\{R\}} = N_{\{R\}} \cup \{x : R \mid \exists n : N_{\{R\}} \,.\, x = -n\}$
(integers quasitype).

Thm12: TOTAL-ON2$(+, Z_{\{R\}}, Z_{\{R\}}, Z_{\{R\}})$ ($Z_{\{R\}}$ is closed under $+$).

Thm13: TOTAL-ON2$(*, Z_{\{R\}}, Z_{\{R\}}, Z_{\{R\}})$ ($Z_{\{R\}}$ is closed under $*$).

Thm14: TOTAL-ON$(-_{R \to R}, Z_{\{R\}}, Z_{\{R\}})$
($Z_{\{R\}}$ is closed under $-_{R \to R}$).

Def7: $Q_{\{R\}} = \{x : R \mid \exists m, n : Z_{\{R\}} \,.\, n \neq 0 \wedge x = m \,/\, n\}$
(rational numbers quasitype).

Thm15: TOTAL-ON2$(+, Q_{\{R\}}, Q_{\{R\}}, Q_{\{R\}})$ ($Q_{\{R\}}$ is closed under $+$).

Thm16: TOTAL-ON2$(*, Q_{\{R\}}, Q_{\{R\}}, Q_{\{R\}})$ ($Q_{\{R\}}$ is closed under $*$).

Thm17: TOTAL-ON$(-_{R \to R}, Q_{\{R\}}, Q_{\{R\}})$
($Q_{\{R\}}$ is closed under $-_{R \to R}$).

Thm18: TOTAL-ON$(\cdot^{-1}, \{q : Q_{\{R\}} \mid q \neq 0\}, \{q : Q_{\{R\}} \mid q \neq 0\})$
$\qquad\qquad\qquad$ ($Q_{\{R\}}$ is almost closed under $\cdot^{-1}$).

Thm19: INF$(N_{\{R\}}) \wedge$ COUNT$(N_{\{R\}})$ $\qquad$ ($N_{\{R\}}$ is countably infinite).

Thm20: INF$(Z_{\{R\}}) \wedge$ COUNT$(N_{\{R\}})$ $\qquad$ ($Z_{\{R\}}$ is countably infinite).

Thm21: INF$(Q_{\{R\}}) \wedge$ COUNT$(N_{\{R\}})$ $\qquad$ ($Q_{\{R\}}$ is countably infinite).

Thm22: $\neg(\exists f : R \to R \,.\, \text{BIJ-ON}(f, N_{\{R\}}, U_{\{R\}}))$ (R is uncountable).

12.3.3 Iterated Sum and Product Operators

We will continue the development of COF by defining the iterated sum

$$\text{sum}_{R \to R \to (R \to R) \to R}$$

and iterated product

$$\text{prod}_{R \to R \to (R \to R) \to R}$$

operators.

Development Extension 12.9 (COF Development 3)

$\quad$ **Name:** COF-dev-3.

$\quad$ **Extends** COF-dev-2.

$\quad$ **New definitions and theorems:**

$\quad$ Def8: $\text{sum}_{R \to R \to (R \to R) \to R} =$
$\qquad$ I $f : Z_{\{R\}} \to Z_{\{R\}} \to (Z_{\{R\}} \to R) \to R$.
$\qquad$ $\forall m, n : Z_{\{R\}}, g : Z_{\{R\}} \to R \,.\, f\,m\,n\,g \simeq$
$\qquad$ $(m > n \mapsto 0 \mid (f\,m\,(n-1)\,g) + (g\,n))$ $\qquad$ (iterated sum).

$\quad$ Thm23: $\text{sum}_{R \to R \to (R \to R) \to R} \downarrow (Z_{\{R\}} \to Z_{\{R\}} \to (Z_{\{R\}} \to R) \to R)$.
$\qquad\qquad\qquad\qquad\qquad\qquad\qquad$ (iterated sum is defined).

$\quad$ Def9: $\text{prod}_{R \to R \to (R \to R) \to R} =$
$\qquad$ I $f : Z_{\{R\}} \to Z_{\{R\}} \to (Z_{\{R\}} \to R) \to R$.
$\qquad$ $\forall m, n : Z_{\{R\}}, g : Z_{\{R\}} \to R \,.\, f\,m\,n\,g \simeq$
$\qquad$ $(m > n \mapsto 1 \mid (f\,m\,(n-1)\,g) * (g\,n))$ $\qquad$ (iterated product).

$\quad$ Thm24: $\text{prod}_{R \to R \to (R \to R) \to R} \downarrow (Z_{\{R\}} \to Z_{\{R\}} \to (Z_{\{R\}} \to R) \to R)$.
$\qquad\qquad\qquad\qquad\qquad\qquad\qquad$ (iterated product is defined).

12.3.4　Calculus

We will now continue the development of COF by introducing the basic definitions and some of the basic theorems of single-variable calculus. The development contains the definitions of limits, continuity, derivatives, and integrals. The development also contains four major theorems: the intermediate value theorem, the mean value theorem, the fundamental theorem of calculus, and the well-known corollary of the fundamental theorem of calculus.

Development Extension 12.10 (COF Development 4)

Name: COF-dev-4.

Extends　COF-dev-3.

New definitions and theorems:

Def10: $\lim_{(R \to R) \to R \to R} =$
$\lambda f : R \to R \, . \, \lambda a : R \, . \, Ib : R \, .$
$(\forall e : R \, . \, 0 < e \Rightarrow$
$(\exists d : R \, . \, 0 < d \land$
$(\forall x : R \, . \, 0 < |x - a| < d \Rightarrow$
$|f \, x - b| < e)))$　　　　　　　　　　(limit of a function).

Def11: right-$\lim_{(R \to R) \to R \to R} =$
$\lambda f : R \to R \, . \, \lambda a : R \, . \, Ib : R \, .$
$(\forall e : R \, . \, 0 < e \Rightarrow$
$(\exists d : R \, . \, 0 < d \land$
$(\forall x : R \, . \, 0 < x - a < d \Rightarrow$
$|f \, x - b| < e)))$　　　　　(right limit of a function).

Def12: left-$\lim_{(R \to R) \to R \to R} =$
$\lambda f : R \to R \, . \, \lambda a : R \, . \, Ib : R \, .$
$(\forall e : R \, . \, 0 < e \Rightarrow$
$(\exists d : R \, . \, 0 < d \land$
$(\forall x : R \, . \, 0 < a - x < d \Rightarrow$
$|f \, x - b| < e)))$　　　　　(left limit of a function).

Def13: lim-$seq_{(R \to R) \to R} =$
$\lambda s : N_{\{R\}} \to R \, . \, Ib : R \, .$
$(\forall e : R \, . \, 0 < e \Rightarrow$
$(\exists m : N_{\{R\}} \, .$

$$(\forall n : N_{\{R\}} . m < n \Rightarrow$$
$$|s\,n - b| < e)))\qquad\qquad\text{(limit of a sequence).}$$

Def14: cont-at$_{(R\to R)\to R\to o}$ =
$$\lambda f : R \to R . \lambda a : R . \lim_{x\to a} f\,x = f\,a \qquad\text{(continuous at a point).}$$

Def15: right-cont-at$_{(R\to R)\to R\to o}$ =
$$\lambda f : R \to R . \lambda a : R . \lim_{x\to a^+} f\,x = f\,a \quad\text{(right continuous at a point).}$$

Def16: left-cont-at$_{(R\to R)\to R\to o}$ =
$$\lambda f : R \to R . \lambda a : R . \lim_{x\to a^-} f\,x = f\,a \qquad\text{(left continuous at a point).}$$

Def17: cont$_{(R\to R)\to o}$ = $\lambda f : R \to R . \forall x : R .$ cont-at $f\,x$
$$\text{(continuous).}$$

Def18: cont-on-closed-int$_{(R\to R)\to R\to R\to o}$ =
$\lambda f : R \to R . \lambda a : R . \lambda b : R .$
$(\forall x : R . a < x < b \Rightarrow$ cont-at $f\,x) \wedge$
right-cont-at $f\,a \wedge$
left-cont-at $f\,b$ \qquad\qquad (continuous on a closed interval).

Def19: deriv-at$_{(R\to R)\to R\to R}$ =
$$\lambda f : R \to R . \lambda a : R . \lim_{h\to 0} \tfrac{f(a+h)-f\,a}{h} \qquad\text{(derivative at a point).}$$

Def20: right-deriv-at$_{(R\to R)\to R\to R}$ =
$$\lambda f : R \to R . \lambda a : R . \lim_{h\to 0^+} \tfrac{f(a+h)-f\,a}{h} \quad\text{(right derivative at a point).}$$

Def21: left-deriv-at$_{(R\to R)\to R\to R}$ =
$$\lambda f : R \to R . \lambda a : R . \lim_{h\to 0^-} \tfrac{f(a+h)-f\,a}{h} \qquad\text{(left derivative at a point).}$$

Def22: deriv$_{(R\to R)\to(R\to R)}$ = $\lambda f : R \to R . \lambda x : R .$ deriv-at $f\,x$
$$\text{(derivative).}$$

Def23: diff-at$_{(R\to R)\to R\to o}$ = $\lambda f : R \to R . \lambda a : R .$ (deriv-at $f\,a)\!\downarrow$
$$\text{(differentiable at a point).}$$

Def24: diff$_{(R\to R)\to o}$ = $\lambda f : R \to R . \forall x : R .$ diff-at $f\,x$
$$\text{(differentiable).}$$

Def25: diff-on-open-int$_{(R\to R)\to R\to R\to o}$ =
$\lambda f : R \to R . \lambda a : R . \lambda b : R .$
$(\forall x : R . a < x < b \Rightarrow$ diff-at $f\,x)$
$$\text{(differentiable on an open interval).}$$

Def26: integral$_{(R \to R) \to R \to R \to R}$ =
$\lambda f : R \to R . \lambda a : R . \lambda b : R .$

$$\lim_{n \to \infty} \sum_{i=1}^{n} \left(f \left(a + \left(\tfrac{b-a}{n} * i \right) \right) * \tfrac{b-a}{n} \right) \qquad \text{(definite integral)}.$$

Thm25: $\forall f : R \to R, \ a, b, y : R .$
(cont-on-closed-int $f \, a \, b \wedge f \, a < y < f \, b) \Rightarrow$
$(\exists x : R . \ a < x < b \wedge f \, x = y)$ \qquad (intermediate value theorem).

Thm26: $\forall f : R \to R, \ a, b : R .$
(cont-on-closed-int $f \, a \, b \wedge$ diff-on-open-int $f \, a \, b) \Rightarrow$
$(\exists x : R . \ a < x < b \wedge$ deriv-at $f \, x = \tfrac{f \, b - f \, a}{b - a})$
\qquad (mean value theorem).

Thm27: $\forall f, g : R \to R, \ a, b : R .$
(cont-on-closed-int $f \, a \, b \wedge g = \lambda x : R . \int_a^x (f \, s) \, ds) \Rightarrow$
$((\forall x : R . \ a < x < b \Rightarrow$ deriv-at $g \, x = f \, x) \wedge$
right-deriv-at $g \, a = f \, a \wedge$
left-deriv-at $g \, b = f \, b)$ \qquad (fundamental theorem of calculus).[1]

Thm28: $\forall f, g : R \to R, \ a, b : R .$
(cont-on-closed-int $f \, a \, b \wedge f = $ deriv $g) \Rightarrow \int_a^b (f \, x) \, dx = g \, b - g \, a$
\qquad (corollary of fundamental theorem of calculus).

12.3.5   Euclidean Space

For $n \in \mathbb{N}$, $\mathbb{R}^n$ with the standard vector operations on members of $\mathbb{R}^n$ is an inner product space over $\mathbb{R}$ (i.e., a vector space over $\mathbb{R}$ with an inner product) called *n-dimensional Euclidean space*. Using the sequence machinery introduced in Chapter 10, we will extend the development of COF to represent the members of $\mathbb{R}^n$ in COF as lists over R of length n. The development extension will include definitions that introduce constants for the standard vector operations and theorems that show the constants satisfy the properties of an inner product space over $\mathbb{R}$.

Consider the set

$$\mathcal{E}_{\mathsf{nat}} = \{ \mathbf{C}_{\{R\}}^N, \mathbf{C}_R^0, \mathbf{C}_{R \to R}^S, \mathbf{C}_{R \to R}^P, \mathbf{C}_{R \to R \to R}^+, \mathbf{C}_{R \to R \to R}^*, \mathbf{C}_{R \to R \to o}^{\leq} \}$$

of closed expressions where:

---

[1]The fundamental theorem of calculus is often stated without mentioning the right and left derivatives of g at a and b, respectively, which are needed to prove its corollary, Thm28.

1. $\mathbf{C}^N_{\{R\}} \equiv N_{\{R\}}$.

2. $\mathbf{C}^0_R \equiv 0_R$.

3. $\mathbf{C}^S_{R \to R} \equiv \lambda\, x : N_{\{R\}} \,.\, x + 1$.

4. $\mathbf{C}^P_{R \to R} \equiv \lambda\, x : N_{\{R\}} \,.\, x \neq 0 \mapsto x - 1 \mid \perp_R$.

5. $\mathbf{C}^+_{R \to R \to R} \equiv \lambda\, x : N_{\{R\}} \,.\, \lambda\, y : N_{\{R\}} \,.\, x + y$.

6. $\mathbf{C}^*_{R \to R \to R} \equiv \lambda\, x : N_{\{R\}} \,.\, \lambda\, y : N_{\{R\}} \,.\, x * y$.

7. $\mathbf{C}^{\leq}_{R \to R \to o} \equiv \lambda\, x : N_{\{R\}} \,.\, \lambda\, y : N_{\{R\}} \,.\, x \leq y$.

Proposition 12.11 $\mathcal{E}_{\mathsf{nat}}$ *is a system of natural numbers in the top theory of COF-dev-4.*

Proof The proof is left to the reader as an exercise. □

Now using this system of natural numbers we can extend the development of COF as described above.

Development Extension 12.12 (COF Development 5)

Name: COF-dev-5.

Extends COF-dev-4.

New definitions and theorems:

Def27: $\odot_{R \to (R \to R)} =$
$\lambda\, n : N_{\{R\}} \,.\, \mathrm{I}\, v : [R]^n \,.\, \forall m : N_{\{R\}} \,.\, v\, m = 0 \Leftrightarrow m \leq n$
(zero vectors).

Def28: $\oplus_{(R \to R) \to (R \to R) \to (R \to R)} =$
$\lambda\, u : [R] \,.\, \lambda\, v : [R] \,.\, \mathrm{I}\, w : [R] \,.\, |u| = |v| \wedge \forall n : N_{\{R\}} \,.\, w\, n \simeq u\, n + v\, n$
(vector addition).

Def29: $\otimes_{R \to (R \to R) \to (R \to R)} =$
$\lambda\, r : R \,.\, \lambda\, v : [R] \,.\, \mathrm{I}\, w : [R] \,.\, \forall n : N_{\{R\}} \,.\, w\, n \simeq r * (v\, n)$
(scalar multiplication).

Def30: $\ominus_{(R \to R) \to (R \to R)} = \lambda\, v : [R] \,.\, -1 \otimes v$ (vector negation).

Thm29: $\forall n : N_{\{R\}}, u, v, w : [R]^n . u \oplus (v \oplus w) = (u \oplus v) \oplus w$

$$\text{(associativity of } \oplus).$$

Thm30: $\forall n : N_{\{R\}}, u, v : [R]^n . u \oplus v = v \oplus u$ (commutativity of $\oplus$).

Thm31: $\forall n : N_{\{R\}}, v : [R]^n . v \oplus (\odot n) = v$ ($\oplus$ identity).

Thm32: $\forall n : N_{\{R\}}, v : [R]^n . v \oplus (\ominus v) = \odot n$ ($\oplus$ inverse).

Thm33: $\forall r, s : R, v : [R] . (r * s) \otimes v = r \otimes (s \otimes v)$

$$\text{(} \otimes \text{ field multiplication).}$$

Thm34: $\forall v : [R] . 1 \otimes v = v$ ($\otimes$ identity).

Thm35: $\forall n : N_{\{R\}}, r : R, u, v : [R]^n . r \otimes (u \oplus v) = (r \otimes u) \oplus (r \otimes v)$

$$\text{(} \otimes \text{ vector distributivity).}$$

Thm36: $\forall r, s : R, v : [R] . (r + s) \otimes v = (r \otimes v) \oplus (s \otimes v)$

$$\text{(} \otimes \text{ field distributivity).}$$

Def31: $\odot_{(R \to R) \to (R \to R) \to R} =$

$$\lambda u : [R] . \lambda v : [R] . |u| = |v| \mapsto \sum_{i=0_R}^{|u|-1} u \, i * v \, i \mid \perp_R \qquad \text{(dot product).}$$

Thm37: $\forall n : N_{\{R\}}, r, s : R, u, v, w : [R]^n .$
$((r \otimes u) \oplus (s \otimes v)) \odot w = (r * (u \odot w)) + (s * (v \odot w))$

$$\text{(linearity).}$$

Thm38: $\forall n : N_{\{R\}}, u, v : [R]^n . u \odot v = v \odot u.$ (symmetry).

Thm39: $\forall v : [R] . v \odot v \geq 0$ (nonnegativity).

Thm40: $\forall n : N_{\{R\}}, v : [R]^n . v \odot v = 0 \Leftrightarrow v = \odot n.$ (definiteness).

Def32: $\|\cdot\|_{(R \to R) \to R} = \lambda v : [R] . \sqrt{v \odot v}$ (vector norm).

Let $T = (L, \Gamma)$ be the top theory of **COF-dev-5** where

$$L = (R, 0, 1, +, *, -, \cdot^{-1}, \ldots, N_{\{R\}}, \ldots, \odot, \oplus, \otimes, \odot, \|\cdot\|),$$

$M = (\mathcal{D}, I)$ be a model of T which defines the structure

$$S = (\mathbb{R}, \underline{0}, \underline{1}, \underline{+}, \underline{*}, \underline{-}, \underline{\cdot^{-1}}, \ldots, \mathbb{N}, \ldots, \underline{\odot}, \underline{\oplus}, \underline{\otimes}, \underline{\odot}, \underline{\|\cdot\|}),$$

and $\varphi \in \mathsf{assign}(M)$. Then the structure $(\mathbb{R}, \underline{0}, \underline{1}, \underline{+}, \underline{*}, \underline{-}, \underline{\cdot^{-1}})$ is a field since T is a definitional extension of **COF**.

Let $n \in \mathbb{N}$ and $\mathbf{N}_R$ be an expression of L such that $V_\varphi^M(\mathbf{N}_R) = n$. Define $\mathbb{R}^* = V_\varphi^M([R])$ and $\mathbb{R}^n = V_\varphi^M([R]^{\mathbf{N}_R})$. Def27–Def31 and Thm29–Thm40 guarantee that

$$E^n = (\mathbb{R}^n, \mathbb{R}, \underline{0}, \underline{1}, \underline{+}, \underline{*}, \underline{-}, \underline{\cdot^{-1}}, \underline{\odot}(n), \underline{\oplus}^n, \underline{\otimes}^n, \underline{\odot}^n)$$

is an inner product space over $\mathbb{R}$ where $\underline{\oplus}^n, \underline{\otimes}^n, \underline{\odot}^n$ are $\underline{\oplus}, \underline{\otimes}, \underline{\odot}$ with the arguments ranging over $\mathbb{R}^*$ restricted to $\mathbb{R}^n$. That is, E^n is an n-dimensional Euclidean space.

This machinery for Euclidean spaces provides a suitable foundation for developing multiple-variable calculus in COF. However, we will stop our development of COF here and define REAL to be the top theory of COF-dev-5.

12.4 Skolem's Paradox

REAL is satisfiable, so there is a frugal general model M of REAL by the Consistency Theorem (Corollary 8.7) and Henkin's Theorem (Theorem 8.11). Since M is frugal, M is a nonstandard model by Corollary 8.10 and $|D_R^M| \leq \|M\| \leq \|L\|$. The latter implies D_R^M is countable since $\|L\| = \omega$. On the other hand, Thm22 (R is uncountable), which states that there is no bijection from the quasitype $N_{\{R\}}$ to the type R, is a theorem of REAL and thus D_R^M should be uncountable. How is this possible?

This seeming contradiction was first discussed by Thoralf Skolem in the early 1920s and is thus known as *Skolem's paradox*. It is not actually a paradox because it is easily resolved with careful analysis. As we have shown, D_R^M is certainly countable because M is frugal. However, Thm22 does not actually say that D_R^M is uncountable. Instead, it says that there is no bijection from $V_\varphi^M(N_{\{R\}})$ to $V_\varphi^M(U_{\{R\}})$ in $D_{R \to R}^M$.

If M were a standard model, then $D_{R \to R}^M$ would be full, and thus every possible function would be in $D_{R \to R}^M$. Therefore, Thm22 could only be true in a standard model if the domain of R were uncountable. But Thm22 is true in M, not because the domain of R is uncountable, but because M is a nonstandard model for which there are no bijections from $V_\varphi^M(N_{\{R\}})$ to $V_\varphi^M(U_{\{R\}})$ in $D_{R \to R}^M$. In summary, in a standard model Thm22 says that D_R^M is uncountable, but in a frugal nonstandard model, like M, Thm22 says something weaker: there is no bijection from the denotation of $N_{\{R\}}$ to the denotation of R in $D_{R \to R}^M$ — which enables D_R^M to actually be countable.

12.5 Conclusions

COF is a theory of complete ordered fields that is categorical in the standard sense. Its unique standard model (up to isomorphism) is the complete ordered field of the real numbers. REAL is the top theory of a development D of COF that includes definitions and theorems about the subsets $\mathbb{N}$, $\mathbb{Z}$, and $\mathbb{Q}$ of $\mathbb{R}$; the iterated sum and product operators; single-variable calculus; and n-dimensional Euclidean spaces for all $n \in \mathbb{N}$. REAL is the beginning of a theory of real number mathematics. The development D illustrates how mathematical ideas can be readily and naturally expressed in Alonzo with the machinery of Church's type theory plus undefinedness, notational definitions, quasitypes, and sequences. Skolem's paradox is the observation that there are countable nonstandard models of REAL even though the set of reals numbers is uncountable.

12.6 Exercises

1. Let $\mathbb{R}$ be the construction of the real numbers as Dedekind cuts (which is presented, for example, in [117]). Assuming the properties of the rational numbers, prove that the structure

$$(\mathbb{R}, 0, 1, +, *, -, \cdot^{-1}, \mathsf{pos}, \leq, \mathsf{ub}, \mathsf{lub})$$

is a complete ordered field (Proposition 12.1).

2. Prove Proposition 12.11.

3. Supply all the proofs that were omitted from the definition and theorem packages of REAL.

4. Define the following notions of limits in REAL:

   a. The limit of a function $f : \mathbb{R} \to \mathbb{R}$ as x goes to positive infinity.

   b. The limit of a function $f : \mathbb{R} \to \mathbb{R}$ as x goes to negative infinity.

   c. The limit of a function $f : \mathbb{R} \to \mathbb{R}$ as x goes to a is positive infinite.

   d. The limit of a function $f : \mathbb{R} \to \mathbb{R}$ as x goes to a is negative infinite.

   e. The limit of an infinite sequence $s : \mathbb{N} \to \mathbb{R}$ is positive infinite.

   f. The limit of an infinite sequence $s : \mathbb{N} \to \mathbb{R}$ is negative infinite.

5. Write an extension of the theory COF that includes the theory of polynomials with a single indeterminant x whose coefficients are members of $\mathbb{R}$. Define an expression that denotes the function that maps a polynomial p and real number r to the value of p at r. For example, the function should map the polynomial $x^2 - 1$ and 3 to $3^2 - 1 = 8$.

Chapter 13

Morphisms

This chapter introduces theory and development morphisms and shows how they can be used to construct libraries of mathematical knowledge.

13.1 A Motivating Example

Suppose that we are working on a development of COF and need to prove the sentence

$$\forall x : R . (\forall y : R . x + y = y = y + x) \Rightarrow x = 0$$

which says 0 is the unique additive identity, and which we will call $\mathbf{X}_o^+$. So we create a theorem package for $\mathbf{X}_o^+$ with a very simple proof. And suppose later we need to prove the sentence

$$\forall x : R . (\forall y : R . x * y = y = y * x) \Rightarrow x = 1$$

which says 1 is the unique multiplicative identity, and which we will call $\mathbf{X}_o^*$. So we create a theorem package for $\mathbf{X}_o^*$ with essentially the same proof as for $\mathbf{X}_o^+$. $\mathbf{X}_o^+$ and $\mathbf{X}_o^*$ are instances of the theorem

$$\forall x : S . (\forall y : S . x \cdot y = y = y \cdot x) \Rightarrow x = e,$$

of MON, which we will call $\mathbf{X}_o$. Having to create two theorem packages in COF for instances of the same theorem is redundant and unnecessarily increases both the cost of developing COF and the cost of maintaining the development.

A much better approach is to create a theorem package for $\mathbf{X}_o$ in a development of MON and then automatically transport the theorem $\mathbf{X}_o$ to

W. M. Farmer, *Simple Type Theory*, Computer Science Foundations and Applied Logic, https://doi.org/10.1007/978-3-031-21112-6_13

the development of COF using two different translations to obtain theorem packages for $\mathbf{X}_o^+$ and $\mathbf{X}_o^*$. With this approach, only the theorem package for $\mathbf{X}_o$ needs to be developed and maintained. Definitions in the development of MON can be transported to the development of COF in the same manner as theorems.

So what are the translations? The additive translation Φ_+ maps an expression $\mathbf{A}_\alpha$ of MON to an expression $\mathbf{B}_\beta$ of COF by replacing each occurrence in $\mathbf{A}_\alpha$ of the base type S with the base type R, the constant $\cdot_{S \to S \to S}$ with the constant $+_{R \to R \to R}$, and the constant e_S with the constant 0_R. Similarly, the multiplicative translation Φ_* maps an expression $\mathbf{A}_\alpha$ of MON to an expression $\mathbf{B}_\beta$ of COF by replacing each occurrence in $\mathbf{A}_\alpha$ of S with R, $\cdot_{S \to S \to S}$ with $*_{R \to R \to R}$, and e_S with 1_R. Φ_+ and Φ_* are both meaning preserving in the sense that they map theorems of MON to theorems of COF. (This is because the substructures $(\mathbb{R}, 0, +)$ and $(\mathbb{R}, 1, *)$ of $(\mathbb{R}, 0, 1, +, *, -, \cdot^{-1}, \mathsf{pos}, \leq, \mathsf{ub}, \mathsf{lub})$ are monoids.) In particular, Φ_+ and Φ_* map $\mathbf{X}_o$, which is a theorem of MON, to $\mathbf{X}_o^+$ and $\mathbf{X}_o^*$, respectively, which are theorems of COF.

A meaning-preserving translation Φ from a theory T_1 (like MON) to a theory T_2 (like Φ_+ or Φ_*) is called a *theory morphism* (or *morphism* for short) [41].[1] Can the morphisms Φ_+ and Φ_* be extended to morphisms Ψ_+ and Ψ_* from GRP to COF by replacing $\cdot^{-1}{}_{S \to S}$ with $-_{R \to R}$ and $\cdot^{-1}{}_{R \to R}$, respectively? Ψ_+ is a morphism from GRP to COF (because $(\mathbb{R}, 0, +, -)$ is a group), but Ψ_* is not a morphism from GRP to COF (because $(\mathbb{R}, 1, *, \cdot^{-1})$ is not a group). However, consider the translation Ψ'_* from GRP to COF that replaces S with the quasitype

$$\{r : R \mid r \neq 0_R\},$$

$\cdot_{S \to S \to S}$ with the expression

$$\lambda x : \{r : R \mid r \neq 0_R\} . \lambda y : \{r : R \mid r \neq 0_R\} . *_{R \to R \to R} x\, y,$$

e_S with 1_R, and $\cdot^{-1}{}_{S \to S}$ with the expression

$$\lambda x : \{r : R \mid r \neq 0_R\} . \cdot^{-1}{}_{R \to R} x.$$

Ψ'_* is a morphism from GRP to COF (because $(\mathbb{R} \setminus \{0\}, 1, *, \cdot^{-1})$ is a group).

Thus we see that a translation from a theory T_1 to a theory T_2 is defined by how it maps the base types of T_1 to the types and quasitypes of T_2 and maps the constants of T_1 to the expressions of T_2. We also see that not all translations are morphisms.

---

[1] Morphisms from one theory to another are also known as *interpretations, views, translations, realizations,* and *immersions*.

13.2   The Little Theories Method

Mathematical knowledge is highly interconnected. The same structure and
the results that go with it can appear in many different guises. For example,
the structure of an associative binary operation having an identity element
— that is, a monoid — is found in countless places in mathematics and computing. (We saw two such instances of a monoid in the motivating example
in Section 13.1.) As a result, it is easy for a body of mathematical knowledge to be heavy with redundancy and for a mathematics practitioner to
find themself defining essentially the same concepts and proving essentially
the same facts over and over again.

The *little theories method* [52] is a method for organizing mathematical
knowledge as a network of theories called a *theory graph* [77, 78]. The nodes
of the graph are *theories* and the directed edges are *theory morphisms*, i.e.,
mappings from the language of one theory to the language of another theory
that preserve theorems. Each mathematical topic is developed in the "little
theory" in the theory graph that has the most convenient level of abstraction and the most convenient vocabulary. Then definitions and theorems
produced in the development of the theory are transported, as needed, to
other theories via the morphisms in the theory graph. For example, we
saw in the motivating example in Section 13.1 that definitions and theorems
about a monoid can be developed in a theory of an abstract monoid and then
transported to both the additive and multiplicative contexts in a theory of
the real numbers (such as COF).

The theories in a theory graph are separate units of mathematical knowledge. The morphisms are unidirectional conduits through which information
in the form of concepts (established by definitions) and facts (established by
theorems) flows from a usually more abstract theory T_1 to a usually more
concrete theory T_2 [10]. Although T_1 and T_2 may have very different languages and axioms, the morphism from T_1 to T_2 guarantees that the logical
structure specified by T_1 sits within T_2 but is specified by possibly different
vocabulary and axioms. Concepts defined in T_1 and facts proved in T_1 can
be transported to any other theory T_2 for which there is a morphism from
T_1 to T_2. This is the power of the little theories method: A definition or
theorem established in a little theory T can be utilized in every other theory
that contains the same logical structure as T.

A theory graph is well suited to be the architecture for a *library of
mathematical knowledge*. Libraries of mathematical knowledge organized as
a theory graph are employed in several proof assistants and logical frame-
works including Ergo [91], IMPS [52], Isabelle [9], MMT [104], and PVS [96].

A theory graph is also well suited to be the architecture for a library of software specifications. The nodes are theories that specify software systems and the edges are theory morphisms that indicate how one software system can be implemented by another. Libraries of software specifications organized as a theory graph are employed in several software specification and development systems including CASL [7], EHDM [109], IOTA [87], KIDS [114], OBJ [57], and Specware [118].

13.3 Theory Morphisms

Let $L_i = (\mathcal{B}_i, \mathcal{C}_i)$ be a language, $T_i = (L_i, \Gamma_i)$ be a theory, $\mathcal{T}_i = \mathcal{T}(L_i)$, $\mathcal{T}_i^- = \{\alpha \in \mathcal{T}_i \mid \alpha \neq o\}$, $\mathcal{E}_i = \mathcal{E}(L_i)$, $\mathcal{Q}_i$ be the set of closed quasitypes in $\mathcal{E}_i$, and $\mathcal{Q}_i^- = \{\mathbf{Q}_{\{\alpha\}} \in \mathcal{Q}_i \mid \alpha \neq o\}$ for $i \in \{1, 2, 3\}$. Also, let $\tau(\alpha) = \alpha$ for all types $\alpha \in \mathcal{T}$ and $\tau(\mathbf{Q}_{\{\alpha\}}) = \alpha$ for all quasitypes $\mathbf{Q}_{\{\alpha\}} \in \mathcal{E}$.

13.3.1 Theory Translations

Given a total function $\mu : \mathcal{B}_1 \to \mathcal{T}_2^- \cup \mathcal{Q}_2^-$, let $\overline{\mu} : \mathcal{T}_1 \to \mathcal{T}_2 \cup \mathcal{Q}_2$ be the canonical extension of μ defined as follows by recursion and pattern matching:

1. $\overline{\mu}(o) \equiv o$.

2. $\overline{\mu}(\mathbf{a}) \equiv \mu(\mathbf{a})$.

3. $\overline{\mu}(\alpha \to \beta) \equiv \overline{\mu}(\alpha) \to \overline{\mu}(\beta)$.

4. $\overline{\mu}(\alpha \times \beta) \equiv \overline{\mu}(\alpha) \times \overline{\mu}(\beta)$.

Recall that $\overline{\mu}(\alpha) \to \overline{\mu}(\beta)$ and $\overline{\mu}(\alpha) \times \overline{\mu}(\beta)$ are types if α and β are types and are quasitypes if one or both of α and β are quasitypes.

Lemma 13.1 *Let $\mu : \mathcal{B}_1 \to \mathcal{T}_2^- \cup \mathcal{Q}_2^-$ be total.*

1. *$\overline{\mu} : \mathcal{T}_1 \to \mathcal{T}_2 \cup \mathcal{Q}_2$ is well-defined and total.*

2. *$\tau(\overline{\mu}(\alpha)) = o$ iff $\alpha = o$ for all $\alpha \in \mathcal{T}_1$.*

3. *If $\alpha \in \mathcal{T}(L_{\mathsf{min}})$, then $\overline{\mu}(\alpha) = \alpha$.*

Proof The proof is left to the reader as an exercise. $\square$

A *(theory) translation from T_1 to T_2* of Alonzo is a pair $\Phi = (\mu, \nu)$, where $\mu : \mathcal{B}_1 \to \mathcal{T}_2^- \cup \mathcal{Q}_2^-$ and $\nu : \mathcal{C}_1 \to \mathcal{E}_2$ are total, such that $\nu(\mathbf{c}_\alpha)$ is a closed expression of type $\tau(\overline{\mu}(\alpha))$ for all $\mathbf{c}_\alpha \in \mathcal{C}_1$. T_1 and T_2 are called the *source theory* and the *target theory* of Φ, respectively. Let $\overline{\nu} : \mathcal{E}_1 \to \mathcal{E}_2$ be the canonical extension of ν defined as follows by recursion and pattern matching:

1. $\overline{\nu}((\mathbf{x} : \alpha)) \equiv (\mathbf{x} : \tau(\overline{\mu}(\alpha)))$.

2. $\overline{\nu}(\mathbf{c}_\alpha) \equiv \nu(\mathbf{c}_\alpha)$.

3. $\overline{\nu}(\mathbf{A}_\alpha = \mathbf{B}_\alpha) \equiv (\overline{\nu}(\mathbf{A}_\alpha) = \overline{\nu}(\mathbf{B}_\alpha))$.

4. $\overline{\nu}(\mathbf{F}_{\alpha \to \beta} \, \mathbf{A}_\alpha) \equiv \overline{\nu}(\mathbf{F}_{\alpha \to \beta}) \, \overline{\nu}(\mathbf{A}_\alpha)$.

5. $\overline{\nu}(\lambda \, \mathbf{x} : \alpha . \, \mathbf{B}_\beta) \equiv \lambda \, \mathbf{x} : \overline{\mu}(\alpha) . \, \overline{\nu}(\mathbf{B}_\beta)$.

6. $\overline{\nu}(\mathrm{I} \, \mathbf{x} : \alpha . \, \mathbf{A}_o) \equiv \mathrm{I} \, \mathbf{x} : \overline{\mu}(\alpha) . \, \overline{\nu}(\mathbf{A}_o)$.

7. $\overline{\nu}((\mathbf{A}_\alpha, \mathbf{B}_\beta)) \equiv (\overline{\nu}(\mathbf{A}_\alpha), \overline{\nu}(\mathbf{B}_\beta))$.

We say that, if $\alpha \in \mathcal{T}_1$ and $\mathbf{A}_\alpha \in \mathcal{E}_1$, then Φ *maps* α *to* $\overline{\mu}(\alpha)$ and *maps* $\mathbf{A}_\alpha$ *to* $\overline{\nu}(\mathbf{A}_\alpha)$. We also say that $\overline{\mu}(\alpha)$ is the *image of* α *under* Φ and $\overline{\nu}(\mathbf{A}_\alpha)$ is the *image of* $\mathbf{A}_\alpha$ *under* Φ. Φ is an *identity translation* if $\overline{\mu}$ and $\overline{\nu}$ are identity mappings, *injective* if $\overline{\mu}$ and $\overline{\nu}$ are injective mappings, and *normal* if $\mu : \mathcal{B}_1 \to \mathcal{T}_2^-$. An identity translation is clearly injective and normal.

Lemma 13.2 *Let $\Phi = (\mu, \nu)$ be a translation from T_1 to T_2.*

1. *$\overline{\nu} : \mathcal{E}_1 \to \mathcal{E}_2$ is well-defined and total.*

2. *$\overline{\nu}(\mathbf{A}_\alpha)$ is of type $\tau(\overline{\mu}(\alpha))$ for all $\mathbf{A}_\alpha \in \mathcal{E}_1$.*

3. *If $\mathbf{A}_\alpha$ is closed, then $\overline{\nu}(\mathbf{A}_\alpha)$ is closed.*

4. *If $\mathbf{A}_\alpha \in \mathcal{E}(L_{\mathsf{min}})$, then $\overline{\nu}(\mathbf{A}_\alpha) = \mathbf{A}_\alpha$.*

5. *If μ and ν are identity mappings, then Φ is an identity translation.*

6. *If μ and ν are are injective mappings, then Φ is injective.*

Proof The proof is left to the reader as an exercise. □

A *theory translation definition module* is used to present a translation from one theory to another. It has the following form:

Theory Translation Definition X.Y

Name: Name.

Source theory: Name-of-source-theory.

Target theory: Name-of-target-theory.

Base type mapping:

$\mathbf{a}_1 \mapsto \zeta_1$ where ζ_1 is a type or closed quasitype with $\tau(\zeta_1) \neq o$.

$\vdots$

$\mathbf{a}_m \mapsto \zeta_m$ where ζ_m is a type or closed quasitype with $\tau(\zeta_m) \neq o$.

Constant mapping:

$\mathbf{c}^1_{\alpha_1} \mapsto \mathbf{B}^1_{\beta_1}.$

$\vdots$

$\mathbf{c}^n_{\alpha_n} \mapsto \mathbf{B}^n_{\beta_n}.$

The theory translation definition module is well-formed if Name-of-source-theory is the name of a theory $T_1 = ((\mathcal{B}_1, \mathcal{C}_1), \Gamma_1)$, Name-of-target-theory is the name of a theory T_2, $\mathcal{B}_1 = \{\mathbf{a}_1, \ldots, \mathbf{a}_m\}$, $\mathcal{C}_1 = \{\mathbf{c}^1_{\alpha_1}, \ldots, \mathbf{c}^n_{\alpha_n}\}$, and $\Phi = (\mu, \nu)$ is a translation from T_1 to T_2 where $\mu(\mathbf{a}_i) = \zeta_i$ for all i with $1 \leq i \leq m$ and $\nu(\mathbf{c}^j_{\alpha_j}) = \mathbf{B}^j_{\beta_j}$ for all j with $1 \leq j \leq n$. Name is a name of the translation Φ.

Let $\Phi = (\mu, \nu)$ be a translation from T_1 to T_2. A translation $\Phi' = (\mu', \nu')$ from T'_1 to T'_2 is an *extension* of Φ (or Φ is a *subtranslation* of Φ'), written $\Phi \leq \Phi'$, if $T_1 \leq T'_1$, $T_2 \leq T'_2$, $\mu \sqsubseteq \mu'$, and $\nu \sqsubseteq \nu'$.

A *theory translation extension module* is used to present an extension of a translation from one theory to another. It has the following form:

Theory Translation Extension X.Y

Name: Name.

Extends Name-of-subtranslation.

New source theory: Name-of-new-source-theory.

New target theory: Name-of-new-target-theory.

New base type mapping:

$\mathbf{a}_1 \mapsto \zeta_1$ where ζ_1 is a type or closed quasitype with $\tau(\zeta_1) \neq o$.

$\vdots$

$\mathbf{a}_m \mapsto \zeta_m$ where ζ_m is a type or closed quasitype with $\tau(\zeta_m) \neq o$.

New constant mapping:

$\mathbf{c}_{\alpha_1}^1 \mapsto \mathbf{B}_{\beta_1}^1.$

$\vdots$

$\mathbf{c}_{\alpha_n}^n \mapsto \mathbf{B}_{\beta_n}^n.$

The theory translation extension module is well-formed if Name-of-subtranslation is the name of a translation $\Phi = (\mu, \nu)$ from $T_1 = (\mathcal{B}_1, \mathcal{C}_1)$ to T_2, Name-of-new-source-theory is the name of an extension $T_1' = (\mathcal{B}_1', \mathcal{C}_1')$ of T_1, Name-of-new-target-theory is the name of an extension T_2' of T_2, $\mathcal{B}_1' = \mathcal{B}_1 \cup \{\mathbf{a}_1, \ldots, \mathbf{a}_m\}$ with $\mathcal{B}_1 \cap \{\mathbf{a}_1, \ldots, \mathbf{a}_m\} = \emptyset$, $\mathcal{C}_1' = \mathcal{C}_1 \cup \{\mathbf{c}_{\alpha_1}^1, \ldots, \mathbf{c}_{\alpha_n}^n\}$ with $\mathcal{C}_1 \cap \{\mathbf{c}_{\alpha_1}^1, \ldots, \mathbf{c}_{\alpha_n}^n\} = \emptyset$, $\Phi' = (\mu', \nu')$ is a translation from T_1' to T_2' where $\mu \sqsubseteq \mu'$, $\nu \sqsubseteq \nu'$, $\mu'(\mathbf{a}_i) = \zeta_i$ for all i with $1 \leq i \leq m$ and $\nu'(\mathbf{c}_{\alpha_j}^j) = \mathbf{B}_{\beta_j}^j$ for all j with $1 \leq j \leq n$. Thus $\Phi \leq \Phi'$. Name is a name of the translation Φ'.

Let $\Phi = (\mu, \nu)$ be a translation from T_1 to T_2. Φ is a *(theory) morphism from T_1 to T_2* if $T_1 \vDash \mathbf{A}_o$ implies $T_2 \vDash \overline{\nu}(\mathbf{A}_o)$ for all sentences $\mathbf{A}_o$ of L_1. That is, a morphism is a translation that is meaning preserving in the sense that it maps theorems to theorems. Φ is an *inclusion* if it is an identity translation that is a morphism. Φ is an *embedding* if it is a normal injective translation that is a morphism. An inclusion is clearly an embedding.

Proposition 13.3 $T_1 \preceq T_2$ *iff there is an inclusion from T_1 to T_2.*

Proof Let $T_1 \preceq T_2$. Define $\mu : \mathcal{B}_1 \to \mathcal{B}_2$ and $\nu : \mathcal{C}_1 \to \mathcal{C}_2$ to be the identity mappings. Then $\Phi = (\mu, \nu)$ is an identity translation from T_1 to T_2 by part 5 of Lemma 13.2. It follows from $T_1 \preceq T_2$ that Φ is also a morphism. Thus Φ is an inclusion from T_1 to T_2.

Now let Φ be an inclusion from T_1 to T_2. Then Φ is (a) an identity translation and (b) a morphism. Then (a) and (b) imply $T_1 \preceq T_2$. $\qquad\square$

Corollary 13.4 *If $T_1 \leq T_2$, then the identity translation from T_1 to T_2 is an inclusion.*

Proof Follows immediately from Proposition 9.12 and the proof of Proposition 13.3. □

Remark 13.5 (Extensions and Morphisms) *The notion of a morphism from a theory T is a generalization of the notion of an extension of T. As shown by Corollary 13.4, a theory extension can be viewed as an inclusion.*

Remark 13.6 (Embeddings) *A normal injective translation from T_1 to T_2 embeds the structure of the closed expressions of T_1 into the structure of the closed expressions of T_2. An embedding (i.e., a normal injective morphism) from T_1 to T_2, in addition, embeds the structure of the valid sentences of T_1 into the structure of the valid sentences of T_2.*

In the next subsection, we will prove a theorem (called the Morphism Theorem) that gives a useful sufficient condition for a translation to be a morphism. We will define in Subsection 13.3.3 several examples of translations and then show that they are morphisms using this theorem.

Let $\Phi_1 = (\mu_1, \nu_1)$ be a translation from T_1 to T_2 and $\Phi_2 = (\mu_2, \nu_2)$ be a translation from T_2 to T_3. The *composition* of Φ_1 and Φ_2, written $\Phi_1 \circ \Phi_2$, is the pair (μ, ν) where:

1. $\mu : \mathcal{B}_1 \to \mathcal{T}_3^- \cup \mathcal{Q}_3^-$ such that $\mu(\mathbf{a}) = \overline{\mu_2}(\mu_1(\mathbf{a}))$ if $\mu_1(\mathbf{a}) \in \mathcal{T}_2^-$ and $\mu(\mathbf{a}) = \overline{\nu_2}(\mu_1(\mathbf{a}))$ if $\mu_1(\mathbf{a}) \in \mathcal{Q}_2^-$ for all $\mathbf{a} \in \mathcal{B}_1$.

2. $\nu : \mathcal{C}_1 \to \mathcal{E}_3$ such that $\nu(\mathbf{c}_\alpha) = \overline{\nu_2}(\nu_1(\mathbf{c}_\alpha))$ for all $\mathbf{c}_\alpha \in \mathcal{C}_1$.

Proposition 13.7 *Let $\Phi_1 = (\mu_1, \nu_1)$ be a translation from T_1 to T_2 and $\Phi_2 = (\mu_2, \nu_2)$ be a translation from T_2 to T_3.*

1. *$\Phi_1 \circ \Phi_2$ is a translation from T_1 to T_3.*

2. *If Φ_1 and Φ_2 are morphisms, then $\Phi_1 \circ \Phi_2$ is also a morphism.*

3. *If Φ_1 and Φ_2 are normal morphisms, then $\Phi_1 \circ \Phi_2$ is also a normal morphism.*

Proof The proof is left to the reader as an exercise. □

13.3.2   Morphism Theorem

Let $\Phi = (\mu, \nu)$ be a translation from T_1 to T_2. An *obligation* of Φ is any one of the following sentences of L_2:

1. $\overline{\nu}(U_{\{\mathbf{a}\}} \neq \emptyset_{\{\mathbf{a}\}})$ where $\mathbf{a} \in \mathcal{B}_1$ and $\mu(\mathbf{a}) \in \mathcal{Q}_2$.

2. $\overline{\nu}(\mathbf{c}_\alpha \downarrow U_{\{\alpha\}})$ where $\mathbf{c}_\alpha \in \mathcal{C}_1$.

3. $\overline{\nu}(\mathbf{A}_o)$ where $\mathbf{A}_o \in \Gamma_1$.

Thus Φ has three kinds of obligations corresponding, respectively, to (1) the base types of T_1 mapped to quasitypes, (2) the constants of T_1, and (3) the axioms of T_1. The following proposition, which is easy to prove, states that the obligations of Φ are images of sentences valid in T_1:

Proposition 13.8 *Let $\Phi = (\mu, \nu)$ be a translation from T_1 to T_2. If $\overline{\nu}(\mathbf{B}_o)$ is an obligation of Φ, then $T_1 \vDash \mathbf{B}_o$.*

Lemma 13.9 *Let $\Phi = (\mu, \nu)$ be a translation from T_1 to T_2.*

1. *If $\alpha \in \mathcal{T}(L_1)$, $\overline{\mu}(\alpha) \in \mathcal{Q}_2$, and $T_2 \vDash \overline{\mu}(\alpha)\!\downarrow$, then $T_2 \vDash \overline{\nu}(U_{\{\alpha\}}) = \overline{\mu}(\alpha)$.*

2. *If $\mathbf{a} \in \mathcal{B}_1$ and $\mu(\mathbf{a}) \in \mathcal{Q}_2$, then $T_2 \vDash \overline{\nu}(U_{\{\mathbf{a}\}} \neq \emptyset_{\{\mathbf{a}\}}) \Leftrightarrow \mu(\mathbf{a}) \neq \emptyset_{\{\tau(\mu(\mathbf{a}))\}}$.*

3. *If $\mathbf{c}_\alpha \in \mathcal{C}_1$ and $\overline{\mu}(\alpha) \in \mathcal{T}_2$, then $T_2 \vDash \overline{\nu}(\mathbf{c}_\alpha \downarrow U_{\{\alpha\}}) \Leftrightarrow \nu(\mathbf{c}_\alpha)\!\downarrow$.*

4. *If $\mathbf{c}_\alpha \in \mathcal{C}_1$ and $\overline{\mu}(\alpha) \in \mathcal{Q}_2$, then $T_2 \vDash \overline{\nu}(\mathbf{c}_\alpha \downarrow U_{\{\alpha\}}) \Leftrightarrow \nu(\mathbf{c}_\alpha) \downarrow \overline{\mu}(\alpha)$.*

Proof Let $\alpha \in \mathcal{T}(L_1)$, $\overline{\mu}(\alpha) \in \mathcal{Q}_2$, and $(\star)$ $T_2 \vDash \overline{\mu}(\alpha)\!\downarrow$. Then

$$\overline{\nu}(U_{\{\alpha\}})$$
$$\equiv \overline{\nu}(\lambda\, x : \alpha \,.\, T_o)$$
$$\equiv \lambda\, x : \overline{\mu}(\alpha) \,.\, T_o$$
$$\equiv \lambda\, x : \alpha \,.\, (x \in \overline{\mu}(\alpha) \mapsto T_o \mid F_o)$$

by the definition of $\overline{\nu}$ and notational definitions. The last expression is clearly equal to $\lambda\, x : \alpha \,.\, \overline{\mu}(\alpha)\, x$, which is equal to $\overline{\mu}(\alpha)$ in T_2 by $(\star)$. This proves part 1. Parts 2 and 4 follow immediately from the definition of $\overline{\nu}$ and part 1. Part 3 follows the definition of $\overline{\nu}$ and the notational definition for the defined-in-quasitype operator. $\qquad\square$

Lemma 13.10 *A normal translation has no obligations of the first kind for base types mapped to quasitypes.*

Proof The lemma follows from the fact that a normal translation maps all base types to types. □

Lemma 13.11 *Let* $\Phi = (\mu, \nu)$ *be a translation from* T_1 *to* T_2. *If* $\nu(\mathbf{c}_\alpha) = \mathbf{d}_\beta$, *then the obligation of* Φ *of the second kind for* $\mathbf{c}_\alpha$ *is valid in* T_2.

Proof Let $\nu(\mathbf{c}_\alpha) = \mathbf{d}_\beta$. This implies $\overline{\mu}(\alpha) = \beta \in \mathcal{T}_2$. Hence the obligation of Φ of the second kind for $\mathbf{c}_\alpha$ is logically equivalent to $\mathbf{d}_\beta{\downarrow}$ by part 3 of Lemma 13.9. $\mathbf{d}_\beta{\downarrow}$ is valid by part 2 of Lemma 5.4 and part 3 of Lemma 6.5, and so the obligation is valid in T_2. □

Let $\Phi = (\mu, \nu)$ be a translation from T_1 to T_2,

$$M_2 = (\{D_\alpha^2 \mid \alpha \in \mathcal{T}_2\}, I_2)$$

be a model of T_2, and $\varphi \in \mathsf{assign}(M_2)$. We will extract a structure M_1 from M_2 using Φ as follows. Define

$$D_\alpha^1 = \begin{cases} D_{\overline{\mu}(\alpha)}^2 & \text{if } \overline{\mu}(\alpha) \in \mathcal{T}_2 \\ \{d \in D_{\tau(\overline{\mu}(\alpha))}^2 \mid V_\varphi^{M_2}(\overline{\mu}(\alpha))(d) = \mathsf{T}\} & \text{if } \overline{\mu}(\alpha) \in \mathcal{Q}_2 \end{cases}$$

for $\alpha \in \mathcal{T}_1$. Define $I_1(\mathbf{c}_\alpha) \simeq V_\varphi^{M_2}(\nu(\mathbf{c}_\alpha))$ for $\mathbf{c}_\alpha \in \mathcal{C}_1$. Finally, define

$$M_1 = (\{D_\alpha^1 \mid \alpha \in \mathcal{T}_1\}, I_1).$$

Notice that we are not claiming M_1 is a general model; $\{D_\alpha^1 \mid \alpha \in \mathcal{T}_1\}$ is not necessarily a frame and I_1 is not necessarily an interpretation function.

Lemma 13.12 *Let* $\Phi = (\mu, \nu)$ *be a translation from* T_1 *to* T_2, M_2 *be a model of* T_2, *and* M_1 *be the structure extracted from* M_2 *using* Φ *as defined above. If each obligation of* Φ *is true in* M_2, *then* M_1 *is a model of* T_1.

Proof Assume that $(\star)$ the obligations of Φ are true in M_2.

Let $\alpha \in \mathcal{T}_1$. If $\overline{\mu}(\alpha) \in \mathcal{T}_2$, then D_α^1 obviously satisfies the conditions of a frame. So let $\overline{\mu}(\alpha) \in \mathcal{Q}_2$. Then $\alpha \neq o$ by part 2 of Lemma 13.1. If $\alpha = \mathbf{a}$, then D_α^1 is nonempty by $(\star)$ and part 2 of Lemma 13.9. If $\alpha = \beta \to \gamma$ and D_β^1 and D_γ^1 are nonempty, then D_α^1 is nonempty since it contains the empty function and is a subset of $D_\beta^1 \to D_\gamma^1$ by the semantics of $\to$. Moreover, D_α^1 is a subset of the total functions in $D_\beta^1 \to D_\gamma^1$ if $\tau(\overline{\mu}(\gamma)) = o$ since $\tau(\overline{\mu}(\gamma)) = o$ iff $\gamma = o$ by part 2 of Lemma 13.1. If $\alpha = \beta \times \gamma$ and D_β^1 and D_γ^1 are nonempty, then D_α^1 is nonempty since it equals $D_\beta^1 \times D_\gamma^1$. Therefore, $\{D_\alpha^1 \mid \alpha \in \mathcal{T}_1\}$ is a frame by induction on the syntactic structure of types.

Let $\mathbf{c}_\alpha \in \mathcal{C}_1$. Then $\nu(\mathbf{c}_\alpha)$ is of type $\tau(\overline{\mu}(\alpha))$ by the definition of a translation. If $\overline{\mu}(\alpha) \in \mathcal{T}_2$, then $I_1(\mathbf{c}_\alpha) = V_\varphi^{M_2}(\nu(\mathbf{c}_\alpha)) \in D^2_{\overline{\mu}(\alpha)} = D^1_\alpha$ by $(\star)$ and part 3 of Lemma 13.9. If $\overline{\mu}(\alpha) \in \mathcal{Q}_2$, then $I_1(\mathbf{c}_\alpha) = V_\varphi^{M_2}(\nu(\mathbf{c}_\alpha)) \in \{d \in D^2_{\tau(\overline{\mu}(\alpha))} \mid V_\varphi^{M_2}(\overline{\mu}(\alpha))(d) = \mathrm{T}\} = D^1_\alpha$ by $(\star)$ and part 4 of Lemma 13.9. Therefore, I_1 is an interpretation function and M_1 is an interpretation of L_1.

For all $\mathbf{A}_\alpha \in \mathcal{E}_1$ and $\varphi \in \mathsf{assign}(M_1)$, define

$$(\star\star) \quad V_\varphi^{M_1}(\mathbf{A}_\alpha) \simeq V_{\overline{\nu}(\varphi)}^{M_2}(\overline{\nu}(\mathbf{A}_\alpha)),{}^2$$

where $\overline{\nu}(\varphi)$ is any $\psi \in \mathsf{assign}(M_2)$ such that $\psi(\overline{\nu}((\mathbf{x} : \beta))) = \varphi((\mathbf{x} : \beta))$ for all variables $(\mathbf{x} : \beta) \in \mathcal{E}_1$. By induction on the syntactic structure of expressions, $V_\varphi^{M_1}$ satisfies the seven conditions of the definition of a general model. Therefore, M_1 is a general model of L_1.

$(\star\star)$ implies

$$(\star\star\star) \quad M_1 \vDash \mathbf{A}_o \text{ iff } M_2 \vDash \overline{\nu}(\mathbf{A}_o) \text{ for all sentences } \mathbf{A}_o \in \mathcal{E}_1.$$

If $\mathbf{A}_o \in \Gamma_1$, then $(\star)$ implies $M_2 \vDash \overline{\nu}(\mathbf{A}_o)$. Hence $M_1 \vDash \mathbf{A}_o$ for all $\mathbf{A}_o \in \Gamma_1$ by $(\star\star\star)$. Therefore, M_1 is a model of T_1. $\square$

Theorem 13.13 (Morphism Theorem) *Let $\Phi = (\mu, \nu)$ be a translation from T_1 to T_2. If each obligation of Φ is valid in T_2, then Φ is a morphism from T_1 to T_2.*

Proof Let $\Phi = (\mu, \nu)$ be a translation from T_1 to T_2 and suppose (a) each obligation of Φ is valid in T_2. Let $\mathbf{A}_o$ be a sentence that is valid in T_1. We must show that $\overline{\nu}(\mathbf{A}_o)$ is valid in every model of T_2. Let M_2 be a model of T_2. (We are done if T_2 is unsatisfiable.) Let M_1 be extracted from M_2 and Φ as defined above. Each obligation of Φ is clearly true in M_2 by (a), and so M_1 is a model of T_1 by Lemma 13.12. Therefore, $M_1 \vDash \mathbf{A}_o$, and so $M_2 \vDash \overline{\nu}(\mathbf{A}_o)$ by $(\star\star\star)$ in the proof of Lemma 13.12. $\square$

Theorem 13.14 (Relative Satisfiability) *Let Φ be a morphism from T_1 to T_2. If T_2 is satisfiable, then T_1 is satisfiable.*

Proof Let Φ be a morphism from T_1 to T_2, M_2 be a model of T_2, and M_1 be extracted from M_2 and Φ as defined above. Since Φ is a morphism, each of its obligations is valid in T_2 by Proposition 13.8 and hence true in M_2. Therefore, M_1 is a model of T_1 by Lemma 13.12. $\square$

---

[2]That is, $V_\varphi^{M_1}(\mathbf{A}_\alpha) = V_{\overline{\nu}(\varphi)}^{M_2}(\overline{\nu}(\mathbf{A}_\alpha))$ if $V_{\overline{\nu}(\varphi)}^{M_2}(\overline{\nu}(\mathbf{A}_\alpha))$ is defined and $V_\varphi^{M_1}(\mathbf{A}_\alpha)$ is undefined otherwise.

13.3.3 Examples of Theory Morphisms

In this section, we use theory translation definition and extension modules to present several examples of theory translations. We prove that each of these translations is a morphism using the Morphism Theorem (Theorem 13.13).

Example 13.15 (MON to GRP) Recall from Example 9.14 that MON $\leq$ GRP. The translation MON-to-GRP is defined as:

Theory Translation Definition 13.16 (MON to GRP)

Name: MON-to-GRP.

Source theory: MON.

Target theory: GRP.

Base type mapping:

1. $S \mapsto S$.

Constant mapping:

1. $\cdot_{S \to S \to S} \mapsto \cdot_{S \to S \to S}$.

2. $e_S \mapsto e_S$.

MON-to-GRP is an identity translation by part 5 of Lemma 13.2. So the three kinds of obligations of MON-to-GRP are valid in GRP by Lemma 13.10, Lemma 13.11, and MON $\leq$ GRP, respectively. Therefore, MON-to-GRP is a morphism by the Morphism Theorem. MON-to-GRP is an example of an inclusion. This example illustrates Corollary 13.4. □

Example 13.17 (PA-with-1 to PA) Recall from Example 9.24 that PA $\trianglelefteq_{sd}$ PA-with-1. The translation PA-with-1-to-PA is defined as:

Theory Translation Definition 13.18 (PA-with-1 to PA)

Name: PA-with-1-to-PA.

Source theory: PA-with-1.

Target theory: PA.

Base type mapping:

1. $N \mapsto N$.

Constant mapping:

1. $0_N \mapsto 0_N$.

2. $S_{N \to N} \mapsto S_{N \to N}$.

3. $1_N \mapsto S_{N \to N} 0_N$.

PA-with-1-to-PA is clearly a normal translation. It thus has no obligations of the first kind by Lemma 13.10. The obligations of the second kind for 0 and S are valid in PA by Lemma 13.11. The obligation $(S\,0)\!\downarrow$ of the second kind for 1 is valid in PA since PA $\trianglelefteq_{\mathsf{sd}}$ PA-with-1. The obligation $S\,0 = S\,0$ of the third kind for the axiom $1 = S\,0$ of PA-with-1 is valid in PA since $(S\,0)\!\downarrow$ is valid in PA. Now let $\mathbf{A}_o$ be one of the remaining four axioms of PA-with-1. The obligation for $\mathbf{A}_o$ is $\mathbf{A}_o$ itself which is valid in PA since $\mathbf{A}_o$ is also an axiom of PA. Hence all the obligations of PA-with-1-to-PA are valid in PA, and so PA-with-1-to-PA is a morphism by the Morphism Theorem. Therefore, PA-with-1-to-PA is normal morphism. This example illustrates the next proposition.   □

Proposition 13.19 $T_1 \trianglelefteq_{\mathsf{d}} T_2$ *implies there is a normal morphism from* T_2 *to* T_1.

Proof   Let $T_1 \trianglelefteq_{\mathsf{sd}} T_2$. Then $\mathcal{B}_2 = \mathcal{B}_1$ and there is a constant $\mathbf{c}_\alpha$ and a closed expression $\mathbf{A}_\alpha$ such that $\mathcal{C}_2 = \mathcal{C}_1 \cup \{\mathbf{c}_\alpha\}$ and $\Gamma_2 = \Gamma_1 \cup \{\mathbf{c}_\alpha = \mathbf{A}_\alpha\}$. Define $\mu : \mathcal{B}_2 \to \mathcal{B}_1$ to be the identity mapping and $\nu : \mathcal{C}_2 \to \mathcal{E}_1$ to be the mapping such that $\nu{\restriction}_{\mathcal{C}_1}$ is the identity mapping and $\nu(\mathbf{c}_\alpha) = \mathbf{A}_\alpha$. $\Phi = (\mu, \nu)$ is a normal translation from T_2 to T_1, and thus it has no obligations of the first kind by Lemma 13.10. It has an obligation $\mathbf{A}_\alpha\!\downarrow$ of the second kind for $\mathbf{c}_\alpha$ which is valid in T_1 since $T_1 \trianglelefteq_{\mathsf{sd}} T_2$. Its other obligations of the second kind are valid in T_1 by Lemma 13.11. It has an obligation $\overline{\nu}(\mathbf{c}_\alpha = \mathbf{A}_\alpha) \equiv (\mathbf{A}_\alpha = \mathbf{A}_\alpha)$ of the third kind which is valid in T_1 since $\mathbf{A}_\alpha\!\downarrow$ is valid in T_1. Its other obligations of the third kind are valid in T_1 since $T_1 \leq T_2$. Hence all the obligations of Φ are valid in T_1, and so Φ is a morphism from T_2 to T_1 by the Morphism Theorem. Therefore, Φ is normal morphism from T_2 to T_1.

Now let $T_1 \trianglelefteq_{\mathsf{d}} T_2$. Then there is a sequence of theories $U_1, \ldots, U_n$ with $n \leq 2$ such that $T_1 = U_1 \trianglelefteq_{\mathsf{sd}} \cdots \trianglelefteq_{\mathsf{sd}} U_n = T_2$. By repeatedly applying the argument above, there is a normal morphism Φ_{i+1} from U_{i+1} to U_i for all i with $1 \leq i \leq n - 1$. Then $\Phi_n \circ \Phi_{n-1} \circ \cdots \circ \Phi_2$ is a normal morphism from T_2 to T_1 by part 3 of Proposition 13.7.   □

Example 13.20 (**DWTOWE to COF**) Recall the theory DWTOWE from Example 9.18. The translation DWTOWE-to-COF is defined as:

Theory Translation Definition 13.21 (DWTOWE to COF)

Name: DWTOWE-to-COF.

Source theory: DWTOWE.

Target theory: COF.

Base type mapping:

1. $S \mapsto R$.

Constant mapping:

1. $\leq_{S \to S \to o} \mapsto \leq_{R \to R \to o}$.

DWTOWE-to-COF is clearly a normal injective translation that maps constants to constants by part 6 of Lemma 13.2. So the first and second kind of obligations of DWTOWE-to-COF are valid in COF by Lemma 13.10 and Lemma 13.11, respectively. For each axiom $\mathbf{A}_o$ of DWTOWE, the image of $\mathbf{A}_o$ under DWTOWE-to-COF is valid in COF since a complete ordered field is a dense weak total order without endpoints. Hence, DWTOWE-to-COF is a morphism by the Morphism Theorem. Therefore, DWTOWE-to-COF is an embedding. DWTOWE-to-COF certifies that each sentence valid in every dense weak total order without endpoints is valid in COF. □

Lemma 13.22 $T_1 \preceq T_2'$ *and* $T_2 \trianglelefteq_\mathsf{d} T_2'$ *imply there is a normal morphism from* T_1 *to* T_2.

Proof Assume $T_1 \preceq T_2'$ and $T_2 \trianglelefteq_\mathsf{d} T_2'$. $T_1 \preceq T_2'$ implies there is an inclusion Φ_1 from T_1 to T_2' by Proposition 13.3. $T_2 \trianglelefteq_\mathsf{d} T_2'$ implies there is a normal morphism Φ_2 from T_2' to T_2 by Proposition 13.19. Therefore, $\Phi_1 \circ \Phi_2$ is a normal morphism from T_1 to T_2 by part 3 of Proposition 13.7. □

Example 13.23 (**RA to PA**) This example illustrates Lemma 13.22. The following is a translation from Robinson's arithmetic to Peano arithmetic:

Theory Translation Definition 13.24 (RA to PA)

Name: RA-to-PA.

Source theory: RA.

Target theory: PA.

Base type mapping:

1. $N \mapsto N$.

Constant mapping:

1. $0_N \mapsto 0_N$.

2. $S_{N \to N} \mapsto S_{N \to N}$.

3. $+_{N \to N \to N} \mapsto$
   $I f : N \to N \to N . \forall x, y : N . f x 0 = x \wedge f x (S y) = S (f x y)$.

4. $*_{N \to N \to N} \mapsto$
   $I g : N \to N \to N .$
   $\forall x, y : N . g x 0 = 0 \wedge g x (S y) = \mathbf{F}_{N \to N \to N} (g x y) x$

   where $\mathbf{F}_{N \to N \to N}$ is
   $I f : N \to N \to N . \forall x, y : N . f x 0 = x \wedge f x (S y) = S (f x y)$.

RA-to-PA is clearly a normal translation. It is a normal morphism by Proposition 9.32 (which says RA $\preceq$ PA$'$), Example 9.30 (which says PA $\trianglelefteq_d$ PA$'$), and the proof of Lemma 13.22. $\square$

Example 13.25 (MON to COF) The following are the two translations from MON to COF that we saw in the motivating example in Section 13.1:

Theory Translation Definition 13.26 (MON to COF with $+$)

   Name: MON-to-COF-+.

   Source theory: MON.

   Target theory: COF.

   Base type mapping:

   1. $S \mapsto R$.

   Constant mapping:

1. $\cdot_{S \to S \to S} \mapsto +_{R \to R \to R}$.

2. $e_S \mapsto 0_R$.

Theory Translation Definition 13.27 (MON to COF with $*$)

Name: MON-to-COF-$*$.

Source theory: MON.

Target theory: COF.

Base type mapping:

1. $S \mapsto R$.

Constant mapping:

1. $\cdot_{S \to S \to S} \mapsto *_{R \to R \to R}$.

2. $e_S \mapsto 1_R$.

Both MON-to-COF-+ and MON-to-COF-$*$ are normal injective translations from MON to COF by part 6 of Lemma 13.2. They are also morphisms by Lemma 13.10, Lemma 13.11, and the Morphism Theorem since $(\mathbb{R}, +, 0)$ and $(\mathbb{R}, *, 1)$ are both monoids. Therefore, MON-to-COF-+ and MON-to-COF-$*$ are embeddings. $\square$

Example 13.28 (GRP to COF) We saw in motivating example in Section 13.1) that the translations MON-to-COF-+ and MON-to-COF-$*$ defined in Example 13.25 can be extended to translations from GRP to COF as follows:

Theory Translation Extension 13.29 (GRP to COF with $+$)

Name: GRP-to-COF-+.

Extends MON-to-COF-+.

New source theory: GRP.

New target theory: COF.

New base type mapping:

New constant mapping:

3. $\cdot^{-1}{}_{S \to S} \mapsto -_{R \to R}.$

Theory Translation Extension 13.30 (GRP to COF with $*$)

    Name: GRP-to-COF-$*$.

    Extends MON-to-COF-$*$.

    New source theory: GRP.

    New target theory: COF.

    New base type mapping:

    New constant mapping:

    3. $\cdot^{-1}{}_{S \to S} \mapsto \cdot^{-1}{}_{R \to R}.$

GRP-to-COF-$+$ and GRP-to-COF-$*$ are normal injective translations by part 6 of Lemma 13.2. GRP-to-COF-$+$ is a morphism by Lemma 13.10, Lemma 13.11, and the Morphism Theorem since $(\mathbb{R}, +, 0, -)$ is a group. Therefore, GRP-to-COF-$+$ is an embedding from GRP to COF.

    GRP-to-COF-$*$ is not a morphism from GRP to COF since $0^{-1}\uparrow$ is an axiom of COF which implies that the image of the axiom

$$\forall x : S . \, x^{-1} \cdot x = e$$

of GRP under GRP-to-COF-$*$ is not valid in COF. However, the translation can be transformed into a morphism by restricting the image of the type S to a quasitype that denotes the set of nonzero values of type R as follows:

Theory Translation Definition 13.31 (GRP to Nonzero COF)

    Name: GRP-to-nonzero-COF.

    Source theory: GRP.

    Target theory: COF.

    Base type mapping:

    1. $S \mapsto \{r : R \mid r \neq 0_R\}.$

Constant mapping:

1. $\cdot_{S \to S \to S} \mapsto \lambda x : \{r : R \mid r \neq 0_R\} \, . \, \lambda y : \{r : R \mid r \neq 0_R\} \, . \, *_{R \to R \to R} x \, y.$

2. $e_S \mapsto 1_R.$

3. $\cdot^{-1}_{S \to S} \mapsto \lambda x : \{r : R \mid r \neq 0_R\} \, . \, \cdot^{-1}_{R \to R} x.$

GRP-to-nonzero-COF is a nonnormal translation. It is a morphism from GRP to COF by the Morphism Theorem since $\{r : R \mid r \neq 0_R\}$ denotes a nonempty set and

$$(\{r \in \mathbb{R} \mid r \neq 0\}, *, 1, \cdot^{-1})$$

is a group. It illustrates the need to sometimes map types to quasitypes and constants that denote total functions to lambda-expressions restricted to the quasitypes that denote partial functions. This is not unusual; partial functions naturally arise from morphisms that map types to quasitypes [41]. $\square$

Example 13.32 (GRP **to Dual** GRP) The following is a translation from GRP to itself that reverses the order of application of $\cdot_{S \to S \to S}$:

Theory Translation Definition 13.33 (GRP **to Dual** GRP)

Name: GRP-to-dual-GRP.

Source theory: GRP.

Target theory: GRP.

Base type mapping:

1. $S \mapsto S.$

Constant mapping:

1. $\cdot_{S \to S \to S} \mapsto \lambda x : S \, . \, \lambda y : S \, . \, \cdot_{S \to S \to S} y \, x.$

2. $e_S \mapsto e_S.$

3. $\cdot^{-1}_{S \to S} \mapsto \cdot^{-1}_{S \to S}.$

GRP-to-dual-GRP is a normal translation. It maps a sentence $\mathbf{A}_o$ of GRP to a sentence $\mathbf{A}'_o$ that is the dual of $\mathbf{A}_o$ with respect to the order of application of $\cdot_{S \to S \to S}$. For example, GRP-to-dual-GRP maps the left cancellation law

$$\forall x, y : S \,.\, x \cdot y = x \cdot z \Leftrightarrow y = z$$

to a sentence that simplifies by beta-reduction to the right cancellation law

$$\forall x, y : S \,.\, y \cdot x = z \cdot x \Leftrightarrow y = z,$$

the dual of the left cancellation law. It is easy to show that GRP-to-dual-GRP is a normal morphism using Lemma 13.10, Lemma 13.11, and the Morphism Theorem. Morphisms like GRP-to-dual-GRP from a theory to itself are useful for formalizing duality and symmetry arguments. □

Example 13.34 (FUN-COMP to COF)
The previous examples of normal morphisms presented in this subsection are injective on types. In this example, we will see a useful morphism that is not injective on types. Let FUN-COMP be the theory $((\{A, B, C, D\}, \emptyset), \emptyset)$. Even though FUN-COMP has no constants or axioms, it is suitable for reasoning about function composition. In particular, the associativity of function composition

$$\forall f : A \to B, \, g : B \to C, \, h : C \to D \,.\, f \circ (g \circ h) = (f \circ g) \circ h$$

can be proved in FUN-COMP. Now consider the following translation:

Theory Translation Definition 13.35 (FUN-COMP to COF)

 Name: FUN-COMP-to-COF.

 Source theory: FUN-COMP.

 Target theory: COF.

 Base type mapping:

 1. $A \mapsto R$.

 2. $B \mapsto R$.

 3. $C \mapsto R$.

 4. $D \mapsto R$.

Constant mapping:

FUN-COMP-to-COF is obviously a normal translation. It is also obviously a morphism by the Morphism Theorem since it has no obligations. The image of the associativity of function composition theorem under FUN-COMP-to-COF is the associativity of function composition theorem for functions of type $R \to R$. This example illustrates that very simple theories like FUN-COMP can be useful in the application of the little theories method. □

13.3.4 Faithful Theory Morphisms

A morphism Φ from T_1 to T_2 is *faithful* if $T_2 \vDash \overline{\nu}(\mathbf{A}_o)$ implies $T_1 \vDash \mathbf{A}_o$ for all sentences $\mathbf{A}_o$ of L_1. In other words, a translation from Φ from T_1 to T_2 is a *faithful morphism* if $T_1 \vDash \mathbf{A}_o$ iff $T_2 \vDash \overline{\nu}(\mathbf{A}_o)$ for all sentences $\mathbf{A}_o$ of L_1. That is, a faithful morphism is a translation that is strongly meaning preserving in the sense that it maps both theorems to theorems and nontheorems to nontheorems. For example, all the morphisms defined in Subsection 13.3.3 are nonfaithful except for PA-with-1-to-PA and GRP-to-dual-GRP.

Proposition 13.36 PA-with-1-to-PA *is a faithful morphism.*

Proof We saw in Example 13.17 that PA-with-1-to-PA is a morphism. Let $\mathbf{A}_o$ be a sentence of PA-with-1 such that PA $\vDash \overline{\nu}(\mathbf{A}_o)$. Then (a) PA-with-1 $\vDash \overline{\nu}(\mathbf{A}_o)$ since PA $\leq$ PA-with-1. (b) PA-with-1 $\vDash \overline{\nu}(\mathbf{A}_o) \Leftrightarrow \mathbf{A}_o$ since PA $\trianglelefteq_{\mathsf{sd}}$ PA-with-1. So (a) and (b) imply PA-with-1 $\vDash \mathbf{A}_o$. Therefore, PA-with-1-to-PA is faithful. □

Proposition 13.37 GRP-to-dual-GRP *is a faithful morphism.*

Proof We saw in Example 13.32 that $(\star)$ GRP-to-dual-GRP is a morphism. Let $\mathbf{A}_o$ be a sentence of GRP such that GRP $\vDash \overline{\nu}(\mathbf{A}_o)$. Then GRP $\vDash \overline{\nu}(\overline{\nu}(\mathbf{A}_o))$ by $(\star)$. This implies GRP $\vDash \mathbf{A}_o$ since GRP $\vDash \overline{\nu}(\overline{\nu}(\mathbf{A}_o)) \Leftrightarrow \mathbf{A}_o$ by the construction of GRP-to-dual-GRP. Therefore, GRP-to-dual-GRP is faithful. □

The following proposition is the analog of Corollary 13.4 for conservative extensions:

Proposition 13.38 *If $T_1 \trianglelefteq T_2$, then the identity translation from T_1 to T_2 is a faithful inclusion.*

Proof Let $T_1 \trianglelefteq T_2$, $\Phi = (\mu, \nu)$ be the identity translation from T_1 to T_2, and $\mathbf{A}_o$ be a sentence of L_1. Then $(\star)$ $\overline{\nu}(\mathbf{A}_o) \equiv \mathbf{A}_o$. $T_1 \trianglelefteq T_2$ implies $T_1 \leq T_2$. It follows from this and $(\star)$ that $T_1 \vDash \mathbf{A}_o$ implies $T_2 \vDash \overline{\nu}(\mathbf{A}_o)$. Hence Φ is an inclusion. By $T_1 \trianglelefteq T_2$ and $(\star)$, $T_2 \vDash \overline{\nu}(\mathbf{A}_o)$ implies $T_1 \vDash \mathbf{A}_o$. Therefore, Φ is faithful.   □

Remark 13.39 (Conservativity and Faithfulness) *The notion of a faithful morphism from a theory T is a generalization of the notion of a conservative extension of T. As shown by Proposition 13.38, a conservative theory extension can be viewed as a faithful inclusion.*

The following theorem is a generalization of Proposition 9.21:

Theorem 13.40 (Faithfulness implies Equisatisfiability) *Let Φ be a faithful morphism from T_1 to T_2. Then T_1 is satisfiable iff T_2 is satisfiable.*

Proof Let $\Phi = (\mu, \nu)$ be a faithful morphism from T_1 to T_2. Then $T_1 \vDash F_o$ iff $T_2 \vDash \overline{\nu}(F_o)$. However, by part 4 of Lemma 13.2, $\overline{\nu}(F_o) = F_o$ since F_o is a sentence of L_{min}. Hence T_1 is unsatisfiable iff T_2 is unsatisfiable, and so T_1 is satisfiable iff T_2 is satisfiable.   □

An *interface* for a theory T is a faithful embedding Φ from some theory T' to T. T' and T are called, respectively, the *front* and *back* of the interface. An interface is intended to be a convenient means for accessing T. The front of a good interface includes a carefully selected set of constants and theorems that denote orthogonal concepts and facts that can be easily combined to express the concepts and facts of the back of the interface.

Proposition 13.41 *If $T_1 \trianglelefteq T_2$, then T_1 is the front of an interface for T_2.*

Proof The identity translation from T_1 to T_2 is a faithful embedding (faithful inclusion) by Proposition 13.38.   □

The morphisms defined in this subsection and their source and target theories form the theory graph shown in Figure 13.1. An inclusion is designated by a $\hookrightarrow$ arrow, and a morphism that is not an inclusion is designated by a $\rightarrow$ arrow. We could have defined a theory morphism from PA to COF (shown by the dashed arrow), but it is more convenient to define a "development morphism" from a development of PA (e.g., PA-dev-4) to a development of COF (e.g., COF-dev-5). The notion of a development morphism is the subject of the next section.

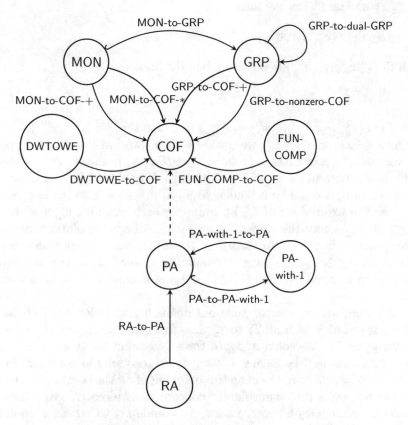

Figure 13.1: An Example Theory Graph

13.4 Development Morphisms

Suppose we work with developments instead of theories. What should a translation between developments be? We will answer this question in this section.

First, we need some definitions. Let $L_i = (\mathcal{B}_i, \mathcal{C}_i)$ and $L_i' = (\mathcal{B}_i, \mathcal{C}_i')$ be languages, $T_i = (L_i, \Gamma_i)$ and $T_i' = (L_i', \Gamma_i')$ be theories, $\mathcal{T}_i = \mathcal{T}(L_i)$, $\mathcal{T}_i^- = \{\alpha \in \mathcal{T}_i \mid \alpha \neq o\}$, $\mathcal{T}_i' = \mathcal{T}(L_i')$, $\mathcal{E}_i = \mathcal{E}(L_i)$, $\mathcal{E}_i' = \mathcal{E}(L_i')$, $\mathcal{Q}_i$ be the set of closed quasitypes in $\mathcal{E}(L_i)$, $\mathcal{Q}_i^- = \{\mathbf{Q}_{\{\alpha\}} \in \mathcal{Q}_i \mid \alpha \neq o\}$, and $D_i = (T_i, \Xi_i)$ be a development where T_i' is the top theory of D_i for $i \in \{1, 2\}$. Then

$T_1 \trianglelefteq_d T_1'$ and so L_1' has the form

$$(\mathcal{B}_1, \mathcal{C}_1 \cup \{\mathbf{c}_{\alpha_1}^1, \ldots, \mathbf{c}_{\alpha_n}^n\})$$

with $\mathcal{C}_1 \cap \{\mathbf{c}_{\alpha_1}^1, \ldots, \mathbf{c}_{\alpha_n}^n\} = \emptyset$ and Γ_1' has the form

$$\Gamma_1 \cup \{\mathbf{c}_{\alpha_1}^1 = \mathbf{A}_{\alpha_1}^1, \ldots, \mathbf{c}_{\alpha_n}^n = \mathbf{A}_{\alpha_n}^n\}$$

with $\Gamma_1 \cap \{\mathbf{c}_{\alpha_1}^1 = \mathbf{A}_{\alpha_1}^1, \ldots, \mathbf{c}_{\alpha_n}^n = \mathbf{A}_{\alpha_n}^n\} = \emptyset$. Let $\mathcal{C} = \{\mathbf{c}_{\alpha_1}^1, \ldots, \mathbf{c}_{\alpha_n}^n\}$.

As a *minimum approach*, we could define a translation from D_1 to D_2 to be, very simply, a translation Ψ from T_1 to T_2, i.e., from the bottom theory of D_1 to the bottom theory of D_2. Although Ψ would not be defined on the constants in $\mathcal{C}$, it could be extended to a full mapping from the expressions of L_1' to the expressions of L_2' by mapping each constant $\mathbf{c}_{\alpha_i}^i \in \mathcal{C}$ to the image of $\mathbf{A}_{\alpha_i}^i$ under the translation. Ψ would be a translation from any development whose bottom theory is T_1 to any other development whose bottom theory is T_2. However, the number of symbols in the image of a constant in $\mathcal{C}$ could explode if $\mathbf{A}_{\alpha_i}^i$ contains instances of other constants in $\mathcal{C}$.

As a *maximum approach*, we could define a translation from D_1 to D_2 to be a translation Ψ' from T_1' to T_2', i.e., from the top theory of D_1 to the top theory of D_2. We could mitigate the explosion of the size of the images of the constants in $\mathcal{C}$ by having Ψ' map these constants to constants in L_2'. However, Ψ' would have to define the mapping of all the constants in $\mathcal{C}$, not just the constants that immediately concern us. Moreover, every time we extend D_1 by defining a new constant, we would have to extend Ψ' to define the mapping of that new constant.

The approach we will actually use provides greater flexibility by compromising between the minimum and maximum approaches. A development translation sits between Ψ of the minimum approach and Ψ' of the maximum approach so that the minimum and maximum approaches are special cases of our approach.

13.4.1   Development Translations

A *(development) translation from D_1 to D_2* is a tuple $\Phi = (\mu, \nu)$ such that:

1. $\mu : \mathcal{B}_1 \to \mathcal{T}_2^- \cup \mathcal{Q}_2^-$ is total.

2. $\nu : \mathcal{C}_1' \to \mathcal{E}_2'$ is partial or total.

3. $\nu\!\restriction_{\mathcal{C}_1}$ is total.

4. $\nu(\mathbf{c}_\alpha)$ is a closed expression of type $\tau(\overline{\mu}(\alpha))$ for all $\mathbf{c}_\alpha \in \mathcal{C}_1'$ for which $\nu(\mathbf{c}_\alpha)$ is defined.

A theory translation is a special case of a development translation since a theory T is identified with the trivial development $(T, [\,])$. Notice that, if $\nu = \nu{\restriction}_{\mathcal{C}_1}$, then Φ is the translation Ψ of the minimum approach, and if ν is total, then Φ is the translation Ψ' of the maximum approach. These observations are encoded in the following proposition:

Proposition 13.42 *Let* $\Phi = (\mu, \nu)$ *be a translation from* D_1 *to* D_2.

1. If $\nu = \nu{\restriction}_{\mathcal{C}_1}$, *then* Φ *is a translation from* T_1 *to* T_2.

2. If ν *is total, then* Φ *is a translation from* T_1' *to* T_2'.

Let $\Phi = (\mu, \nu)$ be a translation from D_1 to D_2. We will now define how to expand the domain of ν to $\mathcal{E}_1'$ by mapping each constant $\mathbf{c}_{\alpha_i}^i \in \mathcal{C}$ outside of the domain of ν to the image of $\mathbf{A}_{\alpha_i}^i$ under the translation as suggested in the discussion of the minimum approach. Notice that this works because $\mathbf{c}_{\alpha_i}^i$ does not occur in $\mathbf{A}_{\alpha_i}^i$. Let $\widetilde{\nu} : \mathcal{E}_1' \to \mathcal{E}_2'$ be the canonical extension of ν defined as follows by recursion and pattern matching:

1. $\widetilde{\nu}((\mathbf{x} : \alpha)) \equiv (\mathbf{x} : \tau(\overline{\mu}(\alpha)))$.

2. $\widetilde{\nu}(\mathbf{c}_\alpha) \equiv \nu(\mathbf{c}_\alpha)$ if $\nu(\mathbf{c}_\alpha)$ is defined.

3. $\widetilde{\nu}(\mathbf{c}_{\alpha_i}^i) \equiv \widetilde{\nu}(\mathbf{A}_{\alpha_i}^i)$ if $\mathbf{c}_{\alpha_i}^i \in \mathcal{C}$ and $\nu(\mathbf{c}_{\alpha_i}^i)$ is undefined.

4. $\widetilde{\nu}(\mathbf{A}_\alpha = \mathbf{B}_\alpha) \equiv (\widetilde{\nu}(\mathbf{A}_\alpha) = \widetilde{\nu}(\mathbf{B}_\alpha))$.

5. $\widetilde{\nu}(\mathbf{F}_{\alpha \to \beta} \, \mathbf{A}_\alpha) \equiv \widetilde{\nu}(\mathbf{F}_{\alpha \to \beta}) \, \widetilde{\nu}(\mathbf{A}_\alpha)$.

6. $\widetilde{\nu}(\lambda \mathbf{x} : \alpha \, . \, \mathbf{B}_\beta) \equiv \lambda \mathbf{x} : \overline{\mu}(\alpha) \, . \, \widetilde{\nu}(\mathbf{B}_\beta)$.

7. $\widetilde{\nu}(\mathrm{I} \mathbf{x} : \alpha \, . \, \mathbf{A}_o) \equiv \mathrm{I} \mathbf{x} : \overline{\mu}(\alpha) \, . \, \widetilde{\nu}(\mathbf{A}_o)$.

8. $\widetilde{\nu}((\mathbf{A}_\alpha, \mathbf{B}_\beta)) \equiv (\widetilde{\nu}(\mathbf{A}_\alpha), \widetilde{\nu}(\mathbf{B}_\beta))$.

Define $\widetilde{\Phi} = (\mu, \widetilde{\nu}{\restriction}_{\mathcal{C}_1'})$.

Proposition 13.43 *Let* $\Phi = (\mu, \nu)$ *be a translation from* D_1 *to* D_2. *Then* $\widetilde{\Phi}$ *is a translation from* T_1' *to* T_2'.

Proof The proof is left to the reader as an exercise.   □

A translation Φ from D_1 to D_2 is a *(development) morphism from D_1 to D_2* if $\widetilde{\Phi}$ is a morphism from T_1' to T_2'. By the Morphism Theorem (Theorem 13.13), Φ is a development morphism if each obligation of the theory translation $\widetilde{\Phi}$ is valid in T_2'. A theory morphism is a special case of a development morphism since a theory T is identified with the trivial development $(T, [\,])$.

A *development translation definition module* is used to present a translation from one development to another. It has the following form:

Development Translation Definition X.Y

   Name: Name.

   Source development: Name-of-source-development.

   Target development: Name-of-target-development.

   Base type mapping:

   $\mathbf{a}_1 \mapsto \zeta_1$ where ζ_1 is a type or closed quasitype with $\tau(\zeta_1) \neq o$.

   $\vdots$

   $\mathbf{a}_m \mapsto \zeta_m$ where ζ_m is a type or closed quasitype with $\tau(\zeta_m) \neq o$.

   Constant mapping:

   $\mathbf{c}_{\alpha_1}^1 \mapsto \mathbf{B}_{\beta_1}^1$.

   $\vdots$

   $\mathbf{c}_{\alpha_n'}^{n'} \mapsto \mathbf{B}_{\beta_{n'}}^{n'}$.

The development translation definition module is well-formed if Name-of-source-development is the name of a development D_1, Name-of-target-theory is the name of a development D_2, $T_1 = ((\mathcal{B}_1, \mathcal{C}_1), \Gamma_1)$, $\mathcal{B}_1 = \{\mathbf{a}_1, \ldots, \mathbf{a}_m\}$, $\mathcal{C}_1 = \{\mathbf{c}_{\alpha_1}^1, \ldots, \mathbf{c}_{\alpha_n}^n\}$, $T_1' = ((\mathcal{B}_1, \mathcal{C}_1'), \Gamma_1')$, $\{\mathbf{c}_{\alpha_1}^1, \ldots, \mathbf{c}_{\alpha_{n'}}^{n'}\} \subseteq \mathcal{C}_1'$ with $n \leq n'$, and $\Phi = (\mu, \nu)$ is a translation from D_1 to D_2 where $\mu(\mathbf{a}_i) = \zeta_i$ for all i with $1 \leq i \leq m$ and $\nu(\mathbf{c}_{\alpha_j}^j) = \mathbf{B}_{\beta_j}^j$ for all j with $1 \leq j \leq n'$. Name is a name of the translation Φ.

Let $\Phi = (\mu, \nu)$ be a translation from D_1 to D_2. A translation $\Phi' = (\mu', \nu')$ from D_1' to D_2' is an *extension* of Φ (or Φ is a *subtranslation* of Φ'), written $\Phi \leq \Phi'$, if $D_1 \leq D_1'$, $D_2 \leq D_2'$, $\mu = \mu'$, and $\nu \sqsubseteq \nu'$.

A *development translation extension module* is used to present an extension of a translation from one development to another. It has the following form:

Development Translation Extension X.Y

Name: Name.

Extends Name-of-subtranslation.

New source development: Name-of-new-source-development.

New target development: Name-of-new-target-development.

New defined constant mapping:

$$\mathbf{c}_{\alpha_1}^1 \mapsto \mathbf{B}_{\beta_1}^1 .$$

$$\vdots$$

$$\mathbf{c}_{\alpha_n}^n \mapsto \mathbf{B}_{\beta_n}^n .$$

The development translation extension module is well-formed if Name-of-subtranslation is the name of a translation $\Phi = (\mu, \nu)$ from D_1 to D_2, Name-of-new-source-development is the name of an extension D_1' of D_1, Name-of-new-target-development is the name of an extension D_2' of D_2, $T_1' = ((\mathcal{B}_1, \mathcal{C}_1), \Gamma_1)$, $T_1'' = ((\mathcal{B}_1, \mathcal{C}_1'), \Gamma_1')$ is the top theory of D_1', $\{\mathbf{c}_{\alpha_1}^1, \ldots, \mathbf{c}_{\alpha_n}^n\} \subseteq \mathcal{C}_1'$ with $\mathrm{dom}(\nu) \cap \{\mathbf{c}_{\alpha_1}^1, \ldots, \mathbf{c}_{\alpha_n}^n\} = \emptyset$, and $\Phi' = (\mu, \nu')$ is a translation from D_1' to D_2' such that $\nu \sqsubseteq \nu'$, $\mathrm{dom}(\nu') = \mathrm{dom}(\nu) \cup \{\mathbf{c}_{\alpha_1}^1, \ldots, \mathbf{c}_{\alpha_n}^n\}$, and $\nu'(\mathbf{c}_{\alpha_j}^j) = \mathbf{B}_{\beta_j}^j$ for all j with $1 \leq j \leq n$. Name is a name of the translation Φ'.

Example 13.44 (PA to COF) The following is a translation from PA-dev-4 (whose top theory is NAT) to COF-dev-5 (whose top theory is REAL):

Development Translation Definition 13.45 (PA to COF 1)

Name: PA-to-COF.

Source development: PA-dev-4.

Target development: COF-dev-5.

Base type mapping:

1. $N \mapsto N_{\{R\}}$.

Constant mapping:

1. $0_N \mapsto 0_R$.

2. $S_{N \to N} \mapsto \lambda\, x : N_{\{R\}} \,.\, x + 1$.

3. $1_N \mapsto 1_R$.

4. $P_{N \to N} \mapsto \lambda\, x : N_{\{R\}} \,.\, x \neq 0 \mapsto x - 1 \mid \perp_R$.

5. $+_{N \to N \to N} \mapsto \lambda\, x : N_{\{R\}} \,.\, \lambda\, y : N_{\{R\}} \,.\, x + y$.

6. $*_{N \to N \to N} \mapsto \lambda\, x : N_{\{R\}} \,.\, \lambda\, y : N_{\{R\}} \,.\, x * y$.

It is easy to check that PA-to-COF $= (\mu, \nu)$ is a development morphism from PA-dev-4 to COF-dev-5. However, PA-to-COF is not a theory morphism from NAT to REAL because ν is not defined on all the constants of NAT, namely, the constants of NAT introduced in the development extensions PA-dev-3 and PA-dev-4.

Suppose $\mathbf{c}_\alpha = \mathbf{A}_\alpha$ is a definition of NAT such that $\nu(\mathbf{c}_\alpha)$ is not defined. There are four ways of dealing with this situation. First, if we do not need to use PA-to-COF to map expressions containing $\mathbf{c}_\alpha$ or if we can tolerate the size of the image of $\mathbf{c}_\alpha$ under $\widetilde{\nu}$, then we can leave PA-to-COF as it is. Second, we can define $\nu(\mathbf{c}_\alpha)$ to be some appropriate constant of REAL provided there is such a constant. Third, we can define $\nu(\mathbf{c}_\alpha)$ to be some appropriate nonconstant expression of REAL provided the size of this expression is tolerable. And fourth, we can transport the definition $\mathbf{c}_\alpha = \mathbf{A}_\alpha$ from PA-dev-4 to COF-dev-5 via PA-to-COF (see the discussion and Example 13.47 in the next subsection).   □

13.4.2 Transportations

We have said that the power of the little theories method is that a definition or theorem established in T can be utilized in every other theory that contains the same logical structure as T. This is done by transporting the definition or theorem in a development D of T to another development D'

that is the target development of a development morphism from D. (It is possible that D and D' are the same development.)

Let $\Phi = (\mu, \nu)$ be a morphism from D_1 to D_2. Assume $\mathbf{c}_\alpha = \mathbf{A}_\alpha$ is a definition of D_1 such that $\mathbf{c}_\alpha \notin \mathrm{dom}(\nu)$. A *transportation of* $\mathbf{c}_\alpha = \mathbf{A}_\alpha$ *from* D_1 *to* D_2 *via* Φ is a pair (D'_2, Φ') where $D'_2 = (T_2, \Xi_2 \mathbin{++} [P])$ is an extension of D_2, P is a definition package of the form $(n, \mathbf{d}_{\tau(\overline{\mu}(\alpha))}, \widetilde{\nu}(\mathbf{A}_\alpha), \pi)$ such that $\mathbf{d}_{\tau(\overline{\mu}(\alpha))} \notin C_2$, and $\Phi' = (\mu, \nu')$ is an extension of Φ such that

$$\nu'(x) = \left\{ \begin{array}{ll} \nu(x) & \text{if } x \in \mathrm{dom}(\nu); \\ \mathbf{d}_{\tau(\overline{\mu}(\alpha))} & \text{if } x = \mathbf{c}_\alpha. \end{array} \right.$$

Hence (1) the definition $\mathbf{c}_\alpha = \mathbf{A}_\alpha$ of D_1 is translated to a new definition $\mathbf{d}_{\tau(\overline{\mu}(\alpha))} = \widetilde{\nu}(\mathbf{A}_\alpha)$, an analog of $\mathbf{c}_\alpha = \mathbf{A}_\alpha$, that is added to the D_2 and (2) Φ is extended to include the mapping $\mathbf{c}_\alpha \mapsto \mathbf{d}_{\tau(\overline{\mu}(\alpha))}$.

Now assume $\mathbf{B}_o$ is a theorem of D_1. A *transportation of* $\mathbf{B}_o$ *from* D_1 *to* D_2 *via* Φ is a development $D'_2 = (T_2, \Xi_2 \mathbin{++} [P])$ where P is a theorem package of the form $(n, \widetilde{\nu}(\mathbf{B}_o), \pi)$. Hence the theorem $\mathbf{B}_o$ of D_1 is translated to a new theorem $\widetilde{\nu}(\mathbf{B}_o)$, an analog of $\mathbf{B}_o$, that is added to the D_2.

Proposition 13.46 *Let* $\Phi = (\mu, \nu)$ *be a morphism from* D_1 *to* D_2,

1. *If* $\mathbf{c}_\alpha = \mathbf{A}_\alpha$ *is a definition of* D_1 *such that* $\mathbf{c}_\alpha \notin \mathrm{dom}(\nu)$, *then there is a transportation of* $\mathbf{c}_\alpha = \mathbf{A}_\alpha$ *from* D_1 *to* D_2 *via* Φ.

2. *If* $\mathbf{B}_o$ *is a theorem of* D_1, *then there is a transportation of* $\mathbf{B}_o$ *from* D_1 *to* D_2 *via* Φ.

Proof

1. We can construct the required development extension of D_2 and development morphism extension of Φ since (1) there is a proof in D_1 that $\mathbf{A}_\alpha\!\downarrow$ is valid in the top theory of D_1 and (2) Φ is a morphism.

2. We can construct the required development extension of D_2 since (1) there is a proof in D_1 that $\mathbf{B}_o$ is valid in the top theory of D_1 and (2) Φ is a morphism. $\qquad\qquad\square$

A *definition transportation module* is used to present a transportation of a definition. It has the following form:

Definition Transportation X.Y

   Name: Name-1.

   Development morphism: Name-2.

   Definition: P.

   Transported definition: P'.

   New target development: Name-3.

   New development translation: Name-4.

The definition transportation is well-formed if Name-2 is the name of a development morphism $\Phi = (\mu, \nu)$ from D_1 to D_2, $P = (n, \mathbf{c}_\alpha, \mathbf{A}_\alpha, \pi)$ is a definition package of D_1 where $\mathbf{c}_\alpha \notin \mathrm{dom}(\nu)$, and

$$P' = (n\text{-via-Name-1}, \mathbf{d}_\beta, \widetilde{\nu}(\mathbf{A}_\alpha), \pi')$$

is a definition package valid in the top theory of D_2 where $\beta = \tau(\overline{\mu}(\alpha))$ and $\mathbf{d}_\beta \notin C_2$. Name-3 is a name for the development $D_2' = (T_2, \Xi_2 ++ [P'])$. Name-4 is a name for the development morphism $\Phi' = (\mu, \nu')$ from D_1 to D_2' where ν' extends ν with $\nu'(\mathbf{c}_\alpha) = \mathbf{d}_\beta$. Name-1 is the name of the definition transportation (D_2', Φ'). Note that π' is a proof that says $\mathbf{A}_\alpha{\downarrow}$ is valid in the top theory of D_1 by π and so $\widetilde{\nu}(\mathbf{A}_\alpha){\downarrow}$ is valid in the top theory of D_2 by the fact that Φ is a development morphism. P' can automatically be constructed from $\mathbf{d}_\beta$, P, and Φ. Thus nothing is lost if we replace P' with $\mathbf{d}_\beta$ in the definition transportation. We have chosen to include P' since it serves as useful documentation.

A *theorem transportation module* is used to present a transportation of a theorem. It has the following form:

Theorem Transportation X.Y

   Name: Name-1.

   Development morphism: Name-2.

   Theorem: P.

   Transported theorem: P'.

   New target development: Name-3.

The theorem transportation is well-formed if Name-2 is the name of a development morphism $\Phi = (\mu, \nu)$ from D_1 to D_2, $P = (n, \mathbf{B}_o, \pi)$ is a theorem package of D_1, and

$$P' = (n\text{-via-Name-1}, \widetilde{\nu}(\mathbf{B}_o), \pi')$$

is a theorem package valid in the top theory of D_2. Name-3 is a name for the development $D_2' = (T_2, \Xi_2 ++ [P'])$. Name-1 is the name of the theorem transportation D_2'. Note that π' is a proof that says $\mathbf{B}_o$ is valid in the top theory of D_1 by π and so $\widetilde{\nu}(\mathbf{B}_o)$ is valid in the top theory of D_2 by the fact that Φ is a development morphism. P' can automatically be constructed from P and Φ. Thus nothing is lost if we remove P' from the theorem transportation. We have chosen to include P' since it serves as useful documentation.

Example 13.47 (Transportation of Divides) The following definition transportation module presents a definition transportation of the definition of $\mid$ (divides) from PA-dev-4 to COF-dev-5 via PA-to-COF:

Definition Transportation 13.48 (Transport of $\mid$ to COF)

    Name: Divides-via-PA-to-COF.

    Development morphism: PA-to-COF.

    Definition:

        Def6: $\mid_{N \to N \to o} = \lambda x : N \,.\, \lambda y : N \,.\, \exists z : N \,.\, x * z = y$ (divides).

    Transported definition:

        Def6-via-PA-to-COF:
        $\mid_{R \to R \to o} = \lambda x : N_{\{R\}} \,.\, \lambda y : N_{\{R\}} \,.\, \exists z : N_{\{R\}} \,.\, x * z = y$ (divides).

    New target development: COF-dev-6.

    New development translation: PA-to-COF-1.

This definition transportation is equivalent to the following development extension and development translation extension:

Development Extension 13.49 (COF Development 6)

Name: COF-dev-6.

Extends COF-dev-5.

New definitions and theorems:

Def6-via-PA-to-COF:
$$|_{R\to R\to o} = \lambda\, x : N_{\{R\}} \,.\, \lambda\, y : N_{\{R\}} \,.\, \exists\, z : N_{\{R\}} \,.\, x * z = y \quad \text{(divides)}.$$

Development Translation Extension 13.50 (PA to COF 2)

Name: PA-to-COF-1.

Extends PA-to-COF.

New source development: PA-dev-4.

New target development: COF-dev-6.

New defined constant mapping:

$$|_{N\to N\to o} \mapsto |_{R\to R\to o}.$$

□

Example 13.51 (Transportation of "0 is Top") The following theorem transportation module presents a theorem transportation of the theorem that says 0 is the top element with respect to | (divides) from **PA-dev-4** to **COF-dev-6** via **PA-to-COF-1**:

Theorem Transportation 13.52 (Transport of "0 is Top" to COF)

Name: 0-is-Top-via-PA-to-COF-1.

Development morphism: PA-to-COF-1.

Theorem:

Thm22: $\forall\, x : N \,.\, x\, |_{N\to N\to o}\, 0$ (0 is top).

Transported theorem:

Thm22-via-PA-to-COF-1: $\forall\, x : N_{\{R\}} \,.\, x\, |_{R\to R\to o}\, 0$ (0 is top).

New target development: COF-dev-7.

□

13.5 Mathematics Libraries

In this subsection, we will define three kinds of graphical structures that are suitable as architectures for libraries of mathematical knowledge.

13.5.1 Theory Graphs

A *theory graph* of Alonzo is a directed graph $G = (\mathcal{N}, \mathcal{V})$ where $\mathcal{N}$ is a set of theories of Alonzo and $\mathcal{V}$ is a set of theory morphisms of Alonzo from T_1 to T_2 where $T_1, T_2 \in \mathcal{N}$. The theories are the nodes of the graph and the theory morphisms are the directed edges of the graph. A theory graph is purely a syntactic object; it contains no proofs. However, proofs are needed to show that one theory is a definitional extension of another and that a translation is a morphism. These proofs, however, reside outside of the theory graph. Theory definition and theory extension modules are used to present the theories of a theory graph and theory translation definition and extension modules are used to present the morphisms.

In section 9.1, we defined the notion of equivalent theories for theories having the same language. We can now generalize this definition to theories having possibly different languages: Two theories T_1 and T_2 are *equivalent*, written $T_1 \equiv T_2$, if there is a morphism from T_1 to T_2 and a morphism from T_2 to T_1.[3] We can also now generalize the definition of the axiomatization of a theory: An *axiomatization* of T is any theory T' (whose language may be different from the language of T) that is equivalent (in the more general sense) to T.

Proposition 13.53 $T_1 \trianglelefteq_d T_2$ *implies* $T_1 \equiv T_2$.

Proof $T_1 \trianglelefteq_d T_2$ implies $T_1 \leq T_2$ which implies there is an inclusion from T_1 to T_2 by Corollary 13.4. $T_1 \trianglelefteq_d T_2$ implies there is a normal morphism from T_2 to T_1 by Proposition 13.19. Therefore, $T_1 \equiv T_2$. □

An unsatisfiable theory is almost never a useful component of any kind of theory graph. By Relative Satisfiability (Theorem 13.14), a theory T is satisfiable if there is a morphism from T to a theory that is known to be satisfiable. Thus, if T is a theory newly added to the theory graph, it is good practice to construct a morphism from T to a theory T' that is trusted to be satisfiable. This will reduce the likelihood that the theory graph contains any unsatisfiable theories.

---

[3]Theories equivalent in this sense are often called *mutually interpretable* theories.

13.5.2   Development Graphs

A theory graph is a suitable architecture for building a library of static theories, but it is not suitable for building a library in which theories are dynamically developed by making definitions and proving theorems. A graph of developments is an architecture that is more suitable for the latter kind of mathematics library.

A *development graph* of Alonzo is a directed graph $G = (\mathcal{N}, \mathcal{V})$ where $\mathcal{N}$ is a set of developments of Alonzo and $\mathcal{V}$ is a set of development morphisms of Alonzo from D_1 to D_2 where $D_1, D_2 \in \mathcal{N}$. The developments are the nodes of the graph and the development morphisms are the directed edges of the graph. Since we are identifying each theory T with its trivial development $(T, [\,])$, each theory translation Φ from T_1 to T_2 is also a development translation from $(T_1, [\,])$ to $(T_2, [\,])$. Thus $\mathcal{N}$ includes theories viewed as developments and $\mathcal{V}$ includes theory morphisms viewed as development morphisms. Unlike a theory graph, a development graph contains proofs that definitions and theorems are valid. These proofs can be used to verify metatheorems about the objects in the graph. A metatheorem can assert, for example, that a theory is a conservative extension of another theory or that a translation is a morphism. All the mathematical knowledge modules of Alonzo can be used in the presentation of a development graph.

A development graph is a development space that supports four ways of developing mathematical knowledge. First, new theories can be added to the graph (via theory definition modules) and theories already in the graph can be extended (via theory extension modules). Second, the theories can be developed by defining new concepts and stating and proving new facts (via development definition and extension modules). Third, theories in the graph can be interconnected with theory morphisms (via theory translation definition and extension modules) and developments in the graph can be interconnected with development modules (via development translation definition and extension modules). And fourth, definitions and theorems can be transported from one development to another (via definition and theorem transportation modules).

13.5.3   Realm Graphs

A theory can have axiomatizations that widely differ in vocabulary and axioms. A particular axiomatization may be better for some situations than others. Thus it is useful to have different axiomatizations of a theory in a mathematics library. Since the axiomatizations of a theory are equivalent to each other, definitions and theorems can be transported between them using the morphisms that establish their equivalence. Thus it is also useful to group equivalent theories together in a mathematics library.

A "realm" is a structure for consolidating knowledge about a mathematical theory. A realm contains several separately developed axiomatizations of a theory. Morphisms interconnect these developments and establish that the axiomatizations are equivalent. A realm also contains a theory called the "face" of the realm that is simultaneously the front of an interface for each axiomatization in the realm. The face enables users to apply the concepts and facts of the mathematical theory without delving into the details of how the concepts and facts were developed.

A *realm* of Alonzo is a tuple $R = (F, \mathcal{D}, \mathcal{M}, \mathcal{I})$ where:

1. F is a theory called the *face* of R.

2. $\mathcal{D}$ is a set $\{D_1, D_2, \ldots, D_n\}$ of developments.

3. $\mathcal{M}$ is a set of theory morphisms that establish that the bottom theories $T_1, T_2, \ldots, T_n$ of $D_1, D_2, \ldots, D_n$, respectively, are pairwise equivalent.

4. $\mathcal{I}$ is a set $\{I_1, I_2, \ldots, I_n\}$ of interfaces such that F is the front of I_i and the top theory T_i' of D_i is the back of I_i for each i with $1 \leq i \leq n$.

We identify a development D with bottom theory T and top theory T' with the trivial realm $(T, \{D\}, \{\,\}, \{\Phi\})$ where Φ is the faithful inclusion from T to T'. See [21] for a full presentation of realms.

A *realm graph* of Alonzo is a directed graph $G = (\mathcal{N}, \mathcal{V})$ where $\mathcal{N}$ is a set of realms of Alonzo and $\mathcal{V}$ is a set of development morphisms of Alonzo that connect the theories and developments in a realm to theories and developments in other realms in $\mathcal{N}$. A realm graph is a natural generalization of a development graph.

13.6 Theory Graph Combinators

We have seen that a theory graph of Alonzo can be constructed one theory or morphism at a time by presenting the theories using theory definition and extension modules and presenting the morphisms using translation definition and extension modules. It is more efficient to construct a theory graph by extending the graph in chunks using "combinators" [18, 19, 23, 24, 25, 57, 103, 105, 113, 115, 116].

A *theory graph combinator*[4] is a function that takes a theory graph G plus some optional arguments $A_1, \ldots, A_n$ as input and returns as output a theory graph G' that is an extension of G obtained by adding theories and morphisms to G. We will illustrate this notion by presenting four examples of theory graph combinators. The first three are basic theory graph combinators presented in [25]. Let $\mathbb{T}$ be the set of theories of Alonzo, $\mathbb{TM}$ be the set of theory morphisms of Alonzo, and $\mathbb{G} \subseteq \mathcal{P}(\mathbb{T}) \times \mathcal{P}(\mathbb{TM})$ be the set of theory graphs of Alonzo.

Example 13.54 (Theory Extension) Let

$$\text{thy-ext} : \mathbb{G} \times \mathcal{P}(\mathcal{T}) \times \mathcal{P}(\mathcal{E}) \times \mathcal{P}(\mathcal{E}) \to \mathbb{G}$$

be a function such that, if

- $G = (\{T\}, \{\emptyset\})$ is a theory graph where $T = ((\mathcal{B}, \mathcal{C}), \Gamma)$,

- $\mathcal{B}'$ is a set of base types,

- $\mathcal{C}'$ is a set of constants, and

- Γ' is a set of sentences,

then $\text{thy-ext}(G, \mathcal{B}', \mathcal{C}', \Gamma')$ is the theory graph $(\{T, T'\}, \{\Phi\})$ where:

- $T' = ((\mathcal{B} \cup \mathcal{B}', \mathcal{C} \cup \mathcal{C}'), \Gamma \cup \Gamma')$ is an extension of T.

- Φ is the inclusion from T to T'.

Otherwise, $\text{thy-ext}(G, \mathcal{B}', \mathcal{C}', \Gamma')$ is undefined. thy-ext is a *theory extension combinator* that can be used to extend a theory graph by extending one of the theories in the graph. Figure 13.2 shows how the input graph is extended to produce the output graph.   □

Before presenting the next example of a theory graph combinator, we will give some definitions concerning functions that rename base types and constants. A *renaming* is a pair $R = (\rho, \pi)$ where ρ is a permutation of $\mathcal{S}_{\text{bt}}$ and π is a permutation of $\mathcal{S}_{\text{con}}$. The *application of R to a type α*, written $R(\alpha)$, is the result of replacing each base type $\text{BaseTy}(\mathbf{a})$ occurring in α with the base type $\text{BaseTy}(\rho(\mathbf{a}))$. The *application of R to an expression $\mathbf{A}_\alpha$*, written $R(\mathbf{A}_\alpha)$, is the result of replacing each variable $\text{Var}(\mathbf{x}, \alpha)$ occurring

---

[4]Theory graph combinators are also called *theory presentation combinators* [24, 25] and *diagram combinators* [105].

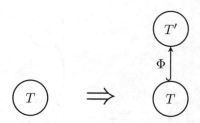

Figure 13.2: Input and Output Graph of thy-ext

in $\mathbf{A}_\alpha$ with the variable $\mathsf{Var}(\mathbf{x}, R(\alpha))$ and each constant $\mathsf{Con}(\mathbf{c}, \alpha)$ occurring in $\mathbf{A}_\alpha$ with the constant $\mathsf{Con}(\pi(\mathbf{c}), R(\alpha))$. The *application of R to a set S of types or expressions*, written $R(S)$, is the set $\{R(x) \mid x \in S\}$. Let $\mathcal{R}$ be the set of renamings.

The *application of R to a theory* $T = ((\mathcal{B},\mathcal{C}),\Gamma)$, written $R(T)$, is the theory

$$T' = ((R(\mathcal{B}), R(\mathcal{C})), R(\Gamma)).$$

T' is called a *renaming* of T. T and T' are isomorphic copies of each other that different only in the names of their base types and constants. The *renaming morphism from T to T' by R* is the translation $\Phi = (\mu, \nu)$ where $\mu : \mathcal{B} \to R(\mathcal{B})$ such that $\mu(\mathbf{a}) = R(\mathbf{a})$ for all $\mathbf{a} \in \mathcal{B}$ and $\nu : \mathcal{C} \to R(\mathcal{C})$ such that $\nu(\mathbf{c}_\alpha) = R(\mathbf{c}_\alpha)$ for all $\mathbf{c}_\alpha \in \mathcal{C}$. $\Phi^{-1} = (\mu^{-1}, \nu^{-1})$ is clearly a renaming morphism from T' to T by the renaming $R^{-1} = (\rho^{-1}, \pi^{-1})$, the inverse of R. Thus Φ is an "isomorphism" from T to T'.

Example 13.55 (Theory Renaming) Let

$$\mathsf{thy\text{-}ren} : \mathbb{G} \times \mathcal{R} \to \mathbb{G}$$

be a function such that, if

- $G = (\{T\}, \{\emptyset\})$ is a theory graph where $T = ((\mathcal{B},\mathcal{C}),\Gamma)$ and

- R is a renaming,

then $\mathsf{thy\text{-}ren}(G, R)$ is the theory graph $(\{T,T'\}, \{\Phi, \Phi^{-1}\})$ where:

- $T' = R(T)$.

- $\Phi = (\mu, \nu)$ is the renaming morphism from T to T' by R.

- $\Phi^{-1} = (\mu^{-1}, \nu^{-1})$.

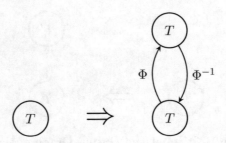

Figure 13.3: Input and Output Graph of thy-ren

Hence T' is a renaming of T whose equivalence to T is shown by Φ and Φ^{-1}. Otherwise, thy-ren(G, R) is undefined. thy-ren is a *theory renaming combinator* that can be used to extend a theory graph by adding to the graph a renaming of one of the theories in the graph. Figure 13.3 shows how the input graph is extended to produce the output graph.   □

Example 13.56 (Theory Combination) Let

$$\text{thy-comb} : \mathbb{G} \times \mathcal{R} \times \mathcal{R} \to \mathbb{G}$$

be a function such that, if

- $G = (\{T_0, T_1, T_2\}, \{\Phi_1, \Phi_2\})$ is a theory graph where $T_i = ((\mathcal{B}_i, \mathcal{C}_i), \Gamma_i)$ for $i \in \{0, 1, 2\}$ and Φ_i is an inclusion from T_0 to T_i for $i \in \{1, 2\}$,

- R_1 is a renaming such that $R_1(T_0) = T_0$,

- R_2 is a renaming such that $R_2(T_0) = T_0$,

- $R_1(\mathcal{B}_1 \setminus \mathcal{B}_0) \cap R_2(\mathcal{B}_2 \setminus \mathcal{B}_0) = \emptyset$,

- $R_1(\mathcal{C}_1 \setminus \mathcal{C}_0) \cap R_2(\mathcal{C}_2 \setminus \mathcal{C}_0) = \emptyset$, and

then thy-comb(G, R_1, R_2) is the theory graph

$$(\{T_0, T_1, T_2, T'\}, \{\Phi_1, \Phi_2, \Phi_1', \Phi_2'\})$$

where:

- $T' = ((\mathcal{B}', \mathcal{C}'), \Gamma')$.

- $\mathcal{B}' = \mathcal{B}_0 \cup R_1(\mathcal{B}_1 \setminus \mathcal{B}_0) \cup R_2(\mathcal{B}_2 \setminus \mathcal{B}_0)$.

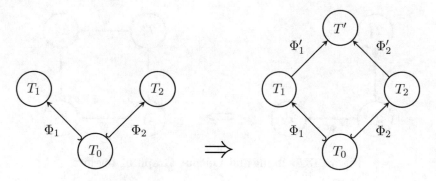

Figure 13.4: Input and Output Graph of thy-comb

- $\mathcal{C}' = \mathcal{C}_0 \cup R_1(\mathcal{C}_1 \setminus \mathcal{C}_0) \cup R_2(\mathcal{C}_2 \setminus \mathcal{C}_0)$.

- $\Gamma' = \Gamma_0 \cup R_1(\Gamma_1 \setminus \Gamma_0) \cup R_2(\Gamma_2 \setminus \Gamma_0)$.

- Φ'_i is the embedding from T_i to T' defined by R_i for $i \in \{1, 2\}$.

Hence T' is, roughly speaking, a disjoint union of T_1 and T_2 over a common subtheory T_0. Otherwise, thy-comb(G, R_1, R_2) is undefined. thy-comb is a *theory combination combinator* that can be used to extend a theory graph by adding to the graph the combination of two theories in the graph over another theory in the graph that is a common subtheory of the two theories. Figure 13.4 shows how the input graph is extended to produce the output graph. $\qquad \square$

Example 13.57 (Theory Instantiation) Let

$$\text{thy-inst} : \mathbb{G} \times \mathcal{R} \to \mathbb{G}$$

be a function such that, if

- $G = (\{T_1, T'_1, T_2\}, \{\Phi_1, \Psi\})$ is a theory graph where $T_i = ((\mathcal{B}_i, \mathcal{C}_i), \Gamma_i)$ for $i \in \{1, 2\}$, $T'_1 = ((\mathcal{B}'_1, \mathcal{C}'_1), \Gamma'_1)$, Φ_1 is an inclusion from T_1 to T'_1, $\Psi = (\mu, \nu)$ is a morphism from T_1 to T_2 and

- $R = (\rho, \pi)$ is a renaming such that $\mathcal{B}_2 \cap R(\mathcal{B}'_1 \setminus \mathcal{B}_1) = \emptyset$ and $\{\mathbf{c} \mid \mathbf{c}_\alpha \in \mathcal{C}_2\} \cap \{\pi(\mathbf{c}) \mid \mathbf{c}_\alpha \in \mathcal{C}'_1 \setminus \mathcal{C}_1\} = \emptyset$,

then thy-inst(G, R) is the theory graph

$$(\{T_1, T'_1, T_2, T'_2\}, \{\Phi_1, \Phi_2, \Psi, \Psi'\})$$

where:

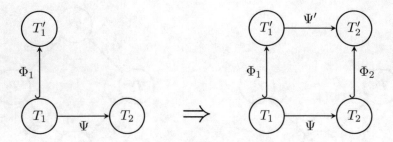

Figure 13.5: Input and Output Graph of thy-inst

- $\Psi' = (\mu', \nu')$ is the morphism from T_1' to T_2' that extends Ψ, $\mu'(\mathbf{a}) = R(\mathbf{a})$ for all $\mathbf{a} \in \mathcal{B}_1' \setminus \mathcal{B}_1$, and $\nu'(\mathbf{c}_\alpha) = \pi(\mathbf{c})_{\overline{\mu'(\alpha)}}$ for all $\mathbf{c}_\alpha \in \mathcal{C}_1' \setminus \mathcal{C}_1$.

- $T_2' = ((\mathcal{B}_2', \mathcal{C}_2'), \Gamma_2')$.

- $\mathcal{B}_2' = \mathcal{B}_2 \cup R(\mathcal{B}_1' \setminus \mathcal{B}_1)$.

- $\mathcal{C}_2' = \mathcal{C}_2 \cup \{\pi(\mathbf{c})_{\overline{\mu'(\alpha)}} \mid \mathbf{c}_\alpha \in \mathcal{C}_1' \setminus \mathcal{C}_1\}$.

- $\Gamma_2' = \Gamma_2 \cup \{\overline{\nu'}(\mathbf{A}_\alpha) \mid \mathbf{A}_\alpha \in \Gamma_1' \setminus \Gamma_1\}$.

- Φ_2 is the inclusion from T_2 to T_2'.

Hence T_2' is, roughly speaking, the result of replacing T_1 with T_2 in T_1'. Otherwise, thy-inst(G, R) is undefined. thy-inst is a *theory instantiation combinator* that can be used to extend a theory graph by adding to the graph the theory that results from instantiating the subtheory of a theory in the graph. Figure 13.5 shows how the input graph is extended to produce the output graph.   □

Remark 13.58 (Other Combinators) *It is also possible to define development graph and realm graph combinators that are analogous to theory graph combinators.*

13.7   Conclusions

A morphism is a mapping from the language of one theory to the language of another theory. It is meaning preserving in the sense that it maps theorems to theorems. Morphisms serve as information conduits through which concepts and facts are transported via definitions and theorems from usually more abstract theories to usually more concrete theories.

The little theories method is a method for organizing mathematical knowledge as a theory graph whose nodes are theories and whose directed edges are morphisms. Each mathematical topic is developed in the "little theory" in the theory graph that has the most convenient level of abstraction and the most convenient vocabulary. Then concepts and facts produced in the development of the theory are transported, as needed, to other theories via the morphisms in the theory graph.

A library of mathematical knowledge can be constructed as graph of theories, developments, or realms. In a theory graph, the nodes are theories and edges are theory morphisms. In a development graph the nodes are developments that include definitions and theorems and the edges are development morphisms. And in a realm graph, the nodes are realms that contain several separately developed axiomatizations of a theory and the edges are development morphisms that connect the theories and developments in a realm to theories and developments in other realms. Each kind of graph can be constructed using either Alonzo modules or graph combinators.

13.8   Exercises

1. Prove Lemma 13.1.

2. Prove Lemma 13.2.

3. Prove Proposition 13.7.

4. Prove that Proposition 13.19 does not hold in all cases if the premise $T_1 \unlhd_d T_2$ is replaced with $T_1 \unlhd_s T_2$.

5. Let $L_i = (\mathcal{B}_i, \mathcal{C}_i)$ be a language and $T_i = (L_i, \Gamma_i)$ be a theory for $i \in \{1, 2\}$. Assume that $T_1 \leq T_2$ and there is a morphism $\Phi = (\mu, \nu)$ from T_2 to T_1 such that $\mu\restriction_{\mathcal{B}_1}$ and $\nu\restriction_{\mathcal{C}_1}$ are identity functions. Prove $T_1 \unlhd T_2$ under these assumptions.

6. Prove that all the morphisms defined in Subsection 13.3.3 other than PA-with-1-to-PA and GRP-to-dual-GRP are nonfaithful.

7. Prove Proposition 13.43.

8. Specify the *Internet Protocol (IP)* as a theory graph in Alonzo.

Chapter 14

Alonzo Variants

We present in this chapter three variants of Alonzo that are obtained by modifying the syntax and semantics of Alonzo. These variants illustrate how an indefinite description operator, sorts (a system of types and subtypes), and quotation and evaluation operators can be added to Alonzo.

14.1 Alonzo with Indefinite Description

14.1.1 Introduction

We saw in Example 3.5 that a *definite description* is an expression of the form $I\,x \in S\,.\,E$ that denotes *the* $a \in S$ that satisfies the property expressed by E. The definite description is undefined if there is no such a or more than one such a. We also saw in Example 3.6 that an *indefinite description* is an expression of the form $\epsilon\,x \in S\,.\,E$ that denotes *some* $a \in S$ that satisfies the property expressed by E. The indefinite description is undefined if there is no such a.

Indefinite description is not as common in mathematics as definite description, but it is more common in computing than definite description. Indefinite description is a stronger principle than definite description. It cannot be defined in terms of definite description, but definite description can be defined in terms of it. Adding definite description to type theory does not increase the theoretical expressivity of the logic, but adding indefinite description does. In fact, the axiom of choice is valid in type theory plus indefinite description, but it is not valid in type theory plus definite description. Indefinite description is a primitive component of the version of type theory underlying the HOL family of proof assistants [58].

© The Author(s), under exclusive license to Springer Nature Switzerland AG 2023
W. M. Farmer, *Simple Type Theory*, Computer Science Foundations
and Applied Logic, https://doi.org/10.1007/978-3-031-21112-6_14

$(\epsilon \mathbf{x} : \alpha \,.\, \mathbf{B}_o)$	stands for	$\mathsf{IndefDes}(\mathsf{Var}(\mathbf{x}, \alpha), \mathbf{B}_o)$	where $\alpha \neq o$.

Table 14.1: Notational Definition for Indefinite Description

We will add indefinite description to Alonzo to obtain a variant of Alonzo called *AlonzoID*. Adding indefinite description to Alonzo requires changes to the definitions of an expression, interpretation, and model.

14.1.2 Syntax

We begin the definition of AlonzoID by adding a new constructor to the definition of an expression of type α:

E8. *Indefinite description*: $\mathsf{IndefDes}(\mathsf{Var}(\mathbf{x}, \alpha), \mathbf{B}_{\mathsf{BoolTy}})$ is an expression of type α where $\alpha \neq \mathsf{BoolTy}$.

The compact notation for indefinite description is given in Table 14.1.

14.1.3 Semantics

The definition of an interpretation of a language is changed to the following definition: An *interpretation* of $L = (\mathcal{B}, \mathcal{C})$ is a tuple $M = (\mathcal{D}, I, \mathcal{F})$ where $\mathcal{D} = \{D_\alpha \mid \alpha \in \mathcal{T}(L)\}$ is a frame for L, I is an *interpretation function* that maps each constant in $\mathcal{C}$ of type α to an element of D_α, and

$$\mathcal{F} = \{f_\alpha \mid f_\alpha \text{ is a choice function on } D_{\{\alpha\}} \text{ and } \alpha \in \mathcal{T}(L) \text{ with } \alpha \neq o\}.$$

(A *choice function* on $B \subseteq \mathcal{P}(A)$ is a function $f : B \to A$ such that, for all $s \in B$, $f(s) \in s$ if s is nonempty and is undefined otherwise.) Notice that different interpretations of L can have different choice functions. Finally, a new condition is added to the definition of a general model:

V8. $V_\varphi^M(\epsilon \mathbf{x} : \alpha \,.\, \mathbf{B}_o) = f_\alpha(D)$ where $D = \{d \in D_\alpha \mid V_{\varphi[(\mathbf{x}:\alpha) \mapsto d]}^M(\mathbf{B}_\beta) = \mathsf{T}\}$, provided D is nonempty. Otherwise, $V_\varphi^M(\epsilon \mathbf{x} : \alpha \,.\, \mathbf{B}_o)$ is undefined.

14.1.4 Proof System

$\mathfrak{A}$ can be extended to a sound and complete proof system for AlonzoID by simply adding the following axioms to it:

A8 (Indefinite Description)

1. $(\exists x : \alpha . \mathbf{A}_o) \Rightarrow (\epsilon x : \alpha . \mathbf{A}_o) \in \{x : \alpha \mid \mathbf{A}_o\}$ (where $\alpha \neq o$).

2. $\neg(\exists x : \alpha . \mathbf{A}_o) \Rightarrow (\epsilon x : \alpha . \mathbf{A}_o)\!\uparrow$ (where $\alpha \neq o$).

14.1.5   Theorems

The following proposition shows how definite description can be defined in terms of indefinite description:

Proposition 14.1 (Definite Description as Indefinite Description)
Let AlonzoID be the logic. Then

$$\mathrm{I}\,\mathbf{x} : \alpha . \mathbf{A}_o \simeq \epsilon\,\mathbf{x} : \alpha . (\mathbf{A}_o \wedge \exists!\,\mathbf{x} : \alpha . \mathbf{A}_o)$$

is valid where $\alpha \neq o$.

Proof Follows immediately from the semantics of definite and indefinite description.   □

Lemma 14.2 (Choice Function) *Let AlonzoID be the logic and $\alpha \neq o$. Define $\mathbf{C}_{\{\alpha\}\to\alpha}$ to be the expression*

$$\lambda s : \{\alpha\} . \epsilon x : \alpha . x \in s.$$

Then $\mathbf{C}_{\{\alpha\}\to\alpha}$ is a choice function on α. More precisely, if M is a general model of L, $\varphi \in \mathsf{assign}(M)$, and $\alpha \in \mathcal{T}(L)$ with $\alpha \neq o$, then $V_\varphi^M(\mathbf{C}_{\{\alpha\}\to\alpha})$ is a choice function on $D_{\{\alpha\}}$.

Proof Follows immediately from the semantics of indefinite description. □

Theorem 14.3 (Axiom of Choice) *Let AlonzoID be the logic. Then*

$$\exists f : \{\alpha\} \to \alpha . \forall s : \{\alpha\} . s \neq \emptyset_{\{\alpha\}} \mapsto f\,s \in s \mid (f\,s)\!\uparrow,$$

which expresses the axiom of choice for α, is valid for all $\alpha \in \mathcal{T}(L)$ with $\alpha \neq o$.

Proof The theorem follows easily from Lemma 14.2.   □

14.2 Alonzo with Sorts

14.2.1 Introduction

We have seen that quasitypes are very useful for reasoning about subtypes. They can be used to restrict the value of a bound variable as, e.g., in a function abstraction of the form $\lambda \mathbf{x} : \mathbf{Q}_{\{\alpha\}} \cdot \mathbf{B}_\beta$. They can also be used to sharpen a definedness statement as in a statement of the form $\mathbf{A}_\alpha \downarrow \mathbf{Q}_{\{\alpha\}}$.

However, quasitypes are not interchangeable with types. A constant symbol $\mathbf{c}$ can be bound to a type α to obtain a constant $\mathbf{c}_\alpha$, but $\mathbf{c}$ cannot be bound to a quasitype $\mathbf{Q}_{\{\alpha\}}$ to obtain a constant $\mathbf{c}_{\mathbf{Q}_{\{\alpha\}}}$. The value of a sentence of the form $\forall \mathbf{x} : \alpha \cdot F_o$ is always F, but the value of a sentence of the form $\forall \mathbf{x} : \mathbf{Q}_{\{\alpha\}} \cdot F_o$ is T if $\mathbf{Q}_{\{\alpha\}}$ denotes the empty set.

Is there a mechanism for subtypes without the limitations of quasitypes? Yes, "sorts" introduced in the IMPS proof assistant [39, 53] is such a mechanism. Sorts are an extension of types in which each sort is a type or a subtype of a type. Sorts can be used wherever types can be used. If α is a sort that is a subtype of the type α', then α denotes a nonempty subset of the domain denoted by α'. In particular, if α and β are sorts that are subtypes of the types α' and β', respectively, then $\alpha \to \beta$ is a sort that is a subtype of the type $\alpha' \to \beta'$. Then $\alpha \to \beta$ denotes a set F of functions f in $D_{\alpha'} \to D_{\beta'}$ such that $\mathsf{dom}(f) \subseteq D_\alpha$ and $\mathsf{ran}(f) \subseteq D_\beta$, $\alpha' \to \beta'$ denotes a set G of functions in $D_{\alpha'} \to D_{\beta'}$, and $F \subseteq G$. This works since functions are allowed to be partial: if $D_\alpha \subset D_{\alpha'}$, then each (partial or total) function in $D_\alpha \to D_\beta$ is represented by a partial function in $D_{\alpha'} \to D_{\beta'}$.

We will add sorts to Alonzo to obtain a variant of Alonzo called *AlonzoS*. This requires small changes to most of the syntax and semantics definitions of Alonzo as well as several new definitions.

14.2.2 Syntax

We begin the definition of AlonzoS by introducing a new set of symbols: $\mathcal{S}_{\mathsf{as}}$, a fixed countably infinite set of symbols that will serve as names of atomic sorts. We assume $\mathcal{S}_{\mathsf{bt}}$ and $\mathcal{S}_{\mathsf{as}}$ are disjoint. We will employ the following syntactic variables for the new symbols and syntactic entities defined for AlonzoS:

1. $\mathbf{s}, \mathbf{t}$, etc. range over members of $\mathcal{S}_{\mathsf{as}}$.

2. σ, τ, etc. range over sorts of AlonzoS.

3. α, β, etc. range over types of AlonzoS.

o	stands for	BoolTy.
$\mathbf{a}$	stands for	BaseTy($\mathbf{a}$).
$\mathbf{s}$	stands for	AtSort($\mathbf{s}$).
$(\sigma \to \tau)$	stands for	FunSort(σ, τ).
$(\sigma \times \tau)$	stands for	ProdSort(σ, τ).

Table 14.2: Notational Definitions for Sorts

A *sort* of AlonzoS is an inductive set of strings of symbols defined by the following constructors:

U1. *Type of truth values*: BoolTy is a sort.

U2. *Base type*: BaseTy($\mathbf{a}$) is a sort.

U3. *Atomic sort*: AtSort($\mathbf{s}$) is a sort.

U4. *Function sort*: FunSort(σ, τ) is a sort.

U5. *Product sort*: ProdSort(σ, τ) is a sort.

A *type* of AlonzoS is a sort of Alonzo that is constructed without using U3. Let $\mathcal{T}$ and $\mathcal{U}$ denote the set of types and sorts, respectively, of AlonzoS. Obviously, $\mathcal{T} \subseteq \mathcal{U}$. Compact notation for sorts is given in Table 14.2.

A *sort system* of AlonzoS is a tuple $S = (\mathcal{B}, \mathcal{A}, \xi)$ where $\mathcal{B}$ is a finite set of base types, $\mathcal{A}$ is a finite set of atomic sorts, and ξ is a function that maps each member of $\mathcal{A}$ to a type α such that all the base types occurring in α are members of $\mathcal{B}$. A sort σ is a *sort of S* if all the base types occurring in σ are members of $\mathcal{B}$ and all the atomic sorts occurring in σ are members of $\mathcal{A}$. A type α is a *type of S* if α is a sort of S. Let $\overline{\xi}$ be the canonical extension of ξ to the sorts of S that is defined as follows by recursion and pattern matching:

1. $\overline{\xi}(o) = o$.

2. $\overline{\xi}(\mathbf{a}) = \mathbf{a}$.

3. $\overline{\xi}(\mathbf{s}) \equiv \xi(\mathbf{s})$.

4. $\overline{\xi}(\sigma \to \tau) \equiv \overline{\xi}(\sigma) \to \overline{\xi}(\tau)$.

5. $\overline{\xi}(\sigma \times \tau) \equiv \overline{\xi}(\sigma) \times \overline{\xi}(\tau)$.

We say that $\overline{\xi}(\sigma)$ is the *type of σ in S*. Notice that, if σ is a type of S, then $\overline{\xi}(\sigma) = \sigma$.

Let $S = (\mathcal{B}, \mathcal{A}, \xi)$ be a sort system of AlonzoS, and let $\mathbf{A}_\sigma, \mathbf{B}_\sigma$, etc. be syntactic variables that range over expressions of S of sort σ (which we define next). An *expression of S of sort σ* of AlonzoS is an inductive set of strings of symbols defined by the following constructors:

E1. *Variable*: $\mathsf{Var}(\mathbf{x}, \sigma)$ is an expression of S of sort σ if σ is a sort of S.

E2. *Constant*: $\mathsf{Con}(\mathbf{c}, \sigma)$ is an expression of S of sort σ if σ is a sort of S.

E3. *Equality*: $\mathsf{Eq}(\mathbf{A}_\sigma, \mathbf{B}_\tau)$ is an expression of S of sort BoolTy if $\overline{\xi}(\sigma) = \overline{\xi}(\tau)$.

E4. *Function application*: $\mathsf{FunApp}(\mathbf{F}_{\sigma \to \tau}, \mathbf{A}_{\sigma'})$ is an expression of S of sort τ if $\overline{\xi}(\sigma) = \overline{\xi}(\sigma')$.

E5. *Function abstraction*: $\mathsf{FunAbs}(\mathsf{Var}(\mathbf{x}, \sigma), \mathbf{B}_\tau)$ is an expression of S of sort $\mathsf{FunSort}(\sigma, \tau)$ if σ is a sort of S.

E6. *Definite description*: $\mathsf{DefDes}(\mathsf{Var}(\mathbf{x}, \sigma), \mathbf{B}_{\mathsf{BoolTy}})$ is an expression of S of sort σ if σ is a sort of S with $\sigma \neq \mathsf{BoolTy}$.

E7. *Ordered pair*: $\mathsf{OrdPair}(\mathbf{A}_\sigma, \mathbf{B}_\tau)$ is an expression of S of sort $\mathsf{ProdSort}(\sigma, \tau)$.

Let $\mathcal{E}(S)$ denote the set of expressions of S of AlonzoS. We employ the same compact notation for the expressions of AlonzoS as for the expressions of Alonzo.

A *language* (or *signature*) of AlonzoS is a pair $L = (S, \mathcal{C})$ where S is sort system of AlonzoS and $\mathcal{C}$ is a set of constants $\mathbf{c}_\sigma$ of S. σ is a *type of L* if α is a type of S, α is a *sort of L* if σ is a sort of S, and $\mathbf{A}_\sigma$ is an *expression of L* if $\mathbf{A}_\sigma$ is an expression of S and all the constants occurring in $\mathbf{A}_\sigma$ are members of $\mathcal{C}$. Let $\mathcal{T}(L) \subseteq \mathcal{T}$ denote the set of types of L, $\mathcal{U}(L) \subseteq \mathcal{U}$ denote the set of sorts of L, and $\mathcal{E}(L) \subseteq \mathcal{E}(S)$ denote the set of expressions of L.

14.2.3   Semantics

Let $L = (S, \mathcal{C})$ be a language of AlonzoS. A *frame* for L is a collection $\mathcal{D} = \{D_\sigma \mid \sigma \in \mathcal{U}(L)\}$ of nonempty domains (sets) of values such that:

F1. *Domain of truth values*: $D_o = \mathbb{B} = \{\mathrm{F}, \mathrm{T}\}$.

F2. *Predicate domain*: $D_{\sigma \to o}$ is a set of *some* total functions f from $D_{\overline{\xi}(\sigma)}$ to D_o such that $\mathsf{dom}(f) \subseteq D_\sigma$ for $\sigma \in \mathcal{U}(L)$.

F3. *Function domain*: $D_{\sigma \to \tau}$ is a set of *some* partial and total functions f from $D_{\overline{\xi}(\sigma)}$ to $D_{\overline{\xi}(\tau)}$ such that $\mathsf{dom}(f) \subseteq D_\sigma$ and $\mathsf{ran}(f) \subseteq D_\tau$ for $\sigma, \tau \in \mathcal{U}(L)$ with $\tau \neq o$.

F4. *Product domain*: $D_{\sigma \times \tau} = D_\sigma \times D_\tau$ for $\sigma, \tau \in \mathcal{U}(L)$.

F5. *Type domain*: $D_\sigma \subseteq D_{\overline{\xi}(\sigma)}$ for $\sigma \in \mathcal{U}(L)$.

An *interpretation* of L is a pair $M = (\mathcal{D}, I)$ where $\mathcal{D} = \{D_\sigma \mid \sigma \in \mathcal{U}(L)\}$ is a frame for L and I is an *interpretation function* that maps each constant in $\mathcal{C}$ of sort σ to an element of D_σ.

Let $\mathcal{D} = \{D_\sigma \mid \sigma \in \mathcal{U}(L)\}$ be a frame for L and $M = (\mathcal{D}, I)$ be an interpretation of L. An *assignment* into $\mathcal{D}$ is defined the same as for Alonzo. M is a *general model* of L if there is a partial binary *valuation function* V^M such that, for all assignments $\varphi \in \mathsf{assign}(M)$ and expressions $\mathbf{C}_\sigma$ of L, (1) either $V^M_\varphi(\mathbf{C}_\sigma) \in D_\sigma$ or $V^M_\varphi(\mathbf{C}_\sigma)$ is undefined and (2) each of the following conditions is satisfied:

V1. $V^M_\varphi((\mathbf{x} : \sigma)) = \varphi((\mathbf{x} : \sigma))$.

V2. $V^M_\varphi(\mathbf{c}_\sigma) = I(\mathbf{c}_\sigma)$.

V3. $V^M_\varphi(\mathbf{A}_\sigma = \mathbf{B}_\tau) = \mathrm{T}$ if $V^M_\varphi(\mathbf{A}_\sigma)$ is defined, $V^M_\varphi(\mathbf{B}_\tau)$ is defined, and $V^M_\varphi(\mathbf{A}_\sigma) = V^M_\varphi(\mathbf{B}_\tau)$. Otherwise, $V^M_\varphi(\mathbf{A}_\sigma = \mathbf{B}_\tau) = \mathrm{F}$.

V4. $V^M_\varphi(\mathbf{F}_{\sigma \to \tau}\, \mathbf{A}_{\sigma'}) = V^M_\varphi(\mathbf{F}_{\sigma \to \tau})(V^M_\varphi(\mathbf{A}_{\sigma'}))$ if $V^M_\varphi(\mathbf{F}_{\sigma \to \tau})$ is defined, $V^M_\varphi(\mathbf{A}_{\sigma'})$ is defined, and $V^M_\varphi(\mathbf{F}_{\sigma \to \tau})$ is defined at $V^M_\varphi(\mathbf{A}_{\sigma'})$. Otherwise, $V^M_\varphi(\mathbf{F}_{\sigma \to \tau}\, \mathbf{A}_{\sigma'}) = \mathrm{F}$ if $\tau = o$ and $V^M_\varphi(\mathbf{F}_{\sigma \to \tau}\, \mathbf{A}_{\sigma'})$ is undefined if $\tau \neq o$.

V5. $V^M_\varphi(\lambda \mathbf{x} : \sigma \,.\, \mathbf{B}_\tau)$ is the (partial or total) function $f \in D_{\sigma \to \tau}$ such that, for each $d \in D_\sigma$, $f(d) = V^M_{\varphi[(\mathbf{x}:\sigma) \mapsto d]}(\mathbf{B}_\tau)$ if $V^M_{\varphi[(\mathbf{x}:\sigma) \mapsto d]}(\mathbf{B}_\tau)$ is defined and $f(d)$ is undefined if $V^M_{\varphi[(\mathbf{x}:\sigma) \mapsto d]}(\mathbf{B}_\tau)$ is undefined.

V6. $V^M_\varphi(\mathrm{I}\mathbf{x} : \sigma \,.\, \mathbf{A}_o)$ is the $d \in D_\sigma$ such that $V^M_{\varphi[(\mathbf{x}:\sigma) \mapsto d]}(\mathbf{A}_o) = \mathrm{T}$ if there is exactly one such d. Otherwise, $V^M_\varphi(\mathrm{I}\mathbf{x} : \sigma \,.\, \mathbf{A}_o)$ is undefined.

V7. $V^M_\varphi((\mathbf{A}_\sigma, \mathbf{B}_\tau)) = (V^M_\varphi(\mathbf{A}_\sigma), V^M_\varphi(\mathbf{B}_\tau))$ if $V^M_\varphi(\mathbf{A}_\sigma)$ and $V^M_\varphi(\mathbf{B}_\tau)$ are defined. Otherwise, $V^M_\varphi((\mathbf{A}_\sigma, \mathbf{B}_\tau))$ is undefined.

Let $\alpha = \overline{\xi}(\sigma) = \overline{\xi}(\tau)$.

$\mathrm{if}_{o \to \sigma \to \tau \to \alpha}$	stands for	$\lambda b : o . \lambda x : \sigma . \lambda y : \tau .$
		$\mathrm{I} z : \alpha . (b \Rightarrow z = x) \wedge (\neg b \Rightarrow z = y).$
$(\mathbf{A}_o \mapsto \mathbf{B}_\sigma \mid \mathbf{C}_\tau)$	stands for	$\mathrm{if}_{o \to \sigma \to \tau \to \alpha} \mathbf{A}_o \mathbf{B}_\sigma \mathbf{C}_\tau.$
$(\mathbf{A}_\sigma \downarrow \tau)$	stands for	$(\lambda x : \tau . \mathrm{T}_o) \mathbf{A}_\sigma.$
$(\mathbf{A}_\sigma \uparrow \tau)$	stands for	$\neg (\mathbf{A}_\sigma \downarrow \tau).$
$(\sigma \subseteq \tau)$	stands for	$\forall x : \sigma . x \downarrow \tau.$
$(\sigma = \tau)$	stands for	$\forall x : \alpha . x \downarrow \sigma \Leftrightarrow x \downarrow \tau.$

Table 14.3: Additional Notational Definitions for AlonzoS

This definition of a general model of L for AlonzoS defines essentially the same semantics for the expressions of L as does the definition for Alonzo.

Nearly all the notational definitions of Alonzo work for AlonzoS, and those that do not can be easily modified so that they do. Table 14.3 contains some additional notational definitions for AlonzoS. An expression

$$(\mathbf{A}_o \mapsto \mathbf{B}_\sigma \mid \mathbf{C}_\tau)$$

denotes the value of $\mathbf{B}_\sigma$ if $\mathbf{A}_o$ holds and otherwise denotes the value of $\mathbf{C}_\tau$, provided σ and τ have a common type. A formula $(\mathbf{A}_\sigma \downarrow \tau)$, read as $\mathbf{A}_\sigma$ *is defined in* τ, asserts that the value of $\mathbf{A}_\sigma$ is defined and is a member of the domain denoted by τ. A formula $(\sigma \subseteq \tau)$, read as "σ is a subsort of τ", asserts that σ denotes a subset of the set denoted by τ. And a formula $(\sigma = \tau)$ asserts that σ and τ denote the same set.

Remark 14.4 (Sorts are Nondependent) *Like the denotation of a type in Alonzo, the denotation of a sort in AlonzoS never depends on an individual value. That is, AlonzoS does not have dependent sorts. Of course, it is still possible to define dependent quasitypes in AlonzoS as in Alonzo.*

14.2.4 Proof System

We present in [39] a version of Church's type theory with undefined expressions and sorts called $\mathbf{PF}^*$ that is similar to AlonzoS. The paper [39] includes a proof system for a language L of $\mathbf{PF}^*$ that is sound and complete (in the general sense). Following the design of the proof system for L, we can modify $\mathfrak{A}$ to obtain a proof system for a language of AlonzoS that is sound and complete.

14.2.5   Theorems

Proposition 14.5 (Subsorts) *Let AlonzoS be the logic and* $L = (S, \mathcal{C})$ *be a language of AlonzoS where* $S = (\mathcal{B}, \mathcal{A}, \xi)$. *Suppose the following are formulas of* L:

1. $\mathbf{A}_\sigma{\downarrow} \Rightarrow \mathbf{A}_\sigma \downarrow \sigma$.

2. $\mathbf{A}_\sigma \downarrow \tau \Rightarrow \mathbf{A}_\sigma{\downarrow}$.

3. $(\mathbf{A}_\sigma \downarrow \tau \wedge \tau \subseteq \tau') \Rightarrow \mathbf{A}_\sigma \downarrow \tau'$.

4. $\sigma \subseteq \overline{\xi}(\sigma)$.

5. $\sigma \subseteq \sigma$.

6. $(\sigma_1 \subseteq \sigma_2 \wedge \sigma_2 \subseteq \sigma_1) \Rightarrow \sigma_1 = \sigma_2$.

7. $(\sigma_1 \subseteq \sigma_2 \wedge \sigma_2 \subseteq \sigma_3) \Rightarrow \sigma_1 \subseteq \sigma_3$.

8. $(\sigma \subseteq \sigma' \wedge \tau \subseteq \tau') \Rightarrow \sigma \to \tau \subseteq \sigma' \to \tau'$.

9. $(\sigma \subseteq \sigma' \wedge \tau \subseteq \tau') \Rightarrow \sigma \times \tau \subseteq \sigma' \times \tau'$.

Then each of these formulas is valid in every general model of L.

Proof This proposition follows easily from the notational definitions of $\mathbf{A}_\sigma{\downarrow}$, $\mathbf{A}_\sigma \downarrow \tau$, $\sigma \subseteq \tau$, and $\sigma = \tau$ and the definition of a general model.   □

Theorem 14.6 (Sort Beta-Reduction) *Let AlonzoS be the logic and* $\mathbf{C}_o$ *be a formula of* L *of the form*

$$\mathbf{A}_{\sigma'} \downarrow \sigma \Rightarrow (\lambda \mathbf{x} : \sigma \,.\, \mathbf{B}_\tau) \, \mathbf{A}_{\sigma'} \simeq \mathbf{B}_\tau[(\mathbf{x} : \sigma) \mapsto \mathbf{A}_{\sigma'}].$$

If $\mathbf{A}_{\sigma'}$ *is free for* $(\mathbf{x} : \sigma)$ *in* $\mathbf{B}_\tau$, *then* $\mathbf{C}_o$ *is valid in every general model of* L.

Proof Similar to the proof of Theorem 7.1.   □

14.3   Alonzo with Quotation and Evaluation

14.3.1   Introduction

It is often useful to be able to reason about the interplay of the syntax and semantics of expressions. For example, a *syntax-based mathematical algorithm (SBMA)* [48], such as a symbolic differentiation algorithm, manipulates mathematical expressions in a mathematically meaningful way. Reasoning about an SBMA requires reasoning about the relationship between how the expressions are manipulated and what the manipulations mean mathematically. In standard predicate logics including Alonzo, it is not possible to directly refer to the syntax of an expression and thus we cannot reason in Alonzo directly about the interplay of syntax and semantics. What is needed to do so are quotation and evaluation operators similar to quote and eval in the Lisp programming language [48, 50].

Quotation is a mechanism for referring to a *syntactic value* (e.g., a syntax tree) that represents the syntactic structure of an expression, while *evaluation* is a mechanism for referring to the value of the expression that a syntactic value represents. Incorporating quotation and evaluation into a traditional logic like first-order logic or type theory is tricky; there are several challenging problems that the logic engineer must overcome [50]. We present in [50] a version of Church's type theory called CTT_{qe} with quotation and evaluation operators that overcomes these problems. We also present in [49] a variant of CTT_{qe} called CTT_{uqe} with undefined expressions as well as quotation and evaluation operators.

Following the design of CTT_{qe} and CTT_{uqe}, we will add quotation and evaluation operators to Alonzo to obtain a variant of Alonzo called *AlonzoQE*. Adding these operators to Alonzo requires changes to the definitions of a type, expression, frame, interpretation, and model as well as a few new definitions. The semantics of AlonzoQE is a bit simpler than the semantics of both CTT_{qe} and CTT_{uqe}, but it is closer to the semantics of CTT_{uqe} since CTT_{uqe} admits undefined expressions like AlonzoQE.

14.3.2 Syntax

We begin the definition of AlonzoQE by adding a new constructor to the definition of a type:

T5. *Type of expressions*: ExprTy is a type.

Next we add two new constructors to the definition of an expression of type α:

E8. *Quotation*: $\text{Quote}(\mathbf{A}_\alpha)$ is an expression of type ExprTy if $\mathbf{A}_\alpha$ is eval-free.

E9. *Evaluation*: $\text{Eval}(\mathbf{A}_{\text{ExprTy}}, \alpha)$ is an expression of type α.

ϵ	stands for	ExprTy.
$\ulcorner \mathbf{A}_\alpha \urcorner$	stands for	$\mathrm{Quote}(\mathbf{A}_\alpha)$.
$[\![\mathbf{A}_\epsilon]\!]_\alpha$	stands for	$\mathrm{Eval}(\mathbf{A}_\epsilon, \alpha)$.

Table 14.4: Notational Definitions for AlonzoQE

q-eq$_{\epsilon \to \epsilon \to \epsilon}$
q-app$_{\epsilon \to \epsilon \to \epsilon}$
q-abs$_{\epsilon \to \epsilon \to \epsilon}$
q-dd$_{\epsilon \to \epsilon \to \epsilon}$
q-op$_{\epsilon \to \epsilon \to \epsilon}$
q-quo$_{\epsilon \to \epsilon}$

Table 14.5: Logical Constants of AlonzoQE

An expression is *eval-free* if it is constructed using only the constructors presented in E1–E8. The compact notation for these three new (type and expression) constructors is given in Table 14.4.

AlonzoQE includes the set of logical constants given in Table 14.5. They correspond to the expression constructors presented in E3–E8. (Recall that Alonzo has no proper logical constants.)

A *construction* is an expression of type ϵ defined inductively as follows:

1. $\ulcorner (\mathbf{x} : \alpha) \urcorner$ is a construction.

2. $\ulcorner \mathbf{c}_\alpha \urcorner$ is a construction.

3. If $\mathbf{A}_\epsilon$ and $\mathbf{B}_\epsilon$ are constructions, then

$$\text{q-eq}_{\epsilon \to \epsilon \to \epsilon} \, \mathbf{A}_\epsilon \, \mathbf{B}_\epsilon,$$
$$\text{q-app}_{\epsilon \to \epsilon \to \epsilon} \, \mathbf{A}_\epsilon \, \mathbf{B}_\epsilon,$$
$$\text{q-abs}_{\epsilon \to \epsilon \to \epsilon} \, \mathbf{A}_\epsilon \, \mathbf{B}_\epsilon,$$
$$\text{q-dd}_{\epsilon \to \epsilon \to \epsilon} \, \mathbf{A}_\epsilon \, \mathbf{B}_\epsilon,$$
$$\text{q-op}_{\epsilon \to \epsilon \to \epsilon} \, \mathbf{A}_\epsilon \, \mathbf{B}_\epsilon, \text{ and}$$
$$\text{q-quo}_{\epsilon \to \epsilon} \, \mathbf{A}_\epsilon$$

are constructions.

Let Ω be the function mapping eval-free expressions to constructions that is defined as follows by recursion and pattern matching:

1. $\Omega((\mathbf{x} : \alpha)) = \ulcorner(\mathbf{x} : \alpha)\urcorner$.

2. $\Omega(\mathbf{c}_\alpha) = \ulcorner\mathbf{c}_\alpha\urcorner$.

3. $\Omega(\mathbf{A}_\alpha = \mathbf{B}_\alpha) = \mathsf{q\text{-}eq}_{\epsilon\to\epsilon\to\epsilon}\,\Omega(\mathbf{A}_\alpha)\,\Omega(\mathbf{B}_\alpha)$.

4. $\Omega(\mathbf{F}_{\alpha\to\beta}\,\mathbf{A}_\alpha) = \mathsf{q\text{-}app}_{\epsilon\to\epsilon\to\epsilon}\,\Omega(\mathbf{F}_{\alpha\to\beta})\,\Omega(\mathbf{A}_\alpha)$.

5. $\Omega(\lambda\,\mathbf{x} : \alpha\,.\,\mathbf{B}_\beta) = \mathsf{q\text{-}abs}_{\epsilon\to\epsilon\to\epsilon}\,\Omega((\mathbf{x} : \alpha))\,\Omega(\mathbf{B}_\beta)$.

6. $\Omega(\mathrm{I}\,\mathbf{x} : \alpha\,.\,\mathbf{A}_o) = \mathsf{q\text{-}dd}_{\epsilon\to\epsilon\to\epsilon}\,\Omega((\mathbf{x} : \alpha))\,\Omega(\mathbf{A}_o)$.

7. $\Omega((\mathbf{A}_\alpha, \mathbf{B}_\alpha)) = \mathsf{q\text{-}op}_{\epsilon\to\epsilon\to\epsilon}\,\Omega(\mathbf{A}_\alpha)\,\Omega(\mathbf{B}_\alpha)$.

8. $\Omega(\ulcorner\mathbf{A}_\alpha\urcorner) = \mathsf{q\text{-}quo}_{\epsilon\to\epsilon}\,\Omega(\mathbf{A}_\alpha)$.

When $\mathbf{A}_\alpha$ is eval-free, $\Omega(\mathbf{A}_\alpha)$ is the unique construction that represents the syntax tree of $\mathbf{A}_\alpha$.

14.3.3 Semantics

A new clause is added to the definition of a frame:

F5. D_ϵ is the set of eval-free expressions in $\mathcal{E}(L)$.

Thus the syntactic values in AlonzoQE are eval-free expressions (while in $\mathrm{CTT}_{\mathrm{qe}}$ and $\mathrm{CTT}_{\mathrm{uqe}}$ they are constructions).

The definition of an interpretation of a language is changed to the following definition: An *interpretation* of $L = (\mathcal{B},\mathcal{C})$ is a pair $M = (\mathcal{D}, I)$ where $\mathcal{D} = \{D_\alpha \mid \alpha \in \mathcal{T}(L)\}$ is a frame for L and I is an *interpretation function* that maps each constant in $\mathcal{C}$ of type α and each logical constant of type α to an element of D_α such that:

1. $I(\mathsf{q\text{-}eq}_{\epsilon\to\epsilon\to\epsilon})(\mathbf{A}_\alpha)(\mathbf{B}_\beta) = (\mathbf{A}_\alpha = \mathbf{B}_\beta)$ (i.e., $I(\mathsf{q\text{-}eq}_{\epsilon\to\epsilon\to\epsilon})$ applied to the expressions $\mathbf{A}_\alpha$ and $\mathbf{B}_\beta$ is the expression $(\mathbf{A}_\alpha = \mathbf{B}_\beta)$) if $\alpha = \beta$ and is undefined otherwise.

2. $I(\mathsf{q\text{-}app}_{\epsilon\to\epsilon\to\epsilon})(\mathbf{A}_\alpha)(\mathbf{B}_\beta) = (\mathbf{A}_\alpha\,\mathbf{B}_\beta)$ if $\alpha = \beta \to \gamma$ and is undefined otherwise.

3. $I(\mathsf{q\text{-}abs}_{\epsilon\to\epsilon\to\epsilon})(\mathbf{A}_\alpha)(\mathbf{B}_\beta) = (\lambda\,\mathbf{x} : \alpha\,.\,\mathbf{B}_\beta)$ if $\mathbf{A}_\alpha$ has the form $(\mathbf{x} : \alpha)$ and is undefined otherwise.

4. $I(\mathsf{q\text{-}dd}_{\epsilon\to\epsilon\to\epsilon})(\mathbf{A}_\alpha)(\mathbf{B}_\beta) = (\mathrm{I}\,\mathbf{x} : \alpha\,.\,\mathbf{B}_\beta)$ if $\mathbf{A}_\alpha$ has the form $(\mathbf{x} : \alpha)$ with $\alpha \neq o$ and $\beta = o$ and is undefined otherwise.

5. $I(\mathsf{q\text{-}op}_{\epsilon\to\epsilon\to\epsilon})(\mathbf{A}_\alpha)(\mathbf{B}_\beta) = (\mathbf{A}_\alpha, \mathbf{B}_\beta)$ if $\alpha = \beta$ and is undefined otherwise.

6. $I(\mathsf{q\text{-}quo}_{\epsilon\to\epsilon})(\mathbf{A}_\alpha) = \ulcorner \mathbf{A}_\alpha \urcorner$.

Finally, two new conditions are added to the definition of a general model:

V8. $V_\varphi^M(\ulcorner \mathbf{A}_\alpha \urcorner) = \mathbf{A}_\alpha$ (i.e., $V_\varphi^M(\ulcorner \mathbf{A}_\alpha \urcorner)$ is the expression $\mathbf{A}_\alpha$).

V9. $V_\varphi^M(\llbracket \mathbf{A}_\epsilon \rrbracket_\alpha) = V_\varphi^M(V_\varphi^M(\mathbf{A}_\epsilon))$ if $V_\varphi^M(\mathbf{A}_\epsilon)$ is an expression of type α. Otherwise, $V_\varphi^M(\llbracket \mathbf{A}_\epsilon \rrbracket_\alpha) = \mathrm{F}$ if $\alpha = o$ and is undefined if $\alpha \neq o$.

14.3.4   Proof System

It is not clear how to extend $\mathfrak{A}$ to a sound and complete proof system for AlonzoQE. However, $\mathfrak{A}$ can be extended to a sound proof system for AlonzoQE that is complete for eval-free formulas by following the design of the proof system for CTT$_{\mathsf{qe}}$ in [50]. Beta-reduction and substitution are tricky to define in the proof system for CTT$_{\mathsf{qe}}$, so they will also be tricky to define in the proof system for AlonzoQE.

14.3.5   Theorems

Theorem 14.7 (Denotation of a Construction) *Let AlonzoQE be the logic, $\mathbf{A}_\alpha$ be an eval-free expression of a language L, M be a general model of L, and $\varphi \in \mathsf{assign}(M)$. Then $V_\varphi^M(\Omega(\mathbf{A}_\alpha)) = \mathbf{A}_\alpha$.*

Proof By induction on the syntactic structure of expressions using the definition of Ω, the interpretation of the logical constants, and condition V8 of the definition of a general model.   □

Theorem 14.8 (Law of Quotation) *Let AlonzoQE be the logic. Then $\ulcorner \mathbf{A}_\alpha \urcorner = \Omega(\mathbf{A}_\alpha)$ is valid.*

Proof Let $\mathbf{A}_\alpha$ be an eval-free expression of a language L, M be a general model of L, and $\varphi \in \mathsf{assign}(M)$. Then $V_\varphi^M(\ulcorner \mathbf{A}_\alpha \urcorner) = \mathbf{A}_\alpha$ by condition V8 of the definition of a general model, and $V_\varphi^M(\Omega(\mathbf{A}_\alpha)) = \mathbf{A}_\alpha$ by the Denotation of a Construction (Theorem 14.7). Hence $V_\varphi^M(\ulcorner \mathbf{A}_\alpha \urcorner) = V_\varphi^M(\Omega(\mathbf{A}_\alpha))$ for all general models M that interpret $\mathbf{A}_\alpha$ and all $\varphi \in \mathsf{assign}(M)$, and so $\ulcorner \mathbf{A}_\alpha \urcorner = \Omega(\mathbf{A}_\alpha)$ is valid by part 10 of Lemma 5.4.   □

Theorem 14.9 (Law of Disquotation) *Let AlonzoQE be the logic. Then $\llbracket \ulcorner \mathbf{A}_\alpha \urcorner \rrbracket_\alpha = \mathbf{A}_\alpha$ is valid.*

Proof Let $\mathbf{A}_\alpha$ be an eval-free expression of a language L, M be a general model of L, and $\varphi \in \mathsf{assign}(M)$. Then

$$V_\varphi^M(\llbracket\ulcorner\mathbf{A}_\alpha\urcorner\rrbracket_\alpha)$$
$$= V_\varphi^M(V_\varphi^M(\ulcorner\mathbf{A}_\alpha\urcorner)) \tag{1}$$
$$= V_\varphi^M(\mathbf{A}_\alpha) \tag{2}$$

(a) $V_\varphi^M(\ulcorner\mathbf{A}_\alpha\urcorner) = \mathbf{A}_\alpha$ by condition V8 of the definition of a general model. (a) implies (b) $V_\varphi^M(\ulcorner\mathbf{A}_\alpha\urcorner)$ is an expression of type α. (1) is by (b) and condition V9 of the definition of a general model; and (2) is by (a). Hence $V_\varphi^M(\llbracket\ulcorner\mathbf{A}_\alpha\urcorner\rrbracket_\alpha) = V_\varphi^M(\mathbf{A}_\alpha)$ for all general models M that interpret $\mathbf{A}_\alpha$ and all $\varphi \in \mathsf{assign}(M)$, and so $\llbracket\ulcorner\mathbf{A}_\alpha\urcorner\rrbracket_\alpha = \mathbf{A}_\alpha$ is valid by part 10 of Lemma 5.4. $\square$

14.4 Conclusions

AlonzoID, AlonzoS, and AlonzoQE are three variants of Alonzo that have useful machinery that is not available in Alonzo. However, these variants have a more complex syntax and semantics than Alonzo, and in the case of AlonzoQE, a significantly more elaborate proof system.

Chapter 15

Software Support

Before the Information Age, mathematics was done with the help of fairly simple tools such as pencil and paper, blackboard and chalk, log and trig tables, abaci, slide rules, and mechanical calculators. Today there is a wide range of software tools used in doing mathematics such as software calculators, programming languages, numerical analysis packages, statistical packages, interactive geometry systems, mathematical data bases (such as the On-Line Encyclopedia of Integer Sequences (OEIS) [125]), computer algebra systems, SAT solvers, SMT solvers, automated theorem provers, and proof assistants. So software can support mathematics, but can it support mathematics done using Alonzo?

The answer is clearly yes since, by virtue of being a logic (as defined in section 2.2), the languages of Alonzo have a precise common syntax plus a precise common semantics with a notion of logical consequence. Since Alonzo has a precise syntax, software can be used to process — build, store, analyze, manipulate, and present — Alonzo types and expressions. Since Alonzo has a precise semantics in addition to a precise syntax, there is a basis for verifying the correctness of this software. Hence Alonzo possesses the prerequisites needed for successful software support.

Although Alonzo can provide a rigorous framework for expressing and reasoning about mathematical ideas without software, effective software support can greatly enhance the benefits of Alonzo. This chapter surveys the various kinds of support that software could provide to Alonzo users. It is intended to help the reader imagine what software could do for Alonzo users and to inspire system developers to make what is imagined a reality.

© The Author(s), under exclusive license to Springer Nature Switzerland AG 2023 210
W. M. Farmer, *Simple Type Theory*, Computer Science Foundations
and Applied Logic, https://doi.org/10.1007/978-3-031-21112-6_15

15.1 Basis Support

The most basis kind of software support is software for processing Alonzo types and expressions. Types and expressions, which can be very large, are difficult to process by hand. Software can make the processing much easier and much more reliable. There are three main approaches for using software to do the processing.

The first, and simplest way, is to build LaTeX commands that represent Alonzo types and expressions. More precisely, each LaTeX command represents a type or expression presented with a particular choice of notation. The LaTeX commands can be stored in text files and presented in mathematical documents as LaTeX output.

We have defined a set of LaTeX macros for constructing LaTeX commands that produce in LaTeX presentations of types and expressions in either formal or compact notation. They make it relatively easy to include Alonzo types and expressions in a mathematics document. We have also defined a set of LaTeX environments for constructing LaTeX commands that produce in LaTeX presentations of Alonzo modules. All of the Alonzo types, expressions, and modules given in this book have been written using these LaTeX macros and environments. These LaTeX macros and environments are described in a separate document entitled "LaTeX for Alonzo".[1]

Since mathematics practitioners customarily write their mathematical documents using LaTeX, this approach would be very beneficial to mathematics practitioners who want to express mathematical ideas in Alonzo without doing a lot of work. However, it is certainly possible to build LaTeX commands that do not represent well-formed Alonzo types, expressions, or modules. For example, an (ill-formed) equality between expressions of different type can be easily built. Hence the user is responsible for ensuring the well-formedness of the objects represented by LaTeX commands. This approach also does not provide support for analyzing or manipulating the Alonzo objects represented by the LaTeX commands. For example, if E is the expression represented by a LaTeX command C, there is no support for determining the type of E or for translating C to another LaTeX command C' that represents E with a different choice of notation. The second and third approaches do not have these shortcomings.

The second approach is to write a language L for representing Alonzo

---

[1]The document "LaTeX for Alonzo" is available at `http://imps.mcmaster.ca/d oc/latex-for-alonzo.pdf`, and a LaTeX source file containing just the macros and environments for Alonzo is available at `http://imps.mcmaster.ca/doc/alonzo-notati on.tex` .

types and expressions as strings of characters plus a parser for translating the strings into data structures. A data structure produced by the parsing represents the Alonzo type or expression represented by the string. The parser generates an error on a string that does not represent a well-formed Alonzo type or expression. So an Alonzo type or expression is built, stored, analyzed, manipulated, and presented via the data structure representing it. This approach does not provide support, however, for constructing the strings that represent the Alonzo types and expressions. The user is responsible for understanding how Alonzo types and expressions are represented in L. The third approach provides this missing support for Alonzo types and expressions.

The third approach is produce a *interactive development environment (IDE)* for Alonzo. The IDE interactively assists the user in building Alonzo types and expressions that are represented and stored as data structures. The IDE can also interactively help the user to analyze, manipulate, and present in various formats the represented Alonzo types and expressions.

15.2   Advanced Support

The basic software support discussed in the previous section provides a foundation for more sophisticated forms of software support for Alonzo. We divide these kinds of advanced support into five areas, corresponding to the five aspects of the tetrapod model of mathematical knowledge [22] discussed in Section 3.1.

15.2.1   Organization

Organization is representing, structuring, and interconnecting mathematical knowledge. The key organizational task for Alonzo is to build libraries of mathematical knowledge represented as theory, development, or realm graphs. This requires software support for building nodes (theories, developments, or realms) and edges (morphisms between the nodes). It also requires support to find relevant definitions and theorems in the graph and then transport them to where they are needed. This kind of organizational support, based on the little theories method, is best illustrated in the IMPS proof assistant [52].

15.2.2 Inference

Inference is deriving mathematical statements by deduction and other forms
of reasoning. In Section 9.6, we discussed the fundamental form of a mathematical problem in Alonzo:

$$T \vDash^s \mathbf{A}_o$$

where $T = (L, \Gamma)$ is a theory of Alonzo and $\mathbf{A}_o$ is a sentence of L. We also
described the three principal approaches to solving a problem of this form.
Software can support each of these approaches.

For the model-theoretic approach, software can be used to help the user
to construct a traditional proof that $T \vDash^s \mathbf{A}_o$ holds by, for example, enabling
the user to find relevant previously proved lemmas. Software can also be
used to search for counterexamples to $T \vDash^s \mathbf{A}_o$.

For the proof-theoretic approach, software can be used to (1) search
for or (2) help the user to develop a proof of $\mathbf{A}_o$ from T in a sound proof
system $\mathfrak{P}$ for Alonzo. The software systems that search for formal proofs are
known as *automated theorem provers*, while software systems that help the
user develop formal proofs are known as *interactive theorem provers*. In both
kinds of systems, the correctness of a proof produced in $\mathfrak{P}$ is mechanically
checked. TPS [6] is an example of an automatic theorem prover for Church's
type theory, and HOL [58], HOL Light [60], IMPS [53], Isabelle/HOL [98],
ProofPower [101], and PVS [95] are examples of interactive theorem provers
for Church's type theory.

For the decision procedure approach, decision procedures can be implemented by software. These algorithms can then be integrated into other
software systems such as automatic and interactive theorem provers and
expression simplifiers (see Subsection 15.2.3).

A proof assistant can incorporate all three of these approaches, but
many contemporary proof assistants only incorporate proof methods that
ultimately produce formal, machined-checked proofs.

15.2.3 Computation

Computation is algorithmically manipulating and simplifying mathematical
expressions and other representations of mathematical values. There are
two main kinds of computational support for Alonzo. The first is internal:
support for computing within a theory T the function denoted by an ex-
pression $\mathbf{F}_{\alpha \to \beta}$ in T. The second is external: support for computing within
a theory T a *syntax-based mathematical algorithm (SBMA)* [48], such as a

symbolic differentiation algorithm, that manipulates Alonzo expressions in a mathematically meaningful way.

The first can be performed, for example, using beta-reduction (as shown in Chapter 7) or term rewriting [8], and the second can be done by symbolic computation (i.e., computation on expressions in which variables are treated as symbols). Proof assistants and term rewriting systems provide support for the first kind of computation, while computer algebra systems such as Maple [81], Mathematica [128], and SageMath [111] provide support for the second.

An *expression simplifier*, that reduces a given expression $\mathbf{A}_\alpha$ to a simpler equivalent expression $\mathbf{A}'_\alpha$, is an important application of computation. Using either internal or external computation, it can incorporate decision procedures and algebraic simplification such as collecting like terms. An expression simplifier is especially useful in a logic like Alonzo that admits undefined expressions for verifying the numerous definedness conditions of the form $\mathbf{A}_\alpha{\downarrow}$ that arise in formal proofs. The IMPS simplifier [54] is example of an expression simplifier for a logic like Alonzo that admits undefined expressions.

15.2.4   Concretization

Concretization is generating, collecting, maintaining, and accessing concrete mathematical values. Collections of Alonzo objects — such as types, expressions, theories, traditional and formal proofs, definitions and theorems, morphisms, and general models — would be very useful for "big math" projects done in Alonzo. This requires software support for building and managing databases of Alonzo objects that are integrated into the software for the other four tetrapod aspects.

15.2.5   Narration

Narration is communicating mathematical ideas in a linear fashion using both natural and formal language. It is the main way of communicating mathematical knowledge in mathematical articles, monographs, and textbooks. In its most popular form, it is a story told by a series of definitions, theorems, and proofs. Narration is needed in Alonzo at two levels: theories and theory graphs. A theory narrative presents a mathematical topic as the development of a theory, while a theory graph narrative presents a mathematical library as the development of a theory graph. Most proof assistants support the first level of narration, but the proofs are entirely formal. Software support for narration should include both levels and should provide the option of having proofs that are entirely formal, entirely traditional (nonformal), or a mixture of traditional and formal proofs (what Michael Kohlhase calls *flexiformal mathematics* [76]).

15.3 Fully Integrated Support

The most effective software system for Alonzo would be an *interactive mathematics laboratory (IML)* [43] in which the various kinds of support mentioned above are fully integrated. An IML of this kind is needed for large-scale mathematics research [22]. A comprehensive IML for Alonzo (or some other practical logic) would have the potential to transform how people learn and practice mathematics.

15.4 Conclusions

By virtue of being a logic, Alonzo possesses the prerequisites needed for successful software support. Software can support the use of Alonzo at three levels: *basic* (support for processing types and expressions), *advanced* (support for one or more of the five aspects of the tetrapod model of mathematical knowledge), and *fully integrated* (integrated support for all five aspects within an interactive mathematics laboratory).

Appendix A

Metatheorems of $\mathfrak{A}$

In this appendix we prove the basic *metatheorems of* $\mathfrak{A}$, which are statements about what can be proved in $\mathfrak{A}$ in contrast to *theorems of* $\mathfrak{A}$ which are formulas of Alonzo proved in $\mathfrak{A}$. The metatheorems are very similar to those of $\mathcal{Q}_0$ presented by Peter Andrews in §52 (Elementary Logic in $\mathcal{Q}_0$) of [5]. The development of the metatheorems closely follows Andrews' development in [5].

A.1 Universal Instantiation

We start with two key lemmas that follow immediately from the definition of a proof from a set of premises:

Lemma A.1 (Rule R1′) *If* $\Gamma \vdash_{\mathfrak{A}} \mathbf{A}_o$ *and* $\Gamma \vdash_{\mathfrak{A}} \mathbf{A}_o \Rightarrow \mathbf{B}_o$, *then* $\Gamma \vdash_{\mathfrak{A}} \mathbf{B}_o$.

Lemma A.2 (Rule R2′) *If* $\Gamma \vdash_{\mathfrak{A}} \mathbf{A}_\alpha \simeq \mathbf{B}_\alpha$ *and* $\Gamma \vdash_{\mathfrak{A}} \mathbf{C}_o$, *then* $\Gamma \vdash_{\mathfrak{A}} \mathbf{D}_o$, *where* $\mathbf{D}_o$ *is the result of replacing one occurrence of* $\mathbf{A}_\alpha$ *in* $\mathbf{C}_o$ *by an occurrence of* $\mathbf{B}_\alpha$, *provided that the occurrence of* $\mathbf{A}_\alpha$ *in* $\mathbf{C}_o$ *is not the first argument of a function abstraction or a definite description and not in the second argument of a function abstraction* $\lambda \mathbf{x} : \beta . \mathbf{E}_\gamma$ *or a definite description* $\mathrm{I} \mathbf{x} : \beta . \mathbf{E}_o$ *where* $(\mathbf{x} : \beta)$ *is free in a member of* Γ *and free in* $\mathbf{A}_\alpha \simeq \mathbf{B}_\alpha$.

Proposition A.3 $\vdash_{\mathfrak{A}} \mathbf{A}_\alpha \simeq \mathbf{A}_\alpha$.

W. M. Farmer, *Simple Type Theory*, Computer Science Foundations and Applied Logic, https://doi.org/10.1007/978-3-031-21112-6

Proof

$$\vdash_{\mathfrak{A}} (\mathbf{x} : \beta){\downarrow} \tag{1}$$

$$\vdash_{\mathfrak{A}} (\lambda \mathbf{x} : \beta \,.\, \mathbf{A}_\alpha)\,(\mathbf{x} : \beta) \simeq \mathbf{A}_\alpha \tag{2}$$

$$\vdash_{\mathfrak{A}} \mathbf{A}_\alpha \simeq \mathbf{A}_\alpha \tag{3}$$

(1) is by Axiom A5.1; (2) follows from (1) and Axiom A4 by Rule R1 since $(\mathbf{x} : \beta)$ is free for $(\mathbf{x} : \beta)$ in $\mathbf{A}_\alpha$ and $\mathbf{A}_\alpha[(\mathbf{x} : \beta) \mapsto (\mathbf{x} : \beta)] \equiv \mathbf{A}_\alpha$; and (3) follows from (2) and (2) by Rule R2.   □

Lemma A.4 (Quasi-Equality Rules)

1. *If* $\Gamma \vdash_{\mathfrak{A}} \mathbf{A}_o$ *and* $\Gamma \vdash_{\mathfrak{A}} \mathbf{A}_o \simeq \mathbf{B}_o$, *then* $\Gamma \vdash_{\mathfrak{A}} \mathbf{B}_o$.

2. *If* $\Gamma \vdash_{\mathfrak{A}} \mathbf{A}_\alpha \simeq \mathbf{B}_\alpha$, *then* $\Gamma \vdash_{\mathfrak{A}} \mathbf{B}_\alpha \simeq \mathbf{A}_\alpha$.

3. *If* $\Gamma \vdash_{\mathfrak{A}} \mathbf{A}_\alpha \simeq \mathbf{B}_\alpha$ *and* $\Gamma \vdash_{\mathfrak{A}} \mathbf{B}_\alpha \simeq \mathbf{C}_\alpha$, *then* $\Gamma \vdash_{\mathfrak{A}} \mathbf{A}_\alpha \sim \mathbf{C}_\alpha$.

4. *If* $\Gamma \vdash_{\mathfrak{A}} \mathbf{A}_{\alpha \to \beta} \simeq \mathbf{B}_{\alpha \to \beta}$, *then* $\Gamma \vdash_{\mathfrak{A}} \mathbf{A}_{\alpha \to \beta}\,\mathbf{C}_\alpha \simeq \mathbf{B}_{\alpha \to \beta}\,\mathbf{C}_\alpha$.

5. *If* $\Gamma \vdash_{\mathfrak{A}} \mathbf{C}_\alpha \simeq \mathbf{D}_\alpha$, *then* $\Gamma \vdash_{\mathfrak{A}} \mathbf{A}_{\alpha \to \beta}\,\mathbf{C}_\alpha \simeq \mathbf{A}_{\alpha \to \beta}\,\mathbf{D}_\alpha$.

6. *If* $\Gamma \vdash_{\mathfrak{A}} \mathbf{A}_{\alpha \to \beta} \simeq \mathbf{B}_{\alpha \to \beta}$ *and* $\Gamma \vdash_{\mathfrak{A}} \mathbf{C}_\alpha \simeq \mathbf{D}_\alpha$, *then* $\Gamma \vdash_{\mathfrak{A}} \mathbf{A}_{\alpha \to \beta}\,\mathbf{C}_\alpha \simeq \mathbf{B}_{\alpha \to \beta}\,\mathbf{D}_\alpha$.

7. *If* $\Gamma \vdash_{\mathfrak{A}} \mathbf{A}_\alpha \simeq \mathbf{B}_\alpha$, *then* $\Gamma \vdash_{\mathfrak{A}} (\mathbf{A}_\alpha, \mathbf{C}_\beta) \simeq (\mathbf{B}_\alpha, \mathbf{C}_\beta)$.

8. *If* $\Gamma \vdash_{\mathfrak{A}} \mathbf{C}_\beta \simeq \mathbf{D}_\beta$, *then* $\Gamma \vdash_{\mathfrak{A}} (\mathbf{A}_\alpha, \mathbf{C}_\beta) \simeq (\mathbf{A}_\alpha, \mathbf{D}_\beta)$.

9. *If* $\Gamma \vdash_{\mathfrak{A}} \mathbf{A}_\alpha \simeq \mathbf{B}_\alpha$ *and* $\Gamma \vdash_{\mathfrak{A}} \mathbf{C}_\beta \simeq \mathbf{D}_\beta$, *then* $\Gamma \vdash_{\mathfrak{A}} (\mathbf{A}_\alpha, \mathbf{C}_\beta) \simeq (\mathbf{B}_\alpha, \mathbf{D}_\beta)$.

Proof By Proposition A.3 and Rule R2′.   □

The next lemma is a special case of Rule R2′ in which the quasi-equality $\mathbf{A}_\alpha \simeq \mathbf{B}_\alpha$ is a theorem of $\mathfrak{A}$. As a result, the restriction involving variables free in both Γ and the quasi-equality is unnecessary.

Lemma A.5 (Rule R2″) *If* $\vdash_{\mathfrak{A}} \mathbf{A}_\alpha \simeq \mathbf{B}_\alpha$ *and* $\Gamma \vdash_{\mathfrak{A}} \mathbf{C}_o$, *then* $\Gamma \vdash_{\mathfrak{A}} \mathbf{D}_o$, *where* $\mathbf{D}_o$ *is the result of replacing one occurrence of* $\mathbf{A}_\alpha$ *in* $\mathbf{C}_o$ *by an occurrence of* $\mathbf{B}_\alpha$, *provided that the occurrence of* $\mathbf{A}_\alpha$ *in* $\mathbf{C}_o$ *is not the first argument of a function abstraction or a definite description.*

Proof

$$\vdash_{\mathfrak{A}} \mathbf{A}_\alpha \simeq \mathbf{B}_\alpha \tag{1}$$

$$\vdash_{\mathfrak{A}} \mathbf{C}_o \simeq \mathbf{C}_o \tag{2}$$

$$\vdash_{\mathfrak{A}} \mathbf{C}_o \simeq \mathbf{D}_o \tag{3}$$

$$\Gamma \vdash_{\mathfrak{A}} \mathbf{C}_o \tag{4}$$

$$\Gamma \vdash_{\mathfrak{A}} \mathbf{C}_o \simeq \mathbf{D}_o \tag{5}$$

$$\Gamma \vdash_{\mathfrak{A}} \mathbf{D}_o \tag{6}$$

(1) and (4) are given; (2) is by Proposition A.3; (3) follows from (1) and (2) by Rule R2; (5) follows from (3) since $\emptyset \subseteq \Gamma$; and (6) follows from (5) and (4) by Rule R2′.   □

Lemma A.6 (Beta-Reduction Rule 1) $\Gamma \ \vdash_{\mathfrak{A}} \ \mathbf{A}_\alpha{\downarrow}$ *and* $\Gamma \ \vdash_{\mathfrak{A}} \ \mathbf{C}_o,$ *then* $\Gamma \ \vdash_{\mathfrak{A}} \ \mathbf{D}_o,$ *where* $\mathbf{D}_o$ *is the result of replacing one occurrence of* $(\lambda \mathbf{x} : \alpha \, . \, \mathbf{B}_\beta) \, \mathbf{A}_\alpha$ *in* $\mathbf{C}_o$ *by an occurrence of* $\mathbf{B}_\beta[(\mathbf{x} : \alpha) \mapsto \mathbf{A}_\alpha],$ *provided that* $\mathbf{A}_\alpha$ *is free for* $(\mathbf{x} : \alpha)$ *in* $\mathbf{B}_\beta$ *and the occurrence of* $(\lambda \mathbf{x} : \alpha \, . \, \mathbf{B}_\beta) \, \mathbf{A}_\alpha$ *is not in the second argument of a function abstraction* $\lambda \mathbf{y} : \gamma \, . \, \mathbf{E}_\delta$ *or a definite description* $I \mathbf{y} : \gamma \, . \, \mathbf{E}_o$ *where* $(\mathbf{y} : \gamma)$ *is free in a member of* Γ *and free in* $(\lambda \mathbf{x} : \alpha \, . \, \mathbf{B}_\beta) \, \mathbf{A}_\alpha \simeq \mathbf{B}_\beta[(\mathbf{x} : \alpha) \mapsto \mathbf{A}_\alpha].$

Proof Let $(\star)$ $\mathbf{A}_\alpha$ be free for $(\mathbf{x} : \alpha)$ in $\mathbf{B}_\beta.$

$$\Gamma \vdash_{\mathfrak{A}} \mathbf{A}_\alpha{\downarrow} \tag{1}$$

$$\Gamma \vdash_{\mathfrak{A}} (\lambda \mathbf{x} : \alpha \, . \, \mathbf{B}_\beta) \, \mathbf{A}_\alpha \simeq \mathbf{B}_\beta[(\mathbf{x} : \alpha) \mapsto \mathbf{A}_\alpha] \tag{2}$$

$$\Gamma \vdash_{\mathfrak{A}} \mathbf{C}_o \tag{3}$$

$$\Gamma \vdash_{\mathfrak{A}} \mathbf{D}_o \tag{4}$$

(1) and (3) are given; (2) follows from (1) and Axiom A4 by Rule R1′ using $(\star)$; and (4) follows from (2) and (3) by Rule R2′.   □

Lemma A.7 (Beta-Reduction Rule 2) $\vdash_{\mathfrak{A}} \mathbf{A}_\alpha{\downarrow}$ *and* $\Gamma \vdash_{\mathfrak{A}} \mathbf{C}_o,$ *then* $\Gamma \ \vdash_{\mathfrak{A}} \ \mathbf{D}_o,$ *where* $\mathbf{D}_o$ *is the result of replacing one occurrence of* $(\lambda \mathbf{x} : \alpha \, . \, \mathbf{B}_\beta) \, \mathbf{A}_\alpha$ *in* $\mathbf{C}_o$ *by an occurrence of* $\mathbf{B}_\beta[(\mathbf{x} : \alpha) \mapsto \mathbf{A}_\alpha],$ *provided that* $\mathbf{A}_\alpha$ *is free for* $(\mathbf{x} : \alpha)$ *in* $\mathbf{B}_\beta.$

Proof Let $(\star)$ $\mathbf{A}_\alpha$ be free for $(\mathbf{x} : \alpha)$ in $\mathbf{B}_\beta.$

$$\vdash_{\mathfrak{A}} \mathbf{A}_\alpha{\downarrow} \tag{1}$$

$$\vdash_{\mathfrak{A}} (\lambda \mathbf{x} : \alpha \, . \, \mathbf{B}_\beta) \, \mathbf{A}_\alpha \simeq \mathbf{B}_\beta[(\mathbf{x} : \alpha) \mapsto \mathbf{A}_\alpha] \tag{2}$$

$$\Gamma \vdash_{\mathfrak{A}} \mathbf{C}_o \tag{3}$$

$$\Gamma \vdash_{\mathfrak{A}} \mathbf{D}_o \tag{4}$$

(1) and (3) are given; (2) follows from (1) and Axiom A4 by Rule R1 using $(\star)$; and (4) follows from (2) and (3) by Rule R2''. $\qquad\square$

Proposition A.8 (Formula Definedness) $\vdash_{\mathfrak{A}} \mathbf{A}_o\!\downarrow$ *for all formulas* $\mathbf{A}_o$.

Proof Every formula is a variable, constant, equality, or function application. Hence the proposition follows immediately from Axioms A5.1, A5.2, A5.3, and A5.6. $\qquad\square$

Lemma A.9

1. *If* $\Gamma \vdash_{\mathfrak{A}} \mathbf{A}_\alpha\!\downarrow$ *or* $\Gamma \vdash_{\mathfrak{A}} \mathbf{B}_\alpha\!\downarrow$, *then*

$$\Gamma \vdash_{\mathfrak{A}} (\mathbf{A}_\alpha \simeq \mathbf{B}_\alpha) \simeq (\mathbf{A}_\alpha = \mathbf{B}_\alpha).$$

2. *If* $\Gamma \vdash_{\mathfrak{A}} \mathbf{A}_\alpha\!\downarrow$ *or* $\Gamma \vdash_{\mathfrak{A}} \mathbf{B}_\alpha\!\downarrow$ *and if* $\Gamma \vdash_{\mathfrak{A}} (\mathbf{A}_\alpha \simeq \mathbf{B}_\alpha)$, *then*

$$\Gamma \vdash_{\mathfrak{A}} \mathbf{A}_\alpha = \mathbf{B}_\alpha.$$

3. *If* $\Gamma \vdash_{\mathfrak{A}} \mathbf{A}_\alpha = \mathbf{B}_\alpha$, *then* $\Gamma \vdash_{\mathfrak{A}} \mathbf{A}_\alpha \simeq \mathbf{B}_\alpha$.

Proof Part 1 follows from Axioms A5.12 and A5.13 by Rule R1'. Part 2 is by part 1 of this lemma and Rule R2'. And part 3 is by Axioms A5.4 or A5.5, part 1 of this lemma and parts 1 and 2 of the Quasi-Equality Rules. $\square$

Remark A.10 (Generalized Rules) *By virtue of part 2 of the Quasi-Equality Rules and part 3 of Lemma A.9, we may assume without loss of generality that the first assumption of Rule R2' is*

$$\Gamma \vdash_{\mathfrak{A}} \mathbf{A}_\alpha \simeq \mathbf{B}_\alpha \ or \ \Gamma \vdash_{\mathfrak{A}} \mathbf{B}_\alpha \simeq \mathbf{A}_\alpha \ or \ \Gamma \vdash_{\mathfrak{A}} \mathbf{A}_\alpha = \mathbf{B}_\alpha \ or \ \Gamma \vdash_{\mathfrak{A}} \mathbf{B}_\alpha = \mathbf{A}_\alpha$$

and the first assumption of Rule R2'' is

$$\vdash_{\mathfrak{A}} \mathbf{A}_\alpha \simeq \mathbf{B}_\alpha \ or \ \vdash_{\mathfrak{A}} \mathbf{B}_\alpha \simeq \mathbf{A}_\alpha \ or \ \vdash_{\mathfrak{A}} \mathbf{A}_\alpha = \mathbf{B}_\alpha \ or \ \vdash_{\mathfrak{A}} \mathbf{B}_\alpha = \mathbf{A}_\alpha$$

Lemma A.11 *If* $\Gamma \vdash_{\mathfrak{A}} \mathbf{A}_\alpha\!\downarrow$, *then* $\Gamma \vdash_{\mathfrak{A}} \mathbf{A}_\alpha = \mathbf{A}_\alpha$.

Proof

$$\Gamma \vdash_{\mathfrak{A}} \mathbf{A}_\alpha\!\downarrow \tag{1}$$

$$\vdash_{\mathfrak{A}} \mathbf{A}_\alpha \simeq \mathbf{A}_\alpha \tag{2}$$

$$\Gamma \vdash_{\mathfrak{A}} \mathbf{A}_\alpha = \mathbf{A}_\alpha \tag{3}$$

(1) is given; (2) is by Proposition A.3; and (3) follows from (1) and (2) by part 2 of Lemma A.9. $\qquad\square$

Corollary A.12 $\vdash_{\mathfrak{A}} T_o$.

Proof

$$\vdash_{\mathfrak{A}} (\lambda x : o . x)\!\downarrow \tag{1}$$
$$\vdash_{\mathfrak{A}} (\lambda x : o . x) = (\lambda x : o . x) \tag{2}$$
$$\vdash_{\mathfrak{A}} T_o \tag{3}$$

(1) is by Axiom A5.11; (2) follows from (1) by Lemma A.11; and (3) follows from (2) by the definition of T_o.   □

Lemma A.13 (Equality Rules)

1. *If* $\Gamma \vdash_{\mathfrak{A}} \mathbf{A}_o$ *and* $\Gamma \vdash_{\mathfrak{A}} \mathbf{A}_o \Leftrightarrow \mathbf{B}_o$, *then* $\Gamma \vdash_{\mathfrak{A}} \mathbf{B}_o$.

2. *If* $\Gamma \vdash_{\mathfrak{A}} \mathbf{A}_\alpha = \mathbf{B}_\alpha$, *then* $\Gamma \vdash_{\mathfrak{A}} \mathbf{B}_\alpha = \mathbf{A}_\alpha$.

3. *If* $\Gamma \vdash_{\mathfrak{A}} \mathbf{A}_\alpha = \mathbf{B}_\alpha$ *and* $\Gamma \vdash_{\mathfrak{A}} \mathbf{B}_\alpha = \mathbf{C}_\alpha$, *then* $\Gamma \vdash_{\mathfrak{A}} \mathbf{A}_\alpha = \mathbf{C}_\alpha$.

4. *If* $\Gamma \vdash_{\mathfrak{A}} \mathbf{A}_{\alpha\to\beta} = \mathbf{B}_{\alpha\to\beta}$, *then* $\Gamma \vdash_{\mathfrak{A}} \mathbf{A}_{\alpha\to\beta} \mathbf{C}_\alpha \simeq \mathbf{B}_{\alpha\to\beta} \mathbf{C}_\alpha$.

5. *If* $\Gamma \vdash_{\mathfrak{A}} \mathbf{C}_\alpha = \mathbf{D}_\alpha$, *then* $\Gamma \vdash_{\mathfrak{A}} \mathbf{A}_{\alpha\to\beta} \mathbf{C}_\alpha \simeq \mathbf{A}_{\alpha\to\beta} \mathbf{D}_\alpha$.

6. *If* $\Gamma \vdash_{\mathfrak{A}} \mathbf{A}_{\alpha\to\beta} = \mathbf{B}_{\alpha\to\beta}$ *and* $\Gamma \vdash_{\mathfrak{A}} \mathbf{C}_\alpha = \mathbf{D}_\alpha$, *then* $\Gamma \vdash_{\mathfrak{A}} \mathbf{A}_{\alpha\to\beta} \mathbf{C}_\alpha \simeq \mathbf{B}_{\alpha\to\beta} \mathbf{D}_\alpha$.

7. *If* $\Gamma \vdash_{\mathfrak{A}} \mathbf{A}_\alpha = \mathbf{B}_\alpha$, *then* $\Gamma \vdash_{\mathfrak{A}} (\mathbf{A}_\alpha, \mathbf{C}_\beta) \simeq (\mathbf{B}_\alpha, \mathbf{C}_\beta)$.

8. *If* $\Gamma \vdash_{\mathfrak{A}} \mathbf{C}_\beta = \mathbf{D}_\beta$, *then* $\Gamma \vdash_{\mathfrak{A}} (\mathbf{A}_\alpha, \mathbf{C}_\beta) \simeq (\mathbf{A}_\alpha, \mathbf{D}_\beta)$.

9. *If* $\Gamma \vdash_{\mathfrak{A}} \mathbf{A}_\alpha = \mathbf{B}_\alpha$ *and* $\Gamma \vdash_{\mathfrak{A}} \mathbf{C}_\beta = \mathbf{D}_\beta$, *then* $\Gamma \vdash_{\mathfrak{A}} (\mathbf{A}_\alpha, \mathbf{C}_\beta) = (\mathbf{B}_\alpha, \mathbf{D}_\beta)$.

Proof By Axioms A5.4 and A5.5, parts 2 and 3 of Lemma A.9, and the Quasi-Equality Rules.   □

Theorem A.14 (Universal Instantiation) *If* $\Gamma \vdash_{\mathfrak{A}} \mathbf{A}_\alpha\!\downarrow$ *and* $\Gamma \vdash_{\mathfrak{A}} \forall \mathbf{x} : \alpha . \mathbf{B}_o$, *then*

$$\Gamma \vdash_{\mathfrak{A}} \mathbf{B}_o[(\mathbf{x} : \alpha) \mapsto \mathbf{A}_\alpha],$$

provided $\mathbf{A}_\alpha$ *is free for* $(\mathbf{x} : \alpha)$ *in* $\mathbf{B}_o$.

Proof Let $(\star)$ $\mathbf{A}_\alpha$ be free for $(\mathbf{x} : \alpha)$ in $\mathbf{B}_o$.

$$\Gamma \vdash_{\mathfrak{A}} \mathbf{A}_\alpha \!\downarrow \tag{1}$$

$$\Gamma \vdash_{\mathfrak{A}} \forall \mathbf{x} : \alpha \,.\, \mathbf{B}_o \tag{2}$$

$$\Gamma \vdash_{\mathfrak{A}} \lambda x : \alpha \,.\, T_o = \lambda \mathbf{x} : \alpha \,.\, \mathbf{B}_o \tag{3}$$

$$\Gamma \vdash_{\mathfrak{A}} \lambda x : \alpha \,.\, T_o \simeq \lambda \mathbf{x} : \alpha \,.\, \mathbf{B}_o \tag{4}$$

$$\Gamma \vdash_{\mathfrak{A}} (\lambda x : \alpha \,.\, T_o) \, \mathbf{A}_\alpha \simeq (\lambda \mathbf{x} : \alpha \,.\, \mathbf{B}_o) \, \mathbf{A}_\alpha \tag{5}$$

$$\Gamma \vdash_{\mathfrak{A}} T_o \simeq \mathbf{B}_o[(\mathbf{x} : \alpha) \mapsto \mathbf{A}_\alpha] \tag{6}$$

$$\Gamma \vdash_{\mathfrak{A}} T_o \tag{7}$$

$$\Gamma \vdash_{\mathfrak{A}} \mathbf{B}_o[(\mathbf{x} : \alpha) \mapsto \mathbf{A}_\alpha] \tag{8}$$

(1) and (2) are given; (3) follows from (2) by the definition $\forall$; (4) follows from (3) by part 3 of Lemma A.9; (5) follows from (4) by part 4 of the Quasi-Equality Rules; (6) follows from (1) and (5) by Beta-Reduction Rule 1 using $(\star)$ and the fact that $\mathbf{A}_\alpha$ is free for $(\mathbf{x} : \alpha)$ in T_o; (7) is by Corollary A.12; and (8) follows from (6) and (7) by Rule R2′. □

A.2 Alpha-Conversion

Lemma A.15 *If $\Gamma \vdash_{\mathfrak{A}} \mathbf{A}_\alpha \!\downarrow$ and $\vdash_{\mathfrak{A}} \mathbf{B}_\beta \simeq \mathbf{C}_\beta$, then*

$$\Gamma \vdash_{\mathfrak{A}} (\mathbf{B}_\beta \simeq \mathbf{C}_\beta)[(\mathbf{x} : \alpha) \mapsto \mathbf{A}_\alpha],$$

provided $\mathbf{A}_\alpha$ is free for $(\mathbf{x} : \alpha)$ in $\mathbf{B}_\alpha$ and $\mathbf{C}_\alpha$.

Proof Let $(\star)$ $\mathbf{A}_\alpha$ be free for $(\mathbf{x} : \alpha)$ in $\mathbf{B}_\alpha$ and $\mathbf{C}_\alpha$.

$$\Gamma \vdash_{\mathfrak{A}} \mathbf{A}_\alpha \!\downarrow \tag{1}$$

$$\vdash_{\mathfrak{A}} \mathbf{B}_\beta \simeq \mathbf{C}_\beta \tag{2}$$

$$\vdash_{\mathfrak{A}} (\lambda \mathbf{x} : \alpha \,.\, \mathbf{B}_\beta) \, \mathbf{A}_\alpha \simeq (\lambda \mathbf{x} : \alpha \,.\, \mathbf{B}_\beta) \, \mathbf{A}_\alpha \tag{3}$$

$$\vdash_{\mathfrak{A}} (\lambda \mathbf{x} : \alpha \,.\, \mathbf{B}_\beta) \, \mathbf{A}_\alpha \simeq (\lambda \mathbf{x} : \alpha \,.\, \mathbf{C}_\beta) \, \mathbf{A}_\alpha \tag{4}$$

$$\Gamma \vdash_{\mathfrak{A}} \mathbf{B}_\beta[(\mathbf{x} : \alpha) \mapsto \mathbf{A}_\alpha] \simeq \mathbf{C}_\beta[(\mathbf{x} : \alpha) \mapsto \mathbf{A}_\alpha] \tag{5}$$

$$\Gamma \vdash_{\mathfrak{A}} (\mathbf{B}_\beta \simeq \mathbf{C}_\beta)[(\mathbf{x} : \alpha) \mapsto \mathbf{A}_\alpha] \tag{6}$$

(1) and (2) are given; (3) is by Proposition A.3; (4) follows from (2) and (3) by Rule R2″; (5) follows from (1) and (4) by Beta-Reduction Rule 1 using $(\star)$; and (6) follows from (5) by the definition of substitution. □

Lemma A.16 *If* $\Gamma \vdash_{\mathfrak{A}} \mathbf{A}_\alpha\downarrow$ *and* $\vdash_{\mathfrak{A}} \mathbf{B}_\beta = \mathbf{C}_\beta$, *then*

$$\Gamma \vdash_{\mathfrak{A}} (\mathbf{B}_\beta = \mathbf{C}_\beta)[(\mathbf{x} : \alpha) \mapsto \mathbf{A}_\alpha],$$

provided $\mathbf{A}_\alpha$ *is free for* $(\mathbf{x} : \alpha)$ *in* $\mathbf{B}_\alpha$ *and* $\mathbf{C}_\alpha$.

Proof Let $(\star)$ $\mathbf{A}_\alpha$ be free for $(\mathbf{x} : \alpha)$ in $\mathbf{B}_\alpha$ and $\mathbf{C}_\alpha$.

$$\Gamma \vdash_{\mathfrak{A}} \mathbf{A}_\alpha\downarrow \tag{1}$$

$$\vdash_{\mathfrak{A}} \mathbf{B}_\beta = \mathbf{C}_\beta \tag{2}$$

$$\vdash_{\mathfrak{A}} \mathbf{B}_\beta \simeq \mathbf{C}_\beta \tag{3}$$

$$\Gamma \vdash_{\mathfrak{A}} (\mathbf{B}_\beta \simeq \mathbf{C}_\beta)[(\mathbf{x} : \alpha) \mapsto \mathbf{A}_\alpha] \tag{4}$$

$$\Gamma \vdash_{\mathfrak{A}} (\lambda\mathbf{x} : \alpha \cdot \mathbf{B}_\beta \simeq \mathbf{C}_\beta)\mathbf{A}_\alpha \simeq (\mathbf{B}_\beta \simeq \mathbf{C}_\beta)[(\mathbf{x} : \alpha) \mapsto \mathbf{A}_\alpha] \tag{5}$$

$$\Gamma \vdash_{\mathfrak{A}} (\lambda\mathbf{x} : \alpha \cdot \mathbf{B}_\beta \simeq \mathbf{C}_\beta)\mathbf{A}_\alpha \tag{6}$$

$$\vdash_{\mathfrak{A}} (\mathbf{B}_\beta \simeq \mathbf{C}_\beta) \simeq (\mathbf{B}_\beta = \mathbf{C}_\beta) \tag{7}$$

$$\Gamma \vdash_{\mathfrak{A}} (\lambda\mathbf{x} : \alpha \cdot \mathbf{B}_\beta = \mathbf{C}_\beta)\mathbf{A}_\alpha \tag{8}$$

$$\Gamma \vdash_{\mathfrak{A}} (\mathbf{B}_\beta = \mathbf{C}_\beta)[(\mathbf{x} : \alpha) \mapsto \mathbf{A}_\alpha] \tag{9}$$

(1) and (2) are given; (3) follows from (2) by part 3 of Lemma A.9; (4) follows from (1) and (3) by Lemma A.15 using $(\star)$; (5) follows (1) and Axiom A4 by R1$'$ using $(\star)$; (6) follows from (5) and (4) by Rule R2$'$; (7) follows from Axioms A5.4 or A5.5 and (2) by part 1 of Lemma A.9; (8) follows from (7) and (6) by Rule R2$''$; and (9) follows from (1) and (8) by Beta-Reduction Rule 1 using $(\star)$. □

Lemma A.17 $\vdash_{\mathfrak{A}} (f : \alpha \to \beta) = \lambda\mathbf{z} : \alpha \cdot (f : \alpha \to \beta)\mathbf{z}$.

Proof

$$\vdash_{\mathfrak{A}} (f : \alpha \to \beta) = \lambda\mathbf{z} : \alpha \cdot (f : \alpha \to \beta)\mathbf{z} \Leftrightarrow$$
$$\forall x : \alpha \cdot (f : \alpha \to \beta)x \simeq (\lambda\mathbf{z} : \alpha \cdot (f : \alpha \to \beta)\mathbf{z})x \tag{1}$$

$$\vdash_{\mathfrak{A}} (f : \alpha \to \beta) = \lambda\mathbf{z} : \alpha \cdot (f : \alpha \to \beta)\mathbf{z} \Leftrightarrow$$
$$\forall x : \alpha \cdot (f : \alpha \to \beta)x \simeq (f : \alpha \to \beta)x \tag{2}$$

$$\vdash_{\mathfrak{A}} (f : \alpha \to \beta) = (f : \alpha \to \beta) \Leftrightarrow$$
$$\forall x : \alpha \cdot (f : \alpha \to \beta)x \simeq (f : \alpha \to \beta)x \tag{3}$$

$$\vdash_{\mathfrak{A}} (f : \alpha \to \beta) = (f : \alpha \to \beta) \tag{4}$$

$$\vdash_{\mathfrak{A}} (f : \alpha \to \beta) = \lambda\mathbf{z} : \alpha \cdot (f : \alpha \to \beta)\mathbf{z} \tag{5}$$

(1) and (3) follow from Axioms A3, A5.1, and A5.11 by Lemma A.16; (2) follows from Axiom A5.1 and (1) by Beta-Reduction Rule 2; (4) follows from Axiom A5.1 and Lemma A.11; and (5) follows from (2), (3), and (4) by the Equality Rules. □

Theorem A.18 (Alpha-Conversion) *If* $(\mathbf{y} : \alpha)$ *is not free in* $\mathbf{B}_\beta$ *and* $(\mathbf{y} : \alpha)$ *is free for* $(\mathbf{x} : \alpha)$ *in* $\mathbf{B}_\beta$, *then*

$$\vdash_{\mathfrak{A}} (\lambda \mathbf{x} : \alpha . \mathbf{B}_\beta) = (\lambda \mathbf{y} : \alpha . \mathbf{B}_\beta[(\mathbf{x} : \alpha) \mapsto \mathbf{y}]).$$

Proof Let $(\star)$ $(\mathbf{y} : \alpha)$ be not free in $\mathbf{B}_\beta$ and $(\mathbf{y} : \alpha)$ be free for $(\mathbf{x} : \alpha)$ in $\mathbf{B}_\beta$. Choose $(\star\star)$ a variable $(\mathbf{z} : \alpha)$ that is distinct from $(\mathbf{x} : \alpha)$, $(\mathbf{y} : \alpha)$, and the variables in $\mathbf{B}_\beta$.

$$\vdash_{\mathfrak{A}} \lambda \mathbf{x} : \alpha . \mathbf{B}_\beta = \lambda \mathbf{z} : \alpha . (\lambda \mathbf{x} : \alpha . \mathbf{B}_\beta) \mathbf{z} \tag{1}$$

$$\vdash_{\mathfrak{A}} \lambda \mathbf{x} : \alpha . \mathbf{B}_\beta = \lambda \mathbf{z} : \alpha . \mathbf{B}_\beta[(\mathbf{x} : \alpha) \mapsto \mathbf{z}] \tag{2}$$

$$\vdash_{\mathfrak{A}} \lambda \mathbf{y} : \alpha . \mathbf{B}_\beta[(\mathbf{x} : \alpha) \mapsto \mathbf{y}] = $$
$$\lambda \mathbf{z} : \alpha . (\lambda \mathbf{y} : \alpha . \mathbf{B}_\beta[(\mathbf{x} : \alpha) \mapsto \mathbf{y}]) \mathbf{z} \tag{3}$$

$$\vdash_{\mathfrak{A}} \lambda \mathbf{y} : \alpha . \mathbf{B}_\beta[(\mathbf{x} : \alpha) \mapsto \mathbf{y}] = $$
$$\lambda \mathbf{z} : \alpha . (\mathbf{B}_\beta[(\mathbf{x} : \alpha) \mapsto (\mathbf{y} : \alpha)])[(\mathbf{y} : \alpha) \mapsto \mathbf{z}] \tag{4}$$

$$\vdash_{\mathfrak{A}} \lambda \mathbf{y} : \alpha . \mathbf{B}_\beta[(\mathbf{x} : \alpha) \mapsto \mathbf{y}] = \lambda \mathbf{z} : \alpha . \mathbf{B}_\beta[(\mathbf{x} : \alpha) \mapsto \mathbf{z}] \tag{5}$$

$$\vdash_{\mathfrak{A}} (\lambda \mathbf{x} : \alpha . \mathbf{B}_\beta) = (\lambda \mathbf{y} : \alpha . \mathbf{B}_\beta[(\mathbf{x} : \alpha) \mapsto \mathbf{y}]) \tag{6}$$

(1) and (3) follow from Axiom A5.11 and Lemma A.17 by Lemma A.16 using $(\star\star)$; (2) follows from Axiom A5.1 and (1) by Beta-Reduction Rule 2 using $(\star\star)$; (4) follows from Axiom A5.1 and (3) by Beta-Reduction Rule 2 using $(\star\star)$; (5) follows from (4) by Exercise 4.5 using $(\star)$; and (6) follows from (2) and (5) by the Equality Rules. □

A.3 Substitution Rule

Lemma A.19 *If* $\Gamma \vdash_{\mathfrak{A}} \mathbf{A}_\alpha{\downarrow}$, *then* $\Gamma \vdash_{\mathfrak{A}} T_o \Leftrightarrow (\mathbf{A}_\alpha \simeq \mathbf{A}_\alpha)$.

Proof

$$\Gamma \vdash_{\mathfrak{A}} \mathbf{A}_\alpha \!\downarrow \tag{1}$$

$$\vdash_{\mathfrak{A}} \lambda y : \alpha \,.\, y = \lambda y : \alpha \,.\, y \Leftrightarrow$$
$$\forall x : \alpha \,.\, (\lambda y : \alpha \,.\, y) x \simeq (\lambda y : \alpha \,.\, y) x \tag{2}$$

$$\vdash_{\mathfrak{A}} \lambda y : \alpha \,.\, y = \lambda y : \alpha \,.\, y \Leftrightarrow \forall x : \alpha \,.\, x \simeq x \tag{3}$$

$$\vdash_{\mathfrak{A}} \lambda y : \alpha \,.\, y = \lambda y : \alpha \,.\, y \tag{4}$$

$$\vdash_{\mathfrak{A}} \forall x : \alpha \,.\, x \simeq x \tag{5}$$

$$\vdash_{\mathfrak{A}} \lambda x : \alpha \,.\, T_o \Leftrightarrow \lambda x : \alpha \,.\, x \simeq x \tag{6}$$

$$\vdash_{\mathfrak{A}} (\lambda x : \alpha \,.\, T_o) \mathbf{A}_\alpha \Leftrightarrow (\lambda x : \alpha \,.\, x \simeq x) \mathbf{A}_\alpha \tag{7}$$

$$\Gamma \vdash_{\mathfrak{A}} T_o \Leftrightarrow (\mathbf{A}_\alpha \simeq \mathbf{A}_\alpha) \tag{8}$$

(1) is given; (2) follows from Axioms A3 and A5.11 by Lemma A.16; (3) follows from Axiom A5.1 and (2) by Beta-Reduction Rule 2; (4) follows from Axiom A5.11 and Lemma A.11; (5) follows from (3) and (4) by Rule R2''; (6) follows from (5) by the definition of $\forall$; (7) follows from (6) by Proposition A.8, part 2 of Lemma A.9, and the Equality Rules; and (8) follows from (1) and (7) by Beta-Reduction Rule 1.   □

Corollary A.20 $\vdash_{\mathfrak{A}} T_o \Leftrightarrow (\mathbf{A}_o \Leftrightarrow \mathbf{A}_o)$.

Proof   Follows from Proposition A.8 and Lemma A.19 by part 1 of Lemma A.9 and Rule R2''.   □

Lemma A.21 $\vdash_{\mathfrak{A}} (T_o \wedge T_o) \Leftrightarrow T_o$.

Proof

$$\vdash_{\mathfrak{A}} ((\lambda y : o \,.\, T_o) T_o \wedge (\lambda y : o \,.\, T_o) F_o) \Leftrightarrow \forall x : o \,.\, (\lambda y : o \,.\, T_o) x \tag{1}$$

$$\vdash_{\mathfrak{A}} (T_o \wedge T_o) \Leftrightarrow \forall x : o \,.\, T_o \tag{2}$$

$$\vdash_{\mathfrak{A}} T_o \Leftrightarrow \lambda x : o \,.\, T_o \simeq \lambda x : o \,.\, T_o \tag{3}$$

$$\vdash_{\mathfrak{A}} T_o \Leftrightarrow \lambda x : o \,.\, T_o = \lambda x : o \,.\, T_o \tag{4}$$

$$\vdash_{\mathfrak{A}} T_o \Leftrightarrow \forall x : o \,.\, T_o \tag{5}$$

$$\vdash_{\mathfrak{A}} (T_o \wedge T_o) \Leftrightarrow T_o \tag{6}$$

(1) follows from Axioms A1 and A5.11 by Lemma A.16; (2) follows from Axiom A5.1, Proposition A.8, and (1) by Beta-Reduction Rule 2; (3) follows from Axiom A5.11 by Lemma A.19; (4) follows from Axiom A5.11 and (3) by part 1 of Lemma A.9 and Rule R2''; (5) follows from (4) by the definition of $\forall$; and (6) follows from (2) and (5) by the Equality Rules.   □

Lemma A.22 $\vdash_{\mathfrak{A}} T_o \wedge T_o$.

Proof Follows from Corollary A.12 and Lemma A.21 by the Equality Rules. □

Lemma A.23 *If* $\vdash_{\mathfrak{A}} \mathbf{A}_o = \mathbf{B}_o$ *and* $\vdash_{\mathfrak{A}} \mathbf{C}_o = \mathbf{D}_o$, *then*

$$\vdash_{\mathfrak{A}} \mathbf{A}_o = \mathbf{B}_o \wedge \mathbf{C}_o = \mathbf{D}_o.$$

Proof

$$\vdash_{\mathfrak{A}} \mathbf{A}_o = \mathbf{B}_o \tag{1}$$
$$\vdash_{\mathfrak{A}} \mathbf{C}_o = \mathbf{D}_o \tag{2}$$
$$\vdash_{\mathfrak{A}} T_o \Leftrightarrow (\mathbf{A}_o = \mathbf{B}_o) \tag{3}$$
$$\vdash_{\mathfrak{A}} T_o \Leftrightarrow (\mathbf{C}_o = \mathbf{D}_o) \tag{4}$$
$$\vdash_{\mathfrak{A}} \mathbf{A}_o = \mathbf{B}_o \wedge \mathbf{C}_o = \mathbf{D}_o \tag{5}$$

(1) and (2) are given; (3) and (4) follow from (1) and (2), respectively, and Corollary A.20 by Rule R2″; and (5) follows (3), (4), and Lemma A.22 by Rule R2″. □

Lemma A.24 $\vdash_{\mathfrak{A}} (T_o \wedge F_o) \Leftrightarrow F_o$.

Proof

$$\vdash_{\mathfrak{A}} ((\lambda y : o \, . \, y) \, T_o \wedge (\lambda y : o \, . \, y) \, F_o) \Leftrightarrow \forall x : o \, . \, (\lambda y : o \, . \, y) \, x \tag{1}$$
$$\vdash_{\mathfrak{A}} (T_o \wedge F_o) \Leftrightarrow \forall x : o \, . \, x \tag{2}$$
$$\vdash_{\mathfrak{A}} (T_o \wedge F_o) \Leftrightarrow \lambda x : o \, . \, T_o = \lambda x : o \, . \, x \tag{3}$$
$$\vdash_{\mathfrak{A}} (T_o \wedge F_o) \Leftrightarrow F_o \tag{4}$$

(1) follows from Axioms A1 and A5.11 by Lemma A.16; (2) follows from Axiom A5.1, Proposition A.8, and (1) by Beta-Reduction Rule 2; (3) follows from (2) by the definition of $\forall$; and (4) follows from (3) by the definition of F_o. □

Lemma A.25 $\vdash_{\mathfrak{A}} (T_o \wedge \mathbf{A}_o) \Leftrightarrow \mathbf{A}_o$.

Proof

$$\vdash_{\mathfrak{A}} ((\lambda x : o \,.\, (T_o \wedge x) \Leftrightarrow x) \, T_o \wedge (\lambda x : o \,.\, (T_o \wedge x) \Leftrightarrow x) \, F_o) \Leftrightarrow$$
$$(\forall x : o \,.\, (\lambda x : o \,.\, (T_o \wedge x) \Leftrightarrow x) \, x) \tag{1}$$
$$\vdash_{\mathfrak{A}} (((T_o \wedge T_o) \Leftrightarrow T_o) \wedge ((T_o \wedge F_o) \Leftrightarrow F_o)) \Leftrightarrow$$
$$(\forall x : o \,.\, (T_o \wedge x) \Leftrightarrow x) \tag{2}$$
$$\vdash_{\mathfrak{A}} (((T_o \wedge T_o) \Leftrightarrow T_o) \wedge ((T_o \wedge F_o) \Leftrightarrow F_o)) \tag{3}$$
$$\vdash_{\mathfrak{A}} \forall x : o \,.\, (T_o \wedge x) \Leftrightarrow x \tag{4}$$
$$\vdash_{\mathfrak{A}} (T_o \wedge \mathbf{A}_o) \Leftrightarrow \mathbf{A}_o \tag{5}$$

(1) follows from Axioms A1 and A5.11 by Lemma A.16; (2) follows from Axiom A5.1, Proposition A.8, and (1) by Beta-Reduction Rule 2; (3) follows from Lemmas A.21 and A.24 by Lemma A.23; (4) follows from (2) and (3) by Rule R2''; and (5) follows from Proposition A.8 and (4) by Universal Instantiation.   □

Lemma A.26 $\vdash_{\mathfrak{A}} (T_o \Leftrightarrow T_o) \Leftrightarrow T_o$.

Proof Follows from Corollary A.20 and the Equality Rules.   □

Lemma A.27 $\vdash_{\mathfrak{A}} (T_o \Leftrightarrow F_o) \Leftrightarrow F_o$.

Proof

$$\vdash_{\mathfrak{A}} ((\lambda x : o \,.\, T_o \Leftrightarrow x) \, T_o \wedge (\lambda x : o \,.\, T_o \Leftrightarrow x) \, F_o) \Leftrightarrow$$
$$(\forall x : o \,.\, (\lambda x : o \,.\, T_o \Leftrightarrow x) \, x) \tag{1}$$
$$\vdash_{\mathfrak{A}} ((T_o \Leftrightarrow T_o) \wedge (T_o \Leftrightarrow F_o)) \Leftrightarrow (\forall x : o \,.\, T_o \Leftrightarrow x) \tag{2}$$
$$\vdash_{\mathfrak{A}} (T_o \wedge (T_o \Leftrightarrow F_o)) \Leftrightarrow (\forall x : o \,.\, T_o \Leftrightarrow x) \tag{3}$$
$$\vdash_{\mathfrak{A}} (T_o \Leftrightarrow F_o) \Leftrightarrow (\forall x : o \,.\, T_o \Leftrightarrow x) \tag{4}$$
$$\vdash_{\mathfrak{A}} (\lambda x : o \,.\, T_o) = (\lambda x : o \,.\, x) \Leftrightarrow$$
$$\forall x : o \,.\, (\lambda x : o \,.\, T_o) \, x \simeq (\lambda x : o \,.\, x) \, x \tag{5}$$
$$\vdash_{\mathfrak{A}} F_o \Leftrightarrow \forall x : o \,.\, (\lambda x : o \,.\, T_o) \, x \simeq (\lambda x : o \,.\, x) \, x \tag{6}$$
$$\vdash_{\mathfrak{A}} F_o \Leftrightarrow \forall x : o \,.\, T_o \simeq x \tag{7}$$
$$\vdash_{\mathfrak{A}} (T_o \Leftrightarrow F_o) \Leftrightarrow F_o \tag{8}$$

(1) follows from Axioms A1 and A5.11 by Lemma A.16; (2) follows from Axiom A5.1, Proposition A.8, and (1) by Beta-Reduction Rule 2; (3) follows from Lemma A.26 and (2) by Rule R2''; (4) follows from Lemma A.25 and

(3) by Rule R2″; (5) follows from Axioms A3 and A5.11 by Lemma A.16; (6) follows from (5) by the definition of F_o; (7) follows from Axiom A5.1 and (6) by Beta-Reduction Rule 2; and (8) follows from (4) and (7) by the Equality Rules. □

Lemma A.28 $\vdash_{\mathfrak{A}} (T_o \Leftrightarrow \mathbf{A}_o) \Leftrightarrow \mathbf{A}_o$.

Proof

$$\vdash_{\mathfrak{A}} ((\lambda x : o \,.\, (T_o \Leftrightarrow x) \Leftrightarrow x)\, T_o \wedge (\lambda x : o \,.\, (T_o \Leftrightarrow x) \Leftrightarrow x)\, F_o) \Leftrightarrow$$
$$(\forall x : o \,.\, (\lambda x : o \,.\, (T_o \Leftrightarrow x) \Leftrightarrow x)\, x) \tag{1}$$
$$\vdash_{\mathfrak{A}} ((T_o \Leftrightarrow T_o) \Leftrightarrow T_o \wedge (T_o \Leftrightarrow F_o) \Leftrightarrow F_o) \Leftrightarrow$$
$$(\forall x : o \,.\, (T_o \Leftrightarrow x) \Leftrightarrow x) \tag{2}$$
$$\vdash_{\mathfrak{A}} ((T_o \Leftrightarrow T_o) \Leftrightarrow T_o) \wedge ((T_o \Leftrightarrow F_o) \Leftrightarrow F_o) \tag{3}$$
$$\vdash_{\mathfrak{A}} \forall x : o \,.\, (T_o \Leftrightarrow x) \Leftrightarrow x \tag{4}$$
$$\vdash_{\mathfrak{A}} (T_o \Leftrightarrow \mathbf{A}_o) \Leftrightarrow \mathbf{A}_o \tag{5}$$

(1) follows from Axioms A1 and A5.11 by Lemma A.16; (2) follows from Axiom A5.1, Proposition A.8, and (1) by Beta-Reduction Rule 2; (3) follows from Lemma A.26 and Lemma A.27 by Lemma A.23; (4) follows from (2) and (3) by the Equality Rules; and (5) follows from Proposition A.8 and (4) by Universal Instantiation. □

Lemma A.29 (Rule T) $\Gamma \vdash_{\mathfrak{A}} \mathbf{A}_o$ *iff* $\Gamma \vdash_{\mathfrak{A}} T_o \Leftrightarrow \mathbf{A}_o$ *iff* $\Gamma \vdash_{\mathfrak{A}} \mathbf{A}_o \Leftrightarrow T_o$.

Proof By Lemma A.28 and the Equality Rules. □

Theorem A.30 (Universal Generalization) *If* $\Gamma \vdash_{\mathfrak{A}} \mathbf{A}_o$, *then*

$$\Gamma \vdash_{\mathfrak{A}} \forall \mathbf{x} : \alpha \,.\, \mathbf{A}_o,$$

provided $(\mathbf{x} : \alpha)$ *is not free in any member of* Γ.

Proof Let $(\star)$ $(\mathbf{x} : \alpha)$ be not free in any member of Γ.

$$\Gamma \vdash_{\mathfrak{A}} \mathbf{A}_o \tag{1}$$
$$\Gamma \vdash_{\mathfrak{A}} T_o \Leftrightarrow \mathbf{A}_o \tag{2}$$
$$\Gamma \vdash_{\mathfrak{A}} (\lambda x : o \,.\, T_o) = (\lambda \mathbf{x} : o \,.\, T_o) \tag{3}$$
$$\Gamma \vdash_{\mathfrak{A}} (\lambda x : o \,.\, T_o) = (\lambda \mathbf{x} : o \,.\, \mathbf{A}_o) \tag{4}$$
$$\Gamma \vdash_{\mathfrak{A}} \forall \mathbf{x} : \alpha \,.\, \mathbf{A}_o \tag{5}$$

(1) is given; (2) follows from (1) by Rule T; (3) is by Alpha-Equivalence; (4) follows (2) and (3) by Rule R2′ using (⋆); and (5) follows from (4) by the definition of ∀.   □

Theorem A.31 (Substitution Rule) *If* $\Gamma \vdash_{\mathfrak{A}} \mathbf{A}_\alpha{\downarrow}$ *and* $\Gamma \vdash_{\mathfrak{A}} \mathbf{B}_o$, *then*

$$\Gamma \vdash_{\mathfrak{A}} \mathbf{B}_o[(\mathbf{x} : \alpha) \mapsto \mathbf{A}_\alpha],$$

provided $(\mathbf{x} : \alpha)$ *is not free in any member of* Γ *and* $\mathbf{A}_\alpha$ *is free for* $(\mathbf{x} : \alpha)$ *in* $\mathbf{B}_o$.

Proof Let (⋆) $(\mathbf{x} : \alpha)$ be not free in any member of Γ and (⋆⋆) $\mathbf{A}_\alpha$ be free for $(\mathbf{x} : \alpha)$ in $\mathbf{B}_o$.

$$\Gamma \vdash_{\mathfrak{A}} \mathbf{A}_\alpha{\downarrow} \tag{1}$$

$$\Gamma \vdash_{\mathfrak{A}} \mathbf{B}_o \tag{2}$$

$$\Gamma \vdash_{\mathfrak{A}} \forall \mathbf{x} : \alpha \, . \, \mathbf{B}_o \tag{3}$$

$$\Gamma \vdash_{\mathfrak{A}} \mathbf{B}_o[(\mathbf{x} : \alpha) \mapsto \mathbf{A}_\alpha] \tag{4}$$

(1) and (2) are given; (3) follows from (2) by Universal Generalization using (⋆); and (4) follows from (1) and (3) by Universal Instantiation using (⋆⋆).   □

A.4   Tautology Theorems

Lemma A.32 (Boolean Cases) *If*

$$\Gamma \vdash_{\mathfrak{A}} \mathbf{A}_o[(\mathbf{x} : o) \mapsto T_o] \ and \ \Gamma \vdash_{\mathfrak{A}} \mathbf{A}_o[(\mathbf{x} : o) \mapsto F_o],$$

then $\Gamma \vdash_{\mathfrak{A}} \mathbf{A}_o$.

Proof

$$\Gamma \vdash_{\mathfrak{A}} \mathbf{A}_o[(\mathbf{x} : o) \mapsto T_o] \tag{1}$$

$$\Gamma \vdash_{\mathfrak{A}} \mathbf{A}_o[(\mathbf{x} : o) \mapsto F_o] \tag{2}$$

$$\Gamma \vdash_{\mathfrak{A}} (\lambda \mathbf{x} : o \, . \, \mathbf{A}_o) \, T_o \tag{3}$$

$$\Gamma \vdash_{\mathfrak{A}} (\lambda \mathbf{x} : o \, . \, \mathbf{A}_o) \, F_o \tag{4}$$

$$\Gamma \vdash_{\mathfrak{A}} T_o \Leftrightarrow (\lambda \mathbf{x} : o \, . \, \mathbf{A}_o) \, T_o \tag{5}$$

$$\Gamma \vdash_{\mathfrak{A}} T_o \Leftrightarrow (\lambda \mathbf{x} : o \, . \, \mathbf{A}_o) \, F_o \tag{6}$$

$$\vdash_{\mathfrak{A}} T_o \land T_o \tag{7}$$

$$\Gamma \vdash_{\mathfrak{A}} ((\lambda \mathbf{x} : o . \mathbf{A}_o) T_o \wedge (\lambda \mathbf{x} : o . \mathbf{A}_o) F_o) \tag{8}$$

$$\vdash_{\mathfrak{A}} ((\lambda \mathbf{x} : o . \mathbf{A}_o) T_o \wedge (\lambda \mathbf{x} : o . \mathbf{A}_o) F_o) \Leftrightarrow$$
$$\forall x : o . (\lambda \mathbf{x} : o . \mathbf{A}_o) x \tag{9}$$

$$\Gamma \vdash_{\mathfrak{A}} \forall x : o . (\lambda \mathbf{x} : o . \mathbf{A}_o) x \tag{10}$$

$$\Gamma \vdash_{\mathfrak{A}} \forall \mathbf{x} : o . \mathbf{A}_o \tag{11}$$

$$\Gamma \vdash_{\mathfrak{A}} \mathbf{A}_o \tag{12}$$

(1) and (2) are given; (3) and (4) follow from Proposition A.8, Axiom A4, and (1) and (2), respectively, by Rules R1$'$ and R2$''$; (5) and (6) follows from (3) and (4), respectively, by Rule T; (7) is by Lemma A.22; (8) follows from (5), (6), and (7) by Rule R2$'$; (9) follows from Axioms A1 and A5.11 by the Substitution Rule; (10) follows from (9) and (8) by Rule R2$''$; (11) follows from Axiom A5.1 and (10) by Beta-Reduction Rule 2; and (12) follows from Axiom A5.1 and (11) by Universal Instantiation. $\qquad\square$

Lemma A.33 $\vdash_{\mathfrak{A}} (T_o \Rightarrow \mathbf{A}_o) \Leftrightarrow \mathbf{A}_o$.

Proof

$$\vdash_{\mathfrak{A}} (T_o \Leftrightarrow (T_o \wedge \mathbf{A}_o)) \Leftrightarrow (T_o \Leftrightarrow (T_o \wedge \mathbf{A}_o)) \tag{1}$$

$$\vdash_{\mathfrak{A}} (T_o \Rightarrow \mathbf{A}_o) \Leftrightarrow (T_o \Leftrightarrow (T_o \wedge \mathbf{A}_o)) \tag{2}$$

$$\vdash_{\mathfrak{A}} (T_o \Rightarrow \mathbf{A}_o) \Leftrightarrow (T_o \wedge \mathbf{A}_o) \tag{3}$$

$$\vdash_{\mathfrak{A}} (T_o \Rightarrow \mathbf{A}_o) \Leftrightarrow \mathbf{A}_o \tag{4}$$

(1) follows from Proposition A.8 and Lemma A.11; (2) follows from (1) and the definition of $\Rightarrow$; (3) follows from Lemma A.28 and (2) by Rule R2$''$; (4) follows from Lemma A.25 and (3) by Rule R2$''$. $\qquad\square$

Lemma A.34 $\vdash_{\mathfrak{A}} (\lambda x : \alpha . T_o) = (f : \alpha \to o) \Rightarrow (f : \alpha \to o) (x : \alpha)$.

Proof

$$\vdash_{\mathfrak{A}} (\lambda x : \alpha . T_o) = (f : \alpha \to o) \Rightarrow$$
$$((\lambda f : \alpha \to o . f x) (\lambda x : \alpha . T_o) \Leftrightarrow$$
$$(\lambda f : \alpha \to o . f x) (f : \alpha \to o)) \tag{1}$$

$$\vdash_{\mathfrak{A}} (\lambda x : \alpha . T_o) = (f : \alpha \to o) \Rightarrow (T_o \Leftrightarrow (f : \alpha \to o) (x : \alpha)) \tag{2}$$

$$\vdash_{\mathfrak{A}} (\lambda x : \alpha . T_o) = (f : \alpha \to o) \Rightarrow (f : \alpha \to o) (x : \alpha) \tag{3}$$

(1) follows Axioms A2, A5.1, and A5.11 by the Substitution Rule; (2) follows from Axioms A5.1 and A5.11 and (1) by Beta-Reduction Rule 2; and (3) follows from Lemma A.28 and (2) by Rule R2$''$. $\qquad\square$

Lemma A.35 *If* $\Gamma \vdash_{\mathfrak{A}} \mathbf{A}_\alpha{\downarrow}$, *then*

$$\Gamma \vdash_{\mathfrak{A}} (\forall \mathbf{x} : \alpha \,.\, \mathbf{B}_o) \Rightarrow \mathbf{B}_o[(\mathbf{x} : \alpha) \mapsto \mathbf{A}_\alpha],$$

provided $\mathbf{A}_\alpha$ *is free for* $(\mathbf{x} : \alpha)$ *in* $\mathbf{B}_o$.

Proof Let $(\star)$ $\mathbf{A}_\alpha$ be free for $(\mathbf{x} : \alpha)$ in $\mathbf{B}_o$ and let $(\mathbf{g} : \alpha \to o)$ and $(\mathbf{y} : \alpha)$ be variables that are not free in Γ.

$$\Gamma \vdash_{\mathfrak{A}} \mathbf{A}_\alpha{\downarrow} \tag{1}$$
$$\vdash_{\mathfrak{A}} (\lambda x : \alpha \,.\, T_o) = (\mathbf{g} : \alpha \to o) \Rightarrow (\mathbf{g} : \alpha \to o)\,(\mathbf{y} : \alpha) \tag{2}$$
$$\Gamma \vdash_{\mathfrak{A}} (\lambda x : \alpha \,.\, T_o) = (\lambda \mathbf{x} : \alpha \,.\, \mathbf{B}_o) \Rightarrow (\lambda \mathbf{x} : \alpha \,.\, \mathbf{B}_o)\,\mathbf{A}_\alpha \tag{3}$$
$$\Gamma \vdash_{\mathfrak{A}} (\forall \mathbf{x} : \alpha \,.\, \mathbf{B}_o) \Rightarrow (\lambda \mathbf{x} : \alpha \,.\, \mathbf{B}_o)\,\mathbf{A}_\alpha \tag{4}$$
$$\Gamma \vdash_{\mathfrak{A}} (\forall \mathbf{x} : \alpha \,.\, \mathbf{B}_o) \Rightarrow \mathbf{B}_o[(\mathbf{x} : \alpha) \mapsto \mathbf{A}_\alpha] \tag{5}$$

(1) is given; (2) follows from Axiom A5.1 and Lemma A.34 by the Substitution Rule; (3) follows from Axiom A5.11, (1), and (2) by the Substitution Rule; (4) follows from (3) by the definition of $\forall$; and (5) follows from (1) and (4) by Beta-Reduction Rule 1 using $(\star)$. $\square$

Corollary A.36 $\vdash_{\mathfrak{A}} F_o \Rightarrow \mathbf{A}_o$.

Proof $\vdash_{\mathfrak{A}} (\forall x : o \,.\, x) \Rightarrow \mathbf{A}_o$ follows from Proposition A.8 and Lemma A.35, and so $\vdash_{\mathfrak{A}} F_o \Rightarrow \mathbf{A}_o$ since F_o is equivalent to $\forall x : o \,.\, x$ by the definitions of F_o and $\forall$. $\square$

Lemma A.37 (Implication)

1. $\vdash_{\mathfrak{A}} (F_o \Rightarrow F_o) \Leftrightarrow T_o$.

2. $\vdash_{\mathfrak{A}} (F_o \Rightarrow T_o) \Leftrightarrow T_o$.

3. $\vdash_{\mathfrak{A}} (T_o \Rightarrow F_o) \Leftrightarrow F_o$.

4. $\vdash_{\mathfrak{A}} (T_o \Rightarrow T_o) \Leftrightarrow T_o$.

Proof Parts 1 and 2 follow from Corollary A.36 and Rule T. Parts 3 and 4 follow from Lemma A.33. $\square$

Lemma A.38 (Conjunction)

1. $\vdash_{\mathfrak{A}} (F_o \wedge F_o) \Leftrightarrow F_o$.

2. $\vdash_{\mathfrak{A}} (F_o \wedge T_o) \Leftrightarrow F_o.$

3. $\vdash_{\mathfrak{A}} (T_o \wedge F_o) \Leftrightarrow F_o.$

4. $\vdash_{\mathfrak{A}} (T_o \wedge T_o) \Leftrightarrow T_o.$

Proof $(\star)$ $\vdash_{\mathfrak{A}} (F_o \wedge \mathbf{A}_o) \Leftrightarrow F_o$ follows from Corollary A.36, the definition of $\Rightarrow$, and the Equality Rules. Parts 1 and 2 follow from $(\star)$. Parts 3 and 4 follow from Lemma A.25.   □

Lemma A.39 $\vdash_{\mathfrak{A}} (F_o \Leftrightarrow T_o) \Leftrightarrow F_o.$

Proof

$$\vdash_{\mathfrak{A}} (F_o \Leftrightarrow T_o) \Rightarrow ((\lambda x : o . x \Leftrightarrow F_o) F_o \Leftrightarrow (\lambda x : o . x \Leftrightarrow F_o) T_o) \tag{1}$$
$$\vdash_{\mathfrak{A}} (F_o \Leftrightarrow T_o) \Rightarrow ((F_o \Leftrightarrow F_o) \Leftrightarrow (T_o \Leftrightarrow F_o)) \tag{2}$$
$$\vdash_{\mathfrak{A}} (F_o \Leftrightarrow T_o) \Rightarrow (T_o \Leftrightarrow (T_o \Leftrightarrow F_o)) \tag{3}$$
$$\vdash_{\mathfrak{A}} (F_o \Leftrightarrow T_o) \Rightarrow (T_o \Leftrightarrow F_o) \tag{4}$$
$$\vdash_{\mathfrak{A}} (F_o \Leftrightarrow T_o) \Rightarrow F_o \tag{5}$$
$$\vdash_{\mathfrak{A}} (F_o \Leftrightarrow T_o) \Leftrightarrow ((F_o \Leftrightarrow T_o) \wedge F_o) \tag{6}$$
$$\vdash_{\mathfrak{A}} ((x : o) \wedge F_o) \Leftrightarrow F_o \tag{7}$$
$$\vdash_{\mathfrak{A}} ((F_o \Leftrightarrow T_o) \wedge F_o) \Leftrightarrow F_o \tag{8}$$
$$\vdash_{\mathfrak{A}} (F_o \Leftrightarrow T_o) \Leftrightarrow F_o \tag{9}$$

(1) follows from Axioms A2 and A5.11 and Proposition A.8 by the Substitution Rule; (2) follows from Proposition A.8 and (1) by Beta-Reduction Rule 2; (3) follows from Corollary A.20 and (2) by Rule R2″; (4) follows from Lemma A.28 and (3) by Rule R2″; (5) follows from Lemma A.28 and (4) by Rule R2″; (6) follows from (5) by the definition of $\Rightarrow$; (7) follows from parts 1 and 3 of Lemma A.38 by Boolean Cases; (8) follows from Proposition A.8 and (7) by the Substitution Rule; and (9) follows from (6) and (8) by the Equality Rules.   □

Lemma A.40 (Logical Equivalence)

1. $\vdash_{\mathfrak{A}} (F_o \Leftrightarrow F_o) \Leftrightarrow T_o.$

2. $\vdash_{\mathfrak{A}} (F_o \Leftrightarrow T_o) \Leftrightarrow F_o.$

3. $\vdash_{\mathfrak{A}} (T_o \Leftrightarrow F_o) \Leftrightarrow F_o.$

4. $\vdash_{\mathfrak{A}} (T_o \Leftrightarrow T_o) \Leftrightarrow T_o.$

Proof Parts (1) and (4) follow from Corollary A.20 by Rule T. Part 2 is Lemma A.39. And part 3 is Lemma A.27.   □

Lemma A.41 (Negation)

1. $\vdash_{\mathfrak{A}} \neg F_o \Leftrightarrow T_o.$

2. $\vdash_{\mathfrak{A}} \neg T_o \Leftrightarrow F_o.$

Proof Parts 1 and 2 follow from parts 1 and 3, respectively, of Lemma A.40 by the definition of $\neg$.   □

Lemma A.42 (Disjunction)

1. $\vdash_{\mathfrak{A}} (F_o \vee F_o) \Leftrightarrow F_o.$

2. $\vdash_{\mathfrak{A}} (F_o \vee T_o) \Leftrightarrow T_o.$

3. $\vdash_{\mathfrak{A}} (T_o \vee F_o) \Leftrightarrow T_o.$

4. $\vdash_{\mathfrak{A}} (T_o \vee T_o) \Leftrightarrow T_o.$

Proof Follows from Lemmas A.38 and A.41 by Rule R2″ and the definitions of $\vee$.   □

A *propositional formula* is a formula of Alonzo defined inductively as follows:

1. T_o and F_o are propositional formulas.

2. Each variable of type o is a propositional formula.

3. If $\mathbf{A}_o$ and $\mathbf{B}_o$ are propositional formulas, then so are $\neg\mathbf{A}_o$, $\mathbf{A}_o \Rightarrow \mathbf{B}_o$, $\mathbf{A}_o \Leftrightarrow \mathbf{B}_o$, $\mathbf{A}_o \vee \mathbf{B}_o$, and $\mathbf{A}_o \wedge \mathbf{B}_o$.

A formula $\mathbf{A}_o$ is a *tautology* if there is a valid propositional formula $\mathbf{P}_o$, distinct variables $(\mathbf{x}_1 : o), \ldots, (\mathbf{x}_n : o)$, and formulas $\mathbf{C}_o^1, \ldots, \mathbf{C}_o^n$ with $n \geq 0$ such that $\mathbf{A}_o$ is obtained from $\mathbf{P}_o$ by simultaneously replacing $(\mathbf{x}_i : o)$ with $\mathbf{C}_o^i$ for all i with $1 \leq i \leq n$. Note that a valid propositional formula is trivially a tautology.

Proposition A.43 *If* $\mathbf{P}_o$ *is a closed propositional formula, then:*

1. $\mathbf{P}_o$ *is either valid or unsatisfiable.*

2. $\vdash_{\mathfrak{A}} \mathbf{P}_o \Leftrightarrow T_o$ *if* $\mathbf{P}_o$ *is valid.*

3. $\vdash_{\mathfrak{A}} \mathbf{P}_o \Leftrightarrow F_o$ *if* $\mathbf{P}_o$ *is unsatisfiable.*

Proof Let $\mathbf{P}_o$ be a closed propositional formula. We will prove this proposition by induction on the syntactic structure of $\mathbf{P}_o$.

Case 1: $\mathbf{P}_o$ is T_o. Then $\mathbf{P}_o$ is clearly valid and $\vdash_{\mathfrak{A}} \mathbf{P}_o \Leftrightarrow T_o$ follows from Propositional A.8 and Lemma A.11.

Case 2: $\mathbf{P}_o$ is F_o. Then $\mathbf{P}_o$ is clearly unsatisfiable and $\vdash_{\mathfrak{A}} \mathbf{P}_o \Leftrightarrow F_o$ follows from Propositional A.8 and Lemma A.11.

Case 3: $\mathbf{P}_o$ is $\neg \mathbf{A}_o$. $\mathbf{A}_o$ is either valid or unsatisfiable by the induction hypothesis. If $\mathbf{A}_o$ is valid, then $\mathbf{P}_o$ is clearly unsatisfiable and (a) $\vdash_{\mathfrak{A}} \mathbf{A}_o \Leftrightarrow T_o$ by the induction hypothesis, and so $\vdash_{\mathfrak{A}} \mathbf{P}_o \Leftrightarrow F_o$ follows from (a), part 2 of Lemma A.41, Proposition A.8, and Lemma A.11 by Rule R2″. If $\mathbf{A}_o$ is unsatisfiable, then $\mathbf{P}_o$ is clearly valid and (b) $\vdash_{\mathfrak{A}} \mathbf{A}_o \Leftrightarrow F_o$ by the induction hypothesis, and so $\vdash_{\mathfrak{A}} \mathbf{P}_o \Leftrightarrow T_o$ follows from (b), part 1 of Lemma A.41, Proposition A.8, and Lemma A.11 by Rule R2″.

Case 4: $\mathbf{P}_o$ is $\mathbf{A}_o \Rightarrow \mathbf{B}_o$. Similar to Case 3 using Lemma A.37.

Case 5: $\mathbf{P}_o$ is $\mathbf{A}_o \Leftrightarrow \mathbf{B}_o$. Similar to Case 3 using Lemma A.40.

Case 6: $\mathbf{P}_o$ is $\mathbf{A}_o \vee \mathbf{B}_o$. Similar to Case 3 using Lemma A.42.

Case 7: $\mathbf{P}_o$ is $\mathbf{A}_o \wedge \mathbf{B}_o$. Similar to Case 3 using Lemma A.38.

$\square$

Theorem A.44 (Propositional Tautology Theorem) *If* $\mathbf{P}_o$ *is a valid propositional formula (and hence a tautology), then* $\vdash_{\mathfrak{A}} \mathbf{P}_o$.

Proof Let $\mathbf{P}_o$ be a valid propositional formula. We will prove the theorem by induction on the number n of free variables in $\mathbf{P}_o$.

Base case: $n = 0$. Then $\mathbf{P}_o$ is closed, so $\vdash_{\mathfrak{A}} \mathbf{P}_o$ by Proposition A.43 and Rule T.

Induction step: $n > 0$. Let $(\mathbf{x} : o)$ be one of the free variables in $\mathbf{P}_o$. Then $\mathbf{P}_o[(\mathbf{x} : o) \mapsto T_o]$ and $\mathbf{P}_o[(\mathbf{x} : o) \mapsto F_o]$ are valid propositional formulas having $n - 1$ free variables, and so $\vdash_{\mathfrak{A}} \mathbf{P}_o[(\mathbf{x} : o) \mapsto T_o]$ and $\vdash_{\mathfrak{A}} \mathbf{P}_o[(\mathbf{x} : o) \mapsto F_o]$ by the induction hypothesis. Therefore, $\vdash_{\mathfrak{A}} \mathbf{P}_o$ by Boolean Cases. $\square$

Corollary A.45 (Tautology Theorem) *If* $\mathbf{A}_o$ *is a tautology, then* $\vdash_{\mathfrak{A}} \mathbf{A}_o$.

Proof Let $\mathbf{A}_o$ be a tautology. Then there is a valid propositional formula $\mathbf{P}_o$ such that $\mathbf{A}_o$ is a substitution instance of $\mathbf{P}_o$. Without loss of generality, we may assume that $(\star)$ there is no variable (of type o) that is free in both $\mathbf{A}_o$ and $\mathbf{P}_o$.

$$\vdash_{\mathfrak{A}} \mathbf{P}_o \tag{1}$$
$$\vdash_{\mathfrak{A}} \mathbf{A}_o \tag{2}$$

(1) is by Theorem A.44; and (2) follows from Proposition A.8 and (1) by the Substitution Rule using $(\star)$. □

Corollary A.46 (Tautology Rule) *If* $(\mathbf{A}_o^1 \wedge \cdots \wedge \mathbf{A}_o^n) \Rightarrow \mathbf{B}_o$ *is a tautology for* $n \geq 0$, *then* $\Gamma \vdash_{\mathfrak{A}} \mathbf{A}_o^1, \ldots, \Gamma \vdash_{\mathfrak{A}} \mathbf{A}_o^n$ *implies* $\Gamma \vdash_{\mathfrak{A}} \mathbf{B}_o$.

Proof Let $(\mathbf{A}_o^1 \wedge \cdots \wedge \mathbf{A}_o^n) \Rightarrow \mathbf{B}_o$ be a tautology for $n \geq 0$. Then

$$\mathbf{A}_o^1 \Rightarrow \cdots \Rightarrow \mathbf{A}_o^n \Rightarrow \mathbf{B}_o$$

is logically equivalent to this formula and so $(\star)$ it is also a tautology.

$$\Gamma \vdash_{\mathfrak{A}} \mathbf{A}_o^1 \tag{1}$$
$$\vdots$$
$$\Gamma \vdash_{\mathfrak{A}} \mathbf{A}_o^n \tag{n}$$
$$\vdash_{\mathfrak{A}} \mathbf{A}_o^1 \Rightarrow \cdots \Rightarrow \mathbf{A}_o^n \Rightarrow \mathbf{B}_o \tag{n+1}$$
$$\Gamma \vdash_{\mathfrak{A}} \mathbf{B}_o \tag{n+2}$$

$(1), \ldots, (n)$ are given; $(n+1)$ follows from $(\star)$ by Lemma A.44; and $(n+2)$ follows from $(1), \ldots, (n)$ and $(n+1)$ by n applications of Rule R1$'$. □

A.5   Deduction Theorem

Lemma A.47 *If* $\Gamma \vdash_{\mathfrak{A}} \mathbf{A}_o \Rightarrow \mathbf{B}_o$, *then* $\Gamma \vdash_{\mathfrak{A}} \mathbf{A}_o \Rightarrow \forall \mathbf{x} : \alpha . \mathbf{B}_o$, *provided that* $(\mathbf{x} : \alpha)$ *is not free in* $\mathbf{A}_o$ *or in any member of* Γ.

Proof Let $(\star)$ $(\mathbf{x} : \alpha)$ not be free in $\mathbf{A}_o$, $(\star\star)$ $(\mathbf{x} : \alpha)$ not be free in any member of Γ, and $(\star\star\star)$ $(\mathbf{y} : o)$ does not occur in $\mathbf{B}_o$.

$$\Gamma \vdash_{\mathfrak{A}} \mathbf{A}_o \Rightarrow \mathbf{B}_o \tag{1}$$

$$\Gamma \vdash_{\mathfrak{A}} \forall \mathbf{x} : \alpha \,.\, (\mathbf{A}_o \Rightarrow \mathbf{B}_o) \tag{2}$$

$$\vdash_{\mathfrak{A}} (\forall \mathbf{x} : \alpha \,.\, (F_o \Rightarrow \mathbf{B}_o)) \Rightarrow (F_o \Rightarrow \forall \mathbf{x} : \alpha \,.\, \mathbf{B}_o) \tag{3}$$

$$\vdash_{\mathfrak{A}} (\forall \mathbf{x} : \alpha \,.\, \mathbf{B}_o) \Rightarrow (T_o \Rightarrow \forall \mathbf{x} : \alpha \,.\, \mathbf{B}_o) \tag{4}$$

$$\vdash_{\mathfrak{A}} \mathbf{B}_o \Leftrightarrow (T_o \Rightarrow \mathbf{B}_o) \tag{5}$$

$$\vdash_{\mathfrak{A}} (\forall \mathbf{x} : \alpha \,.\, (T_o \Rightarrow \mathbf{B}_o)) \Rightarrow (T_o \Rightarrow \forall \mathbf{x} : \alpha \,.\, \mathbf{B}_o) \tag{6}$$

$$\vdash_{\mathfrak{A}} (\forall \mathbf{x} : \alpha \,.\, ((\mathbf{y} : o) \Rightarrow \mathbf{B}_o)) \Rightarrow ((\mathbf{y} : o) \Rightarrow \forall \mathbf{x} : \alpha \,.\, \mathbf{B}_o) \tag{7}$$

$$\vdash_{\mathfrak{A}} (\forall \mathbf{x} : \alpha \,.\, (\mathbf{A}_o \Rightarrow \mathbf{B}_o)) \Rightarrow (\mathbf{A}_o \Rightarrow \forall \mathbf{x} : \alpha \,.\, \mathbf{B}_o) \tag{8}$$

$$\Gamma \vdash_{\mathfrak{A}} \mathbf{A}_o \Rightarrow \forall \mathbf{x} : \alpha \,.\, \mathbf{B}_o \tag{9}$$

(1) is given; (2) follows from (1) by Universal Generalization using $(\star\star)$; (3), (4), and (5) are by the Tautology Theorem; (6) follows from (5) and (4) by Rule R2″; (7) follows from (6) and (3) by Boolean Cases using $(\star\star\star)$; (8) follows from (7) by Proposition A.8 and the Substitution Rule using $(\star)$; (9) follows from (2) and (8) by Rule R1′. $\qquad\square$

Lemma A.48

$$\vdash_{\mathfrak{A}} \lambda \mathbf{x}_1 : \alpha_1 \,.\, \cdots \lambda \mathbf{x}_n : \alpha_n \,.\, \mathbf{A}_\beta = \lambda \mathbf{x}_1 : \alpha_1 \,.\, \cdots \lambda \mathbf{x}_n : \alpha_n \,.\, \mathbf{B}_\beta \Leftrightarrow$$
$$\forall \mathbf{x}_1 : \alpha_1 \,.\, \cdots \forall \mathbf{x}_n : \alpha_n \,.\, \mathbf{A}_\beta \simeq \mathbf{B}_\beta$$

for $n \geq 1$.

Proof We will prove the lemma by induction on n.

Base case: $n = 1$.

$$\vdash_{\mathfrak{A}} (\lambda \mathbf{x} : \alpha \,.\, \mathbf{A}_\beta) = (\lambda \mathbf{x} : \alpha \,.\, \mathbf{B}_\beta) \Leftrightarrow$$
$$\forall x : \alpha \,.\, (\lambda \mathbf{x} : \alpha \,.\, \mathbf{A}_\beta) x \simeq (\lambda \mathbf{x} : \alpha \,.\, \mathbf{B}_\beta) x \tag{1}$$

$$\vdash_{\mathfrak{A}} (\lambda \mathbf{x} : \alpha \,.\, \mathbf{A}_\beta) = (\lambda \mathbf{x} : \alpha \,.\, \mathbf{B}_\beta) \Leftrightarrow$$
$$\forall \mathbf{x} : \alpha \,.\, (\lambda \mathbf{x} : \alpha \,.\, \mathbf{A}_\beta) \mathbf{x} \simeq (\lambda \mathbf{x} : \alpha \,.\, \mathbf{B}_\beta) \mathbf{x} \tag{2}$$

$$\vdash_{\mathfrak{A}} (\lambda \mathbf{x} : \alpha \,.\, \mathbf{A}_\beta) = (\lambda \mathbf{x} : \alpha \,.\, \mathbf{B}_\beta) \Leftrightarrow \forall \mathbf{x} : \alpha \,.\, \mathbf{A}_\beta \simeq \mathbf{B}_\beta \tag{3}$$

(1) follows Axioms A3 and A5.11 by the Substitution Rule; (2) follows from Alpha-Conversion and (1) by Rule R2″; and (3) follows from Axiom A5.1 and (2) by Beta-Reduction Rule 2. This proves the lemma for the $n = 1$ case.

Induction step: $n > 1$.

$$\vdash_{\mathfrak{A}} \lambda\,\mathbf{x}_1 : \alpha_1 . \cdots \lambda\,\mathbf{x}_{n-1} : \alpha_{n-1} . (\lambda\,\mathbf{x}_n : \alpha . \mathbf{A}_\beta) =$$
$$\lambda\,\mathbf{x}_1 : \alpha_1 . \cdots \lambda\,\mathbf{x}_{n-1} : \alpha_{n-1} . (\lambda\,\mathbf{x}_n : \alpha . \mathbf{B}_\beta) \Leftrightarrow$$
$$\forall\,\mathbf{x}_1 : \alpha_1 . \cdots \forall\,\mathbf{x}_{n-1} : \alpha_{n-1} . (\lambda\,\mathbf{x}_n : \alpha . \mathbf{A}_\beta) \simeq (\lambda\,\mathbf{x}_n : \alpha . \mathbf{B}_\beta) \quad (4)$$
$$\vdash_{\mathfrak{A}} \lambda\,\mathbf{x}_1 : \alpha_1 . \cdots \lambda\,\mathbf{x}_{n-1} : \alpha_{n-1} . (\lambda\,\mathbf{x}_n : \alpha . \mathbf{A}_\beta) =$$
$$\lambda\,\mathbf{x}_1 : \alpha_1 . \cdots \lambda\,\mathbf{x}_{n-1} : \alpha_{n-1} . (\lambda\,\mathbf{x}_n : \alpha . \mathbf{B}_\beta) \Leftrightarrow$$
$$\forall\,\mathbf{x}_1 : \alpha_1 . \cdots \forall\,\mathbf{x}_{n-1} : \alpha_{n-1} . (\lambda\,\mathbf{x}_n : \alpha . \mathbf{A}_\beta) = (\lambda\,\mathbf{x}_n : \alpha . \mathbf{B}_\beta) \quad (5)$$
$$\vdash_{\mathfrak{A}} \lambda\,\mathbf{x}_1 : \alpha_1 . \cdots \lambda\,\mathbf{x}_{n-1} : \alpha_{n-1} . (\lambda\,\mathbf{x}_n : \alpha . \mathbf{A}_\beta) =$$
$$\lambda\,\mathbf{x}_1 : \alpha_1 . \cdots \lambda\,\mathbf{x}_{n-1} : \alpha_{n-1} . (\lambda\,\mathbf{x}_n : \alpha . \mathbf{B}_\beta) \Leftrightarrow$$
$$\forall\,\mathbf{x}_1 : \alpha_1 . \cdots \forall\,\mathbf{x}_{n-1} : \alpha_{n-1} . (\forall\,\mathbf{x}_n : \alpha . \mathbf{A}_\beta \simeq \mathbf{B}_\beta) \quad (6)$$

(4) is an instance of the induction hypothesis; (5) follows from (4) by Axiom A5.11, part 1 of Lemma A.9, and Rule R2″; and (6) follows from (3) and (5) by Rule R2″. □

Lemma A.49 *Let the following hold:*

1. $\mathbf{D}_o$ *is obtained from* $\mathbf{A}_\beta \simeq \mathbf{B}_\beta$ *and* $\mathbf{C}_o$ *as in Rule R2.*

2. $(\mathbf{x}_1 : \alpha_1), \dots, (\mathbf{x}_n : \alpha_n)$ *for* $n \geq 0$ *are the variables* $(\mathbf{x} : \alpha)$ *such that* $(\mathbf{x} : \alpha)$ *is free in* $\mathbf{A}_\beta \simeq \mathbf{B}_\beta$ *and the occurrence of* $\mathbf{A}_\beta$ *in* $\mathbf{C}_o$ *that was replaced by* $\mathbf{B}_\beta$ *in* $\mathbf{D}_o$ *is not in the second argument of a function abstraction* $\lambda\,\mathbf{x} : \alpha . \mathbf{E}_\gamma$ *or a definite description* $\mathrm{I}\,\mathbf{x} : \alpha . \mathbf{E}_o$.

Then

$$\vdash_{\mathfrak{A}} (\forall\,\mathbf{x}_1 : \alpha_1 . \cdots \forall\,\mathbf{x}_n : \alpha_n . \mathbf{A}_\beta \simeq \mathbf{B}_\beta) \Rightarrow (\mathbf{C}_o \Leftrightarrow \mathbf{D}_o).$$

Proof Let $(\mathbf{f} : \alpha_1 \to \cdots \to \alpha_n \to o)$ be a variable that does not occur in the formula to be proved, and let $\mathbf{X}_o$ be the result of replacing the occurrence of $\mathbf{A}_\beta$ in $\mathbf{C}_o$ with

$$(\mathbf{f} : \alpha_1 \to \cdots \to \alpha_n \to o) \, (\mathbf{x}_1 : \alpha_1) \cdots (\mathbf{x}_n : \alpha_n).$$

$$\vdash_{\mathfrak{A}} \lambda\,\mathbf{x}_1 : \alpha_1 . \cdots \lambda\,\mathbf{x}_n : \alpha_n . \mathbf{A}_\beta = \lambda\,\mathbf{x}_1 : \alpha_1 . \cdots \lambda\,\mathbf{x}_n : \alpha_n . \mathbf{B}_\beta \Rightarrow$$
$$((\lambda\,\mathbf{f} : \alpha_1 \to \cdots \to \alpha_n \to o . \mathbf{X}_o) \, (\lambda\,\mathbf{x}_1 : \alpha_1 . \cdots \lambda\,\mathbf{x}_n : \alpha_n . \mathbf{A}_\beta) \Leftrightarrow$$
$$(\lambda\,\mathbf{f} : \alpha_1 \to \cdots \to \alpha_n \to o . \mathbf{X}_o) \, (\lambda\,\mathbf{x}_1 : \alpha_1 . \cdots \lambda\,\mathbf{x}_n : \alpha_n . \mathbf{B}_\beta)) \quad (1)$$
$$\vdash_{\mathfrak{A}} (\forall\,\mathbf{x}_1 : \alpha_1 . \cdots \forall\,\mathbf{x}_n : \alpha_n . \mathbf{A}_\beta \simeq \mathbf{B}_\beta) \Rightarrow$$
$$((\lambda\,\mathbf{f} : \alpha_1 \to \cdots \to \alpha_n \to o . \mathbf{X}_o) \, (\lambda\,\mathbf{x}_1 : \alpha_1 . \cdots \lambda\,\mathbf{x}_n : \alpha_n . \mathbf{A}_\beta) \Leftrightarrow$$
$$(\lambda\,\mathbf{f} : \alpha_1 \to \cdots \to \alpha_n \to o . \mathbf{X}_o) \, (\lambda\,\mathbf{x}_1 : \alpha_1 . \cdots \lambda\,\mathbf{x}_n : \alpha_n . \mathbf{B}_\beta)) \quad (2)$$
$$\vdash_{\mathfrak{A}} (\forall\,\mathbf{x}_1 : \alpha_1 . \cdots \forall\,\mathbf{x}_n : \alpha_n . \mathbf{A}_\beta \simeq \mathbf{B}_\beta) \Rightarrow (\mathbf{C}_o \Leftrightarrow \mathbf{D}_o) \quad (3)$$

(1) follows from Axioms A2 and A5.11 by the Substitution Rule; (2) follows from Lemma A.48 and (1) by Rule R2″; and (3) follows from Axioms A5.1 and A5.11 and (2) by Beta-Reduction Rule 2. □

Theorem A.50 (Deduction Theorem) *If* $\Gamma \cup \{\mathbf{X}_o\} \vdash_{\mathfrak{A}} \mathbf{Y}_o$, *then* $\Gamma \vdash_{\mathfrak{A}} \mathbf{X}_o \Rightarrow \mathbf{Y}_o$.

Proof Let (Π_1, Π_2) be a proof of $\mathbf{Y}_o$ from $\Gamma \cup \{\mathbf{X}_o\}$ in $\mathfrak{A}$ where $\Pi_2 = [\mathbf{C}_o^1, \ldots, \mathbf{C}_o^n]$. We will prove $\Gamma \vdash_{\mathfrak{A}} \mathbf{X}_o \Rightarrow \mathbf{C}_o^i$ for all i with $1 \leq i \leq n$ by induction on i. This result will then imply the theorem since $\mathbf{C}_o^n$ is $\mathbf{Y}_o$.

Case 1: $\mathbf{C}_o^i$ is a theorem of $\mathfrak{A}$.

$$\vdash_{\mathfrak{A}} \mathbf{C}_o^i \tag{1}$$
$$\vdash_{\mathfrak{A}} \mathbf{X}_o \Rightarrow \mathbf{C}_o^i \tag{2}$$
$$\Gamma \vdash_{\mathfrak{A}} \mathbf{X}_o \Rightarrow \mathbf{C}_o^i \tag{3}$$

(1) is given; (2) follows from (1) by the Tautology Rule; and (3) follows from (2) since $\emptyset \subseteq \Gamma$.

Case 2: $\mathbf{C}_o^i$ is $\mathbf{X}_o$.

$$\vdash_{\mathfrak{A}} \mathbf{X}_o \Rightarrow \mathbf{X}_o \tag{1}$$
$$\Gamma \vdash_{\mathfrak{A}} \mathbf{X}_o \Rightarrow \mathbf{X}_o \tag{2}$$

(1) is by the Tautology Theorem; and (2) follows from (1) since $\emptyset \subseteq \Gamma$.

Case 3: $\mathbf{C}_o^i$ is a member of Γ.

$$\Gamma \vdash_{\mathfrak{A}} \mathbf{C}_o^i \tag{1}$$
$$\Gamma \vdash_{\mathfrak{A}} \mathbf{X}_o \Rightarrow \mathbf{C}_o^i \tag{2}$$

(1) is given; and (2) follows from (1) by the Tautology Rule.

Case 4: $\mathbf{C}_o^i$ is obtained from $\mathbf{C}_o$ and $\mathbf{C}_o \Rightarrow \mathbf{D}_o$ by Rule R1′.

$$\Gamma \vdash_{\mathfrak{A}} \mathbf{X}_o \Rightarrow \mathbf{C}_o \tag{1}$$
$$\Gamma \vdash_{\mathfrak{A}} \mathbf{X}_o \Rightarrow (\mathbf{C}_o \Rightarrow \mathbf{D}_o) \tag{2}$$
$$\Gamma \vdash_{\mathfrak{A}} \mathbf{X}_o \Rightarrow \mathbf{D}_o \tag{3}$$

(1) and (2) are by the induction hypothesis; and (3) follows from (1) and (2) by the Tautology Rule.

Case 5: $\mathbf{C}_o^i$ is obtained from $\mathbf{A}_\alpha \simeq \mathbf{B}_\alpha$ and $\mathbf{C}_o$ by Rule R2′. Let $(\mathbf{x}_1 : \alpha_1), \ldots, (\mathbf{x}_n : \alpha_n)$ for $n \geq 0$ be the variables $(\mathbf{x} : \alpha)$ such that $(\mathbf{x} : \alpha)$ is free in $\mathbf{A}_\beta \simeq \mathbf{B}_\beta$ and the occurrence of $\mathbf{A}_\beta$ in $\mathbf{C}_o$ that was replaced by $\mathbf{B}_\beta$ in $\mathbf{D}_o$ is not in the second argument of a function abstraction $\lambda \mathbf{x} : \alpha \, . \, \mathbf{E}_\gamma$ or a definite description $\mathrm{I}\mathbf{x} : \alpha \, . \, \mathbf{E}_o$. Since $\mathbf{C}_o^i$ is obtained by Rule R2′, $(\star)$ the $(\mathbf{x}_1 : \alpha_1), \ldots, (\mathbf{x}_n : \alpha_n)$ are not free in $\mathbf{X}_o$ or in any member of Γ.

$$\Gamma \vdash_{\mathfrak{A}} \mathbf{X}_o \Rightarrow \mathbf{A}_\alpha \simeq \mathbf{B}_\alpha \tag{1}$$

$$\Gamma \vdash_{\mathfrak{A}} \mathbf{X}_o \Rightarrow \mathbf{C}_o \tag{2}$$

$$\Gamma \vdash_{\mathfrak{A}} \mathbf{X}_o \Rightarrow \forall \mathbf{x}_1 : \alpha_1 \, . \, \cdots \forall \mathbf{x}_n : \alpha_n \, . \, \mathbf{A}_\beta \simeq \mathbf{B}_\beta \tag{3}$$

$$\Gamma \vdash_{\mathfrak{A}} \mathbf{X}_o \Rightarrow (\mathbf{C}_o \Leftrightarrow \mathbf{C}_o^i) \tag{4}$$

$$\Gamma \vdash_{\mathfrak{A}} \mathbf{X}_o \Rightarrow \mathbf{C}_o^i \tag{5}$$

(1) and (2) are by the induction hypothesis; (3) follows from (1) by Lemma A.47 using $(\star)$; (4) follows from (3) and Lemma A.49 by Rule R1′; and (5) follows from (2) and (4) by the Tautology Rule. $\qquad\square$

A.6   Miscellaneous Metatheorems

Theorem A.51 (Existential Generalization) *If* $\Gamma \vdash_{\mathfrak{A}} \mathbf{A}_\alpha\!\downarrow$ *and* $\Gamma \vdash_{\mathfrak{A}} \mathbf{B}_o[(\mathbf{x} : \alpha) \mapsto \mathbf{A}_\alpha]$, *then*

$$\Gamma \vdash_{\mathfrak{A}} \exists \mathbf{x} : \alpha \, . \, \mathbf{B}_o,$$

provided $\mathbf{A}_\alpha$ *is free for* $(\mathbf{x} : \alpha)$ *in* $\mathbf{B}_o$.

Proof Let $(\star)$ $\mathbf{A}_\alpha$ be free for $(\mathbf{x} : \alpha)$ in $\mathbf{B}_o$.

$$\Gamma \vdash_{\mathfrak{A}} \mathbf{A}_\alpha\!\downarrow \tag{1}$$

$$\Gamma \vdash_{\mathfrak{A}} \mathbf{B}_o[(\mathbf{x} : \alpha) \mapsto \mathbf{A}_\alpha] \tag{2}$$

$$\Gamma \vdash_{\mathfrak{A}} (\forall \mathbf{x} : \alpha \, . \, \neg\mathbf{B}_o) \Rightarrow \neg\mathbf{B}_o[(\mathbf{x} : \alpha) \mapsto \mathbf{A}_\alpha] \tag{3}$$

$$\Gamma \vdash_{\mathfrak{A}} \mathbf{B}_o[(\mathbf{x} : \alpha) \mapsto \mathbf{A}_\alpha] \Rightarrow \exists \mathbf{x} : \alpha \, . \, \mathbf{B}_o \tag{4}$$

$$\Gamma \vdash_{\mathfrak{A}} \exists \mathbf{x} : \alpha \, . \, \mathbf{B}_o \tag{5}$$

(1) and (2) are given; (3) follows from (1) by Lemma A.35 using $(\star)$; (4) follows from (3) by the Tautology Rule and the definition of $\exists$; and (5) follows from (2) and (4) by Rule R1′. $\qquad\square$

Theorem A.52 (Alpha-Conversion for Definite Descriptions)
If $(\mathbf{y} : \alpha)$ *is not free in* $\mathbf{B}_o$ *and* $(\mathbf{y} : \alpha)$ *is free for* $(\mathbf{x} : \alpha)$ *in* $\mathbf{B}_o$, *then*

$$\vdash_{\mathfrak{A}} (\mathrm{I}\mathbf{x} : \alpha \, . \, \mathbf{B}_o) \simeq (\mathrm{I}\mathbf{y} : \alpha \, . \, \mathbf{B}_o[(\mathbf{x} : \alpha) \mapsto \mathbf{y}]).$$

Proof By Axioms A6.1 and A6.2 and Alpha-Conversion (for function abstractions). □

Lemma A.53 (First and Second Pseudoconstants)

1. $\vdash_\mathfrak{A} \forall x : \alpha, y : \beta \,.\, \mathsf{fst}_{(\alpha \times \beta) \to \alpha} (x, y) = x.$

2. $\vdash_\mathfrak{A} \forall x : \alpha, y : \beta \,.\, \mathsf{snd}_{(\alpha \times \beta) \to \beta} (x, y) = y.$

Proof By Axioms A7.1, A7.4, and A7.5. □

Appendix B

Soundness of $\mathfrak{A}$

In this appendix we prove that the proof system $\mathfrak{A}$ for Alonzo is sound by showing that its axioms are valid and its rules of inference preserve validity in all general models. The form of the proof is similar to Peter Andrews' proof that $\mathcal{Q}_0$ is sound [5].

Lemma B.1 (Axiom A1) *Axiom A1 is valid.*

Proof Let $\mathbf{X}_o$ be

$$((g : o \rightarrow o)\, T_o \wedge (g : o \rightarrow o)\, F_o) \Leftrightarrow \forall\, x : o\,.\,(g : o \rightarrow o)\, x,$$

M be a general model (that interprets $\mathbf{X}_o$), and $\varphi \in \mathsf{assign}(M)$. We must show $(\star)$ $V_\varphi^M(\mathbf{X}_o) = \mathrm{T}$.

Case 1: $\varphi((g : o \rightarrow o))(\mathrm{T}) = \varphi((g : o \rightarrow o))(\mathrm{F}) = \mathrm{T}$. Then

$$\text{(a) } V_\varphi^M((g : o \rightarrow o)\, T_o \wedge (g : o \rightarrow o)\, F_o) = \mathrm{T}$$

by parts 1, 2, and 5 of Lemma 6.2. Let $\psi = \varphi[(x : o) \mapsto d]$ where $d \in \mathbb{B}$. Then

$$
\begin{aligned}
&V_\psi^M((g : o \rightarrow o)\,(x : o)) \\
&= \psi((g : o \rightarrow o))(\psi((x : o))) & (1)\\
&= \varphi((g : o \rightarrow o))(\psi((x : o))) & (2)\\
&= \mathrm{T}. & (3)
\end{aligned}
$$

(1) is by conditions V1 and V4 of the definition of a general model and the fact that all members of $D_{o \rightarrow o}^M$ are total; (2) is by the definition of ψ; and

W. M. Farmer, *Simple Type Theory*, Computer Science Foundations and Applied Logic, https://doi.org/10.1007/978-3-031-21112-6

(3) is by assumption. Hence $V_\psi^M((g:o\to o)\,(x:o)) = \text{T}$ for all $d\in\mathbb{B}$, and so

    (b) $V_\varphi^M(\forall x:o\,.\,(g:o\to o)\,x) = \text{T}$

by part 1 of Lemma 6.4. Therefore, $(\star)$ follows from (a) and (b) by part 10 of Lemma 5.4.

Case 2: $\varphi((g:o\to o))(\text{T}) = \text{F}$ or $\varphi((g:o\to o))(\text{T}) = \text{F}$. Then

    (a) $V_\varphi^M((g:o\to o)\,T_o\wedge(g:o\to o)\,F_o) = \text{F}$

by parts 1, 2, and 5 of Lemma 6.2. Let $\psi = \varphi[(x:o)\mapsto d]$ where $\varphi((g:o\to o))(d) = \text{F}$. Then

$$
\begin{aligned}
&V_\psi^M((g:o\to o)\,(x:o))\\
&= \psi((g:o\to o))(\psi((x:o))) &(1)\\
&= \varphi((g:o\to o))(\psi((x:o))) &(2)\\
&= \text{F}. &(3)
\end{aligned}
$$

(1) is by conditions V1 and V4 of the definition of a general model and the fact that all members of $D_{o\to o}^M$ are total; (2) is by the definition of ψ; and (3) is by assumption. Hence $V_\psi^M((g:o\to o)\,(x:o)) = \text{F}$ for some $d\in\mathbb{B}$, and so

    (b) $V_\varphi^M(\forall x:o\,.\,(g:o\to o)\,x) = \text{F}$

by part 1 of Lemma 6.4. Therefore, $(\star)$ follows from (a) and (b) by part 10 of Lemma 5.4. $\qquad\qquad\qquad\qquad\qquad\qquad\qquad\qquad\qquad\qquad\qquad\square$

Lemma B.2 (Axiom A2) *Each instance of Axiom A2 is valid.*

Proof Let $\mathbf{X}_o$ be

    $(x:\alpha) = (y:\alpha) \Rightarrow ((h:\alpha\to o)\,(x:\alpha) \Leftrightarrow (h:\alpha\to o)\,(y:\alpha)),$

M be a general model (that interprets $\mathbf{X}_o$), and $\varphi\in\mathsf{assign}(M)$. We must show $(\star)$ $V_\varphi^M(\mathbf{X}_o) = \text{T}$.

Case 1: $\varphi((x:\alpha)) = \varphi((y:\alpha))$. Then (a) $V_\varphi^M((x:\alpha) = (y:\alpha)) = \text{T}$ by conditions V1 and V3 of the definition of a general model.

$$
\begin{aligned}
&V_\varphi^M((h:\alpha\to o)\,(x:\alpha))\\
&= \varphi((h:\alpha\to o))(\varphi((x:\alpha))) &(1)\\
&= \varphi((h:\alpha\to o))(\varphi((y:\alpha))) &(2)\\
&= V_\varphi^M((h:\alpha\to o)\,(y:\alpha)) &(3)
\end{aligned}
$$

(1) and (3) are by conditions V1 and V4 of the definition of a general model and the fact that all members of $D^M_{\alpha \to o}$ are total; and (2) is by assumption. Hence

$$(b) \quad V^M_\varphi((h : \alpha \to o)\,(x : \alpha) \Leftrightarrow (h : \alpha \to o)\,(y : \alpha)) = \text{T}$$

by part 10 of Lemma 5.4. Therefore, $(\star)$ follows from (a) and (b) by part 6 of Lemma 6.2.

Case 2: $\varphi((x : \alpha)) \neq \varphi((y : \alpha))$. Then $V^M_\varphi((x : \alpha) = (y : \alpha)) = \text{F}$ by conditions V1 and V3 of the definition of a general model, and so $(\star)$ holds by part 6 of Lemma 6.2. □

Lemma B.3 (Axiom A3) *Each instance of Axiom A3 is valid.*

Proof Let $\mathbf{X}_o$ be

$$(f : \alpha \to \beta) = (g : \alpha \to \beta) \Leftrightarrow \forall x : \alpha \,.\, (f : \alpha \to \beta)\,x \simeq (g : \alpha \to \beta)\,x,$$

M be a general model (that interprets $\mathbf{X}_o$), and $\varphi \in \mathsf{assign}(M)$. We must show $(\star)$ $V^M_\varphi(\mathbf{X}_o) = \text{T}$.

Case 1: $\varphi((f : \alpha \to \beta)) = \varphi((g : \alpha \to \beta))$.

$$V^M_\varphi((f : \alpha \to \beta) = (g : \alpha \to \beta)) = \text{T}$$

$$\text{iff} \quad V^M_\varphi((f : \alpha \to \beta)) = V^M_\varphi((g : \alpha \to \beta)) \tag{1}$$

$$\text{iff} \quad \varphi((f : \alpha \to \beta)) = \varphi((g : \alpha \to \beta)) \tag{2}$$

$$\text{iff} \quad \text{T} \tag{3}$$

(1) is by part 10 of Lemma 5.4; (2) is by condition V1 of the definition of a general model; and (3) is by assumption. Hence

$$(a) \quad V^M_\varphi((f : \alpha \to \beta) = (g : \alpha \to \beta)) = \text{T}.$$

Let $\psi = \varphi[(x : \alpha) \mapsto d]$ where $d \in D^M_\alpha$.

$$V^M_\psi((f : \alpha \to \beta)\,(x : \alpha))$$

$$\simeq \psi((f : \alpha \to \beta))(\psi((x : \alpha))) \tag{1}$$

$$\simeq \varphi((f : \alpha \to \beta))(\psi((x : \alpha))) \tag{2}$$

$$\simeq \varphi((g : \alpha \to \beta))(\psi((x : \alpha))) \tag{3}$$

$$\simeq \psi((g : \alpha \to \beta))(\psi((x : \alpha))) \tag{4}$$

$$\simeq V^M_\psi((g : \alpha \to \beta)\,(x : \alpha)) \tag{5}$$

(1) and (5) are by conditions V1 and V4 of the definition of a general model; (2) and (4) are by the definition of ψ; and (3) is by assumption. Hence

(b) $V_\psi^M((f : \alpha \to \beta)\,(x : \alpha) \simeq (g : \alpha \to \beta)\,(x : \alpha)) = \mathrm{T}$

for all $d \in D_\alpha^M$ by part 5 of Lemma 6.5, and so

(c) $V_\varphi^M(\forall x : \alpha \,.\, (f : \alpha \to \beta)\,x \simeq (g : \alpha \to \beta)\,x) = \mathrm{T}$

by part 1 of Lemma 6.4. Therefore, $(\star)$ follows from (a) and (c) by part 10 of Lemma 5.4.

Case 2: $\varphi((f : \alpha \to \beta)) \neq \varphi((g : \alpha \to \beta))$.

$$V_\varphi^M((f : \alpha \to \beta) = (g : \alpha \to \beta)) = \mathrm{F}$$
$$\text{iff} \quad V_\varphi^M((f : \alpha \to \beta)) \neq V_\varphi^M((g : \alpha \to \beta)) \tag{1}$$
$$\text{iff} \quad \varphi((f : \alpha \to \beta)) \neq \varphi((g : \alpha \to \beta)) \tag{2}$$
$$\text{iff} \quad \mathrm{T} \tag{3}$$

(1) is by part 10 of Lemma 5.4; (2) is by condition V1 of the definition of a general model; and (3) is by assumption. Hence

(a) $V_\varphi^M((f : \alpha \to \beta) = (g : \alpha \to \beta)) = \mathrm{F}$.

Let $\psi = \varphi[(x : \alpha) \mapsto d]$ where $d \in D_\alpha^M$ such that $\varphi((f : \alpha \to \beta))(d)$ is defined, $\varphi((g : \alpha \to \beta))(d)$ is defined, and

$$\varphi((f : \alpha \to \beta))(d) \neq \varphi((g : \alpha \to \beta))(d).$$

$$V_\psi^M((f : \alpha \to \beta)\,(x : \alpha))$$
$$\simeq \; \psi((f : \alpha \to \beta))(\psi((x : \alpha))) \tag{1}$$
$$\simeq \; \varphi((f : \alpha \to \beta))(\psi((x : \alpha))) \tag{2}$$
$$\not\simeq \; \varphi((g : \alpha \to \beta))(\psi((x : \alpha))) \tag{3}$$
$$\simeq \; \psi((g : \alpha \to \beta))(\psi((x : \alpha))) \tag{4}$$
$$\simeq \; V_\psi^M((g : \alpha \to \beta)\,(x : \alpha)) \tag{5}$$

(1) and (5) are by conditions V1 and V4 of the definition of a general model; (2) and (4) are by the definition of ψ; and (3) is by assumption and the definition of ψ. Hence

(b) $V_\psi^M((f : \alpha \to \beta)\,(x : \alpha) \simeq (g : \alpha \to \beta)\,(x : \alpha)) = \mathrm{F}$

by part 5 of Lemma 6.5, and so

$$\text{(c)} \ V_\varphi^M(\forall x : \alpha \,.\, (f : \alpha \to \beta)\, x \simeq (g : \alpha \to \beta)\, x) = \text{F}$$

by part 1 of Lemma 6.4. Therefore, $(\star)$ follows from (a) and (c) by part 10 of Lemma 5.4. $\qquad\qquad\qquad\qquad\qquad\qquad\qquad\qquad\qquad\qquad\qquad\qquad\qquad\Box$

Lemma B.4 (Axiom A4) *Each instance of Axiom A4 is valid.*

Proof This lemma is the same as Theorem 7.1 previously proved in Chapter 7. $\qquad\qquad\qquad\qquad\qquad\qquad\qquad\qquad\qquad\qquad\qquad\qquad\qquad\qquad\qquad\Box$

Lemma B.5 (Axiom A5) *Each instance of Axiom A5 is valid.*

Proof Let $\mathbf{X}_o$ be an instance of Axiom A5, M be a general model that interprets $\mathbf{X}_o$, and $\varphi \in \mathsf{assign}(M)$. We must show $(\star)$ $V_\varphi^M(\mathbf{X}_o) = \text{T}$.

Case 1: $\mathbf{X}_o \equiv (\mathbf{x} : \alpha)\!\downarrow$. $(\star)$ follows from part 1 of Lemma 5.4 and part 3 of Lemma 6.5.

Case 2: $\mathbf{X}_o \equiv \mathbf{c}_\alpha\!\downarrow$. $(\star)$ follows from part 2 of Lemma 5.4 and part 3 of Lemma 6.5.

Case 3: $\mathbf{X}_o \equiv (\mathbf{A}_\alpha = \mathbf{B}_\alpha)\!\downarrow$. $(\star)$ follows from part 3 of Lemma 5.4 and part 3 of Lemma 6.5.

Case 4: $\mathbf{X}_o \equiv (\mathbf{A}_\alpha = \mathbf{B}_\alpha) \Rightarrow \mathbf{A}_\alpha\!\downarrow$. $(\star)$ follows from parts 3 and 10 of Lemma 5.4, part 6 of Lemma 6.2, and part 3 of Lemma 6.5.

Case 5: $\mathbf{X}_o \equiv (\mathbf{A}_\alpha = \mathbf{B}_\alpha) \Rightarrow \mathbf{B}_\alpha\!\downarrow$. Same as Case 4.

Case 6: $\mathbf{X}_o \equiv (\mathbf{F}_{\alpha \to o}\, \mathbf{A}_\alpha)\!\downarrow$. $(\star)$ follows from part 4 of Lemma 5.4 and part 3 of Lemma 6.5.

Case 7: $\mathbf{X}_o \equiv \mathbf{F}_{\alpha \to o}\, \mathbf{A}_\alpha \Rightarrow \mathbf{F}_{\alpha \to o}\!\downarrow$. $(\star)$ follows from parts 4 and 5 of Lemma 5.4; part 6 of Lemma 6.2; and part 3 of Lemma 6.5.

Case 8: $\mathbf{X}_o \equiv \mathbf{F}_{\alpha \to o}\, \mathbf{A}_\alpha \Rightarrow \mathbf{A}_\alpha\!\downarrow$. Similar to Case 7.

Case 9: $\mathbf{X}_o \equiv (\mathbf{F}_{\alpha \to \beta}\, \mathbf{A}_\alpha)\!\downarrow \Rightarrow \mathbf{F}_{\alpha \to \beta}\!\downarrow$ where $\beta \neq o$. $(\star)$ follows from part 5 of Lemma 5.4, part 6 of Lemma 6.2, and part 3 of Lemma 6.5.

Case 10: $\mathbf{X}_o \equiv (\mathbf{F}_{\alpha \to \beta}\, \mathbf{A}_\alpha)\!\downarrow \Rightarrow \mathbf{A}_\alpha\!\downarrow$ where $\beta \neq o$. Similar to Case 9.

Case 11: $\mathbf{X}_o \equiv (\lambda \mathbf{x} : \alpha \,.\, \mathbf{B}_\beta)\!\downarrow$. $(\star)$ follows from part 6 of Lemma 5.4 and part 3 of Lemma 6.5.

Case 12: $\mathbf{X}_o \equiv \mathbf{A}_\alpha\!\downarrow \Rightarrow (\mathbf{A}_\alpha \simeq \mathbf{B}_\alpha) \simeq (\mathbf{A}_\alpha = \mathbf{B}_\alpha)$. ($\star$) follows part 3 of Lemma 5.4, part 6 of Lemma 6.2, and parts 3, 5, and 6 of Lemma 6.5.

Case 13: $\mathbf{X}_o \equiv \mathbf{B}_\alpha\!\downarrow \Rightarrow (\mathbf{A}_\alpha \simeq \mathbf{B}_\alpha) \simeq (\mathbf{A}_\alpha = \mathbf{B}_\alpha)$. Same as Case 12.

$\square$

Lemma B.6 (Axiom A6) *Each instance of Axiom A6 is valid.*

Proof Let $\mathbf{X}_o$ be an instance of Axiom A6, M be a general model that interprets $\mathbf{X}_o$, and $\varphi \in \mathsf{assign}(M)$. We must show ($\star$) $V_\varphi^M(\mathbf{X}_o) = \mathrm{T}$.

Case 1: $\mathbf{X}_o \equiv (\exists! x : \alpha . \mathbf{A}_o) \Rightarrow (\mathrm{I}\, x : \alpha . \mathbf{A}_o) \in \{x : \alpha \mid \mathbf{A}_o\}$ (where $\alpha \neq o$).
(a) $V_\varphi^M(\exists! x : \alpha . \mathbf{A}_o) = \mathrm{F}$ implies $V_\varphi^M(\mathbf{X}_o) = \mathrm{T}$ by part 6 of Lemma 6.2.

$$V_\varphi^M(\exists! x : \alpha . \mathbf{A}_o) = \mathrm{T}$$

implies $V_{\varphi[(x:\alpha)\mapsto d]}^M(\mathbf{A}_o)$ for exactly one $d \in D_\alpha^M$ \hfill (1)

implies $V_\varphi^M(\mathrm{I}\, x : \alpha . \mathbf{A}_o)$ is defined \hfill (2)

implies $V_{\varphi[(x:\alpha)\mapsto V_\varphi^M(\mathrm{I}\,x:\alpha.\mathbf{A}_o)]}^M(\mathbf{A}_o) = \mathrm{T}$ \hfill (3)

implies $V_\varphi^M((\lambda x : \alpha . \mathbf{A}_o)(\mathrm{I}\, x : \alpha . \mathbf{A}_o)) = \mathrm{T}$ \hfill (4)

implies $V_\varphi^M((\mathrm{I}\, x : \alpha . \mathbf{A}_o) \in \{x : \alpha \mid \mathbf{A}_o\}) = \mathrm{T}$ \hfill (5)

implies $V_\varphi^M(\mathbf{X}_o) = \mathrm{T}$ \hfill (6)

(1) is by part 3 of Lemma 6.4; (2) is by part 7 of Lemma 5.4; (3) is by part 12 of Lemma 5.4; (4) is by part 11 of Lemma 5.4; (5) is by notational definitions; and (6) is part 6 of Lemma 6.2. Hence (b) $V_\psi^M(\exists! x : \alpha . \mathbf{A}_o) = \mathrm{T}$ implies $V_\psi^M(\mathbf{X}_o) = \mathrm{T}$. Therefore, (a) and (b) imply that ($\star$) holds.

Case 2: $\mathbf{X}_o \equiv \neg(\exists! x : \alpha . \mathbf{A}_o) \Rightarrow (\mathrm{I}\, x : \alpha . \mathbf{A}_o)\!\uparrow$ (where $\alpha \neq o$).
(a) $V_\varphi^M(\neg(\exists! x : \alpha . \mathbf{A}_o)) = \mathrm{F}$ implies $V_\varphi^M(\mathbf{X}_o) = \mathrm{T}$ by part 6 of Lemma 6.2.

$$V_\varphi^M(\neg(\exists! x : \alpha . \mathbf{A}_o)) = \mathrm{T}$$

implies $V_{\varphi[(x:\alpha)\mapsto d]}^M(\mathbf{A}_o)$ for not exactly one $d \in D_\alpha^M$ \hfill (1)

implies $V_\varphi^M(\mathrm{I}\, x : \alpha . \mathbf{A}_o)$ is undefined \hfill (2)

implies $V_\varphi^M((\mathrm{I}\, x : \alpha . \mathbf{A}_o)\!\uparrow) = \mathrm{T}$ \hfill (3)

implies $V_\varphi^M(\mathbf{X}_o) = \mathrm{T}$ \hfill (4)

(1) is by part 3 of Lemma 6.4; (2) is by part 7 of Lemma 5.4; (3) is by part 4 of Lemma 6.5; and (4) is part 6 of Lemma 6.2. Hence (b) $V_\psi^M(\neg(\exists! x : \alpha . \mathbf{A}_o)) = \mathrm{T}$ implies $V_\psi^M(\mathbf{X}_o) = \mathrm{T}$. Therefore, (a) and (b) imply that ($\star$) holds. $\square$

Lemma B.7 (Axiom A7) *Each instance of Axiom A7 is valid.*

Proof Let $\mathbf{X}_o$ be an instance of Axiom A7, M be a general model that interprets $\mathbf{X}_o$, and $\varphi \in \mathsf{assign}(M)$. We must show $(\star)$ $V_\varphi^M(\mathbf{X}_o) = \mathrm{T}$.

Case 1: $\mathbf{X}_o \equiv \forall x : \alpha, \, y : \beta \, . \, (x, y)\!\downarrow$. Let

$$\psi = \varphi[(x : \alpha) \mapsto d_1][(y : \beta) \mapsto d_2].$$

$$V_\varphi^M(\forall x : \alpha, \, y : \beta \, . \, (x, y)\!\downarrow) = \mathrm{T}$$

$$\text{iff} \quad V_\psi^M(((x : \alpha), (y : \alpha))\!\downarrow) = \mathrm{T} \text{ for all } d_1 \in D_\alpha^M, d_2 \in D_\beta^M \tag{1}$$

$$\text{iff} \quad V_\psi^M(((x : \alpha), (y : \alpha))) \text{ is defined for all } d_1 \in D_\alpha^M, d_2 \in D_\beta^M \tag{2}$$

$$\text{iff} \quad (d_1, d_2) \in D_{\alpha \times \beta}^M \text{ for all } d_1 \in D_\alpha^M, d_2 \in D_\beta^M \tag{3}$$

(1) is by part 1 of Lemma 6.4; (2) is by part 3 of Lemma 6.5; and (3) is by conditions V1 and V7 of the definition of a general model. Therefore, $(\star)$ holds since (3) is true by the definition of $D_{\alpha \times \beta}^M$.

Case 2: $\mathbf{X}_o \equiv (\mathbf{A}_\alpha, \mathbf{B}_\beta)\!\downarrow \Rightarrow \mathbf{A}_\alpha\!\downarrow$. $(\star)$ follows from part 8 of Lemma 5.4, part 6 of Lemma 6.2, and part 3 of Lemma 6.5.

Case 3: $\mathbf{X}_o \equiv (\mathbf{A}_\alpha, \mathbf{B}_\beta)\!\downarrow \Rightarrow \mathbf{B}_\alpha\!\downarrow$. Same as Case 2.

Case 4: $\mathbf{X}_o \equiv \forall p : \alpha \times \beta \, . \, p = (\mathsf{fst}_{(\alpha \times \beta) \to \alpha} \, p, \mathsf{snd}_{(\alpha \times \beta) \to \beta} \, p)$. Let

$$\psi = \varphi[(p : \alpha \times \beta) \mapsto (d_1, d_2)].$$

$$V_\varphi^M(\forall p : \alpha \times \beta \, . \, p = (\mathsf{fst}_{(\alpha \times \beta) \to \alpha} \, p, \mathsf{snd}_{(\alpha \times \beta) \to \beta} \, p)) = \mathrm{T}$$

$$\text{iff} \quad V_\psi^M((p : \alpha \times \beta) =$$
$$(\mathsf{fst}_{(\alpha \times \beta) \to \alpha} \, (p : \alpha \times \beta), \mathsf{snd}_{(\alpha \times \beta) \to \beta} \, (p : \alpha \times \beta))) = \mathrm{T}$$
$$\text{for all } d_1 \in D_\alpha^M, d_2 \in D_\beta^M \tag{1}$$

$$\text{iff} \quad V_\psi^M((p : \alpha \times \beta)) =$$
$$V_\psi^M((\mathsf{fst}_{(\alpha \times \beta) \to \alpha} \, (p : \alpha \times \beta), \mathsf{snd}_{(\alpha \times \beta) \to \beta} \, (p : \alpha \times \beta)))$$
$$\text{for all } d_1 \in D_\alpha^M, d_2 \in D_\beta^M \tag{2}$$

$$\text{iff} \quad (d_1, d_2) = (d_1, d_2) \text{ for all } d_1 \in D_\alpha^M, d_2 \in D_\beta^M \tag{3}$$

(1) is by part 1 of Lemma 6.4; (2) is by part 10 of Lemma 5.4; and (3) is by conditions V1 and V7 and the definitions of $\mathsf{fst}_{(\alpha \times \beta) \to \alpha}$ and $\mathsf{snd}_{(\alpha \times \beta) \to \beta}$. Therefore, $(\star)$ holds since (3) is true.

Case 5: $\mathbf{X}_o \equiv \forall\, x, x' : \alpha,\, y, y' : \beta\,.\,(x,y) = (x',y') \Rightarrow (x = x' \wedge y = y')$. Let

$$\psi = \varphi[(x:\beta) \mapsto d_1][(x':\beta) \mapsto d_1'][(y:\beta) \mapsto d_2][(y':\beta) \mapsto d_2'],$$

$$\mathbf{A}_o \equiv ((x:\alpha),(y:\beta)) = ((x':\alpha),(y':\beta)),\text{ and}$$

$$\mathbf{B}_o \equiv (x:\alpha) = (x':\alpha) \wedge (y:\beta) = (y':\beta).$$

(a) $V_\psi^M(\mathbf{A}_o) = \mathrm{F}$ implies $V_\psi^M(\mathbf{A}_o \Rightarrow \mathbf{B}_o) = \mathrm{T}$ by part 6 of Lemma 6.2.

$$V_\psi^M(\mathbf{A}_o) = \mathrm{T}$$

implies $(d_1,d_2) = (d_1',d_2')$ (1)

implies $d_1 = d_1'$ and $d_2 = d_2'$ (2)

implies $V_\psi^M(\mathbf{B}_o) = \mathrm{T}$ (3)

implies $V_\psi^M(\mathbf{A}_o \Rightarrow \mathbf{B}_o) = \mathrm{T}$ (4)

(1) is by conditions V1, V3, and V7 of the definition of a general model; (2) is by the injectivity of the ordered pair operation; (3) is by conditions V1 and V3 of the definition of a general model and part 5 of Lemma 6.2; and (4) is by part 6 of Lemma 6.2. Hence (b) $V_\psi^M(\mathbf{A}_o) = \mathrm{T}$ implies $V_\psi^M(\mathbf{A}_o \Rightarrow \mathbf{B}_o) = \mathrm{T}$. Therefore, (a) and (b) imply $V_\psi^M(\mathbf{A}_o \Rightarrow \mathbf{B}_o) = \mathrm{T}$, and so $(\star)$ holds by part 1 of Lemma 6.4. $\qquad\square$

Lemma B.8 (Rule R1) *Rule R1 preserves validity in all general models.*

Proof Let M be a general model and $\mathbf{A}_o$ and $\mathbf{A}_o \Rightarrow \mathbf{B}_o$ be formulas such that (a) $M \vDash \mathbf{A}_o$ and (b) $M \vDash \mathbf{A}_o \Rightarrow \mathbf{B}_o$. We must show (c) $M \vDash \mathbf{B}_o$. (a) and (b) imply $V_\varphi^M(\mathbf{A}_o) = \mathrm{T}$ and $V_\varphi^M(\mathbf{A}_o \Rightarrow \mathbf{B}_o) = \mathrm{T}$, respectively, for all $\varphi \in \mathsf{assign}(M)$. Hence $V_\varphi^M(\mathbf{B}_o) = \mathrm{T}$ for all $\varphi \in \mathsf{assign}(M)$ by part 6 of Lemma 6.2. Therefore, (c) holds. $\qquad\square$

Lemma B.9 *Let M be a general model that interprets $\mathbf{A}_\alpha$, $\mathbf{B}_\alpha$, $\mathbf{C}_\beta$, and $\mathbf{C}_\beta'$ such that $V_\varphi^M(\mathbf{A}_\alpha) \simeq V_\varphi^M(\mathbf{B}_\alpha)$ for all $\varphi \in \mathsf{assign}(M)$ and $\mathbf{C}_\beta'$ is obtained from $\mathbf{C}_\beta$ by replacing at most one occurrence of $\mathbf{A}_\alpha$ in $\mathbf{C}_\beta$, that is not in the first argument of a function abstraction or a definite description, by an occurrence of $\mathbf{B}_\alpha$. Then $V_\varphi^M(\mathbf{C}_\beta) \simeq V_\varphi^M(\mathbf{C}_\beta')$ for all $\varphi \in \mathsf{assign}(M)$.*

Proof Under the assumptions of the lemma, we must show $(\star)$ $V_\varphi^M(\mathbf{C}_\beta) \simeq V_\varphi^M(\mathbf{C}_\beta')$ for all $\varphi \in \mathsf{assign}(M)$. The proof is by induction on the syntactic structure of $\mathbf{C}_\beta$.

Case 1: $\mathbf{C}_\beta \equiv (\mathbf{x} : \beta)$. If $\mathbf{C}_\beta \equiv \mathbf{A}_\alpha$, then $\mathbf{C}'_\beta \equiv \mathbf{B}_\alpha$ and $(\star)$ is obvious. If $\mathbf{C}_\beta \not\equiv \mathbf{A}_\alpha$, then $\mathbf{C}'_\beta \equiv \mathbf{C}_\beta$ and $(\star)$ is again obvious.

Case 2: $\mathbf{C}_\beta \equiv \mathbf{c}_\beta$. Similar to the proof of Case 1.

Case 3: $\mathbf{C}_\beta \equiv \mathbf{D}_\gamma = \mathbf{E}_\gamma$. Then $\mathbf{C}'_\beta$ has the form $\mathbf{D}'_\gamma = \mathbf{E}'_\gamma$, and so by the induction hypothesis $V_\varphi^M(\mathbf{D}_\gamma) \simeq V_\varphi^M(\mathbf{D}'_\gamma)$ and $V_\varphi^M(\mathbf{E}_\gamma) \simeq V_\varphi^M(\mathbf{E}'_\gamma)$ for all $\varphi \in \mathsf{assign}(M)$. Therefore, $V_\varphi^M(\mathbf{C}_\beta) \simeq V_\varphi^M(\mathbf{D}_\gamma = \mathbf{E}_\gamma) \simeq V_\varphi^M(\mathbf{D}'_\gamma = \mathbf{E}'_\gamma) \simeq V_\varphi^M(\mathbf{C}'_\beta)$ for all $\varphi \in \mathsf{assign}(M)$ by condition V3 of the definition of a general model, and so $(\star)$ holds.

Case 4: $\mathbf{C}_\beta \equiv \mathbf{D}_{\gamma \to \beta}\,\mathbf{E}_\gamma$. Then $\mathbf{C}'_\beta$ has the form $\mathbf{D}'_{\gamma \to \beta}\,\mathbf{E}'_\gamma$, and so by the induction hypothesis $V_\varphi^M(\mathbf{D}_{\gamma \to \beta}) \simeq V_\varphi^M(\mathbf{D}'_{\gamma \to \beta})$ and $V_\varphi^M(\mathbf{E}_\gamma) \simeq V_\varphi^M(\mathbf{E}'_\gamma)$ for all $\varphi \in \mathsf{assign}(M)$. Therefore, $V_\varphi^M(\mathbf{C}_\beta) \simeq V_\varphi^M(\mathbf{D}_{\gamma \to \beta}\,\mathbf{E}_\gamma) \simeq V_\varphi^M(\mathbf{D}'_{\gamma \to \beta}\,\mathbf{E}'_\gamma) \simeq V_\varphi^M(\mathbf{C}'_\beta)$ for all $\varphi \in \mathsf{assign}(M)$ by condition V4 of the definition of a general model, and so $(\star)$ holds.

Case 5: $\mathbf{C}_\beta \equiv \lambda\,\mathbf{x} : \gamma\,.\,\mathbf{D}_\delta$. Then $\mathbf{C}'_\beta$ has the form $\lambda\,\mathbf{x} : \gamma\,.\,\mathbf{D}'_\delta$, and so by the induction hypothesis $V_\varphi^M(\mathbf{D}_\delta) \simeq V_\varphi^M(\mathbf{D}'_\delta)$ for all $\varphi \in \mathsf{assign}(M)$. Therefore, $V_\varphi^M(\mathbf{C}_\beta) \simeq V_\varphi^M(\lambda\,\mathbf{x} : \gamma\,.\,\mathbf{D}_\delta) \simeq V_\varphi^M(\lambda\,\mathbf{x} : \gamma\,.\,\mathbf{D}'_\delta) \simeq V_\varphi^M(\mathbf{C}'_\beta)$ for all $\varphi \in \mathsf{assign}(M)$ by condition V5 of the definition of a general model, and so $(\star)$ holds.

Case 6: $\mathbf{C}_\beta \equiv \mathrm{I}\,\mathbf{x} : \beta\,.\,\mathbf{D}_o$. Then $\mathbf{C}'_\beta$ has the form $\mathrm{I}\,\mathbf{x} : \beta\,.\,\mathbf{D}'_o$, and so by the induction hypothesis $V_\varphi^M(\mathbf{D}_o) \simeq V_\varphi^M(\mathbf{D}'_o)$ for all $\varphi \in \mathsf{assign}(M)$. Therefore, $V_\varphi^M(\mathbf{C}_\beta) \simeq V_\varphi^M(\mathrm{I}\,\mathbf{x} : \beta\,.\,\mathbf{D}_o) \simeq V_\varphi^M(\mathrm{I}\,\mathbf{x} : \beta\,.\,\mathbf{D}'_o) \simeq V_\varphi^M(\mathbf{C}'_\beta)$ for all $\varphi \in \mathsf{assign}(M)$ by condition V6 of the definition of a general model, and so $(\star)$ holds.

Case 7: $\mathbf{C}_\beta \equiv (\mathbf{D}_\gamma, \mathbf{E}_\gamma)$. Then $\mathbf{C}'_\beta$ has the form $(\mathbf{D}'_\gamma, \mathbf{E}'_\gamma)$, and so by the induction hypothesis $V_\varphi^M(\mathbf{D}_\gamma) \simeq V_\varphi^M(\mathbf{D}'_\gamma)$ and $V_\varphi^M(\mathbf{E}_\gamma) \simeq V_\varphi^M(\mathbf{E}'_\gamma)$ for all $\varphi \in \mathsf{assign}(M)$. Therefore, $V_\varphi^M(\mathbf{C}_\beta) \simeq V_\varphi^M((\mathbf{D}_\gamma, \mathbf{E}_\gamma)) \simeq V_\varphi^M((\mathbf{D}'_\gamma, \mathbf{E}'_\gamma)) \simeq V_\varphi^M(\mathbf{C}'_\beta)$ for all $\varphi \in \mathsf{assign}(M)$ by condition V7 of the definition of a general model, and so $(\star)$ holds.

$\square$

Lemma B.10 (Rule R2) *Rule R2 preserves validity in all general models.*

Proof   Let M be a general model, $\mathbf{A}_\alpha \simeq \mathbf{B}_\alpha$ and $\mathbf{C}_o$ be formulas such that (a) $M \vDash \mathbf{A}_\alpha \simeq \mathbf{B}_\alpha$ and (b) $M \vDash \mathbf{C}_o$, and $\mathbf{C}'_o$ be obtained from $\mathbf{C}_o$ by replacing one occurrence of $\mathbf{A}_\alpha$ in $\mathbf{C}_o$ by an occurrence of $\mathbf{B}_\alpha$, provided that the occurrence of $\mathbf{A}_\alpha$ in $\mathbf{C}_o$ is not the first argument of a function abstraction or a definite description. We must show (c) $M \vDash \mathbf{C}'_o$.

(a) implies (d) $V_\varphi^M(\mathbf{A}_\alpha) \simeq V_\varphi^M(\mathbf{B}_\beta)$ for all $\varphi \in \mathsf{assign}(M)$ by part 5 of Lemma 6.5. (d) implies (e) $V_\varphi^M(\mathbf{C}_o) \simeq V_\varphi^M(\mathbf{C}'_o)$ for all $\varphi \in \mathsf{assign}(M)$ by Lemma B.9. (b) and (e) implies $V_\varphi^M(\mathbf{C}'_o) = \mathrm{T}$ for all $\varphi \in \mathsf{assign}(M)$. Therefore, (c) holds. $\square$

Theorem B.11 (Soundness Theorem) *Let* $\mathbf{A}_o$ *be a formula and* Γ *be a set of formulas.*

1. $\vdash_\mathfrak{A} \mathbf{A}_o$ *implies* $\vDash \mathbf{A}_o$.

2. $\Gamma \vdash_\mathfrak{A} \mathbf{A}_o$ *implies* $\Gamma \vDash \mathbf{A}_o$.

Proof

Part 1 Let Π be a proof of $\mathbf{A}_o$ in $\mathfrak{A}$ and M interprets $\mathbf{A}_o$. We must show that $\mathbf{A}_o$ is valid in M. By Lemmas B.1–B.7, each axiom of $\mathfrak{A}$ is valid in M, and by Lemmas B.8 and B.10, each rule of inference of $\mathfrak{A}$ preserves validity in M. Therefore, each formula in Π is valid in M, and so $\mathbf{A}_o$ is valid in M.

Part 2 Let (Π_1, Π_2) be a proof of $\mathbf{A}_o$ from Γ in $\mathfrak{A}$, M be a model that interprets $\mathbf{A}_o$ and Γ, $\varphi \in \mathsf{assign}(M)$, and $M \vDash_\varphi \Gamma$. We must show $M \vDash_\varphi \mathbf{A}_o$. Then there is a finite set $\{\mathbf{B}_o^1, \ldots, \mathbf{B}_o^n\} \subseteq \Gamma$ such that $\{\mathbf{B}_o^1, \ldots, \mathbf{B}_o^n\} \vdash_\mathfrak{A} \mathbf{A}_o$. Hence

$$\vdash_\mathfrak{A} \mathbf{B}_o^1 \Rightarrow \cdots \Rightarrow \mathbf{B}_o^n \Rightarrow \mathbf{A}_o$$

by the Deduction Theorem (Theorem A.50), and so

$$\text{(a)} \quad \vDash \mathbf{B}_o^1 \Rightarrow \cdots \Rightarrow \mathbf{B}_o^n \Rightarrow \mathbf{A}_o$$

by part 1 of the theorem. $M \vDash_\varphi \Gamma$ implies (b) $M \vDash_\varphi \mathbf{B}_o^i$ for each i with $1 \leq i \leq n$. (a) and (b) imply $M \vDash_\varphi \mathbf{A}_o$ by part 6 of Lemma 6.2. $\square$

Appendix C

Henkin's Theorem for $\mathfrak{A}$

In this appendix we prove Henkin's theorem for $\mathfrak{A}$. The form of the proof is very similar to Peter Andrews' proof of Henkin's theorem for $\mathcal{Q}_0$ [5].

Let $T = (L, \Gamma)$ be a theory of Alonzo and $\mathfrak{P}$ be a proof system for Alonzo. T is *syntactically complete* in $\mathfrak{P}$ if either $T \vdash_{\mathfrak{P}} \mathbf{A}_o$ or $T \vdash_{\mathfrak{P}} \neg\mathbf{A}_o$ holds for all sentences $\mathbf{A}_o$ of L. T is *extensionally complete* in $\mathfrak{P}$ if, for every sentence of the form $\mathbf{A}_{\alpha\to\beta} = \mathbf{B}_{\alpha\to\beta}$, there is a closed expression $\mathbf{C}_\alpha$ such that:

1. $T \vdash_{\mathfrak{P}} \mathbf{C}_\alpha{\downarrow}$.

2. $T \vdash_{\mathfrak{P}}$
$$(\mathbf{A}_{\alpha\to\beta}{\downarrow} \wedge \mathbf{B}_{\alpha\to\beta}{\downarrow}) \Rightarrow (\mathbf{A}_{\alpha\to\beta}\,\mathbf{C}_\alpha \simeq \mathbf{B}_{\alpha\to\beta}\,\mathbf{C}_\alpha \Rightarrow \mathbf{A}_{\alpha\to\beta} = \mathbf{B}_{\alpha\to\beta}).$$

Remark C.1 (Extensionally Complete) *Let T be a theory that is extensionally complete in $\mathfrak{P}$ such that $T \vdash_{\mathfrak{P}} \mathbf{A}_{\alpha\to\beta}{\downarrow}$ and $T \vdash_{\mathfrak{P}} \mathbf{B}_{\alpha\to\beta}{\downarrow}$. Then there is a closed expression $\mathbf{C}_\alpha$ such that $T \vdash_{\mathfrak{P}} \mathbf{C}_\alpha{\downarrow}$ and, by contraposition,*

$$T \vdash_{\mathfrak{P}} \mathbf{A}_{\alpha\to\beta} \neq \mathbf{B}_{\alpha\to\beta} \Rightarrow \mathbf{A}_{\alpha\to\beta}\,\mathbf{C}_\alpha \not\simeq \mathbf{B}_{\alpha\to\beta}\,\mathbf{C}_\alpha.$$

Hence, if $\mathbf{A}_{\alpha\to\beta}$ and $\mathbf{B}_{\alpha\to\beta}$ denote different functions in a model of T, then $\mathbf{C}_\alpha$ denotes a witness in the model that these functions are different.

Now consider a sentence of the form

$$\lambda x : \alpha \,.\, T_o = \lambda \mathbf{x} : \alpha \,.\, \neg\mathbf{D}_o.$$

Then $\lambda x : \alpha \,.\, T_o \neq \lambda \mathbf{x} : \alpha \,.\, \neg\mathbf{D}_o$ is logically equivalent to $\exists \mathbf{x} : \alpha \,.\, \mathbf{D}_o$. By Axiom A5.11, $T \vdash_{\mathfrak{P}} (\lambda x : \alpha \,.\, T_o){\downarrow}$ and $T \vdash_{\mathfrak{P}} (\lambda \mathbf{x} : \alpha \,.\, \neg\mathbf{D}_o){\downarrow}$. Hence there is a closed expression $\mathbf{C}_\alpha$ such that $T \vdash_{\mathfrak{P}} \mathbf{C}_\alpha{\downarrow}$ and, if $\exists \mathbf{x} : \alpha \,.\, \mathbf{D}_o$ is true in a model of T, then $\mathbf{C}_\alpha$ denotes a witness in the model that $\mathbf{D}_o$ is satisfiable, i.e., that the sentence $(\lambda x : \alpha \,.\, T_o)\,\mathbf{C}_\alpha \not\simeq (\lambda \mathbf{x} : \alpha \,.\, \neg\mathbf{D}_o)\mathbf{C}_\alpha$, which is logically equivalent to $(\lambda \mathbf{x} : \alpha \,.\, \mathbf{D}_o)\,\mathbf{C}_\alpha$, is true in the model.

© The Editor(s) (if applicable) and The Author(s), under exclusive license to Springer Nature Switzerland AG 2023
W. M. Farmer, *Simple Type Theory*, Computer Science Foundations and Applied Logic, https://doi.org/10.1007/978-3-031-21112-6

Lemma C.2 (Extension Lemma) *Let* $T = (L, \Gamma)$ *be a theory of Alonzo. If* T *is consistent in* $\mathfrak{A}$, *then there is a theory* $T' = (L', \Gamma')$ *such that:*

1. $T \leq T'$.

2. T' *is consistent in* $\mathfrak{A}$.

3. T' *is syntactically complete in* $\mathfrak{A}$.

4. T' *is extensionally complete in* $\mathfrak{A}$.

5. $\|L'\| = \|L\|$.

Proof Let $L = (\mathcal{B}, \mathcal{C})$ and $\kappa = \|L\|$. For each $\alpha \in \mathcal{T}(L)$, let $\mathcal{C}_\alpha$ be a well-ordered set of new constants of type α such that $|\mathcal{C}_\alpha| = \kappa$. Define $L' = (\mathcal{B}, \mathcal{C} \cup \mathcal{C}')$ where

$$\mathcal{C}' = \bigcup_{\alpha \in \mathcal{T}(L)} \mathcal{C}_\alpha.$$

Clearly, $|\mathcal{C}'| = \kappa$, so $\|L'\| = \kappa$, and so $\|L'\| = \|L\|$. Therefore, condition 5 is satisfied.

Well-order the sentences in $\mathcal{E}(L')$ and, for each ordinal $\xi < \kappa$, let $\mathbf{S}_o^\xi$ be the ξ-th sentence of L' in this well-order.

For each ordinal $\xi \leq \kappa$, we will define a set Γ_ξ of sentences of L' by transfinite recursion so that (A) $\zeta \leq \xi$ implies $\Gamma_\zeta \subseteq \Gamma_\xi$ and (B) the cardinality of the set of constants in $\mathcal{C}'$ occurring in the sentences of Γ_ξ is finite if ξ is finite and is the cardinality of ξ if ξ is infinite.

Case 1: $\xi = 0$. The $\Gamma_0 = \Gamma$.

Case 2: ξ is a successor ordinal $\zeta + 1$. There are three subcases:

Subcase 2a: $\Gamma_\zeta \cup \{\mathbf{S}_o^\zeta\}$ is consistent in $\mathfrak{A}$. Then $\Gamma_{\zeta+1} = \Gamma_\zeta \cup \{\mathbf{S}_o^\zeta\}$.

Subcase 2b: $\Gamma_\zeta \cup \{\mathbf{S}_o^\zeta\}$ is inconsistent in $\mathfrak{A}$ and $\mathbf{S}_o^\zeta$ does not have the form $\mathbf{A}_{\alpha \to \beta} = \mathbf{B}_{\alpha \to \beta}$. Then $\Gamma_{\zeta+1} = \Gamma_\zeta$.

Subcase 2c: $\Gamma_\zeta \cup \{\mathbf{S}_o^\zeta\}$ is inconsistent in $\mathfrak{A}$ and $\mathbf{S}_o^\zeta$ has the form $\mathbf{A}_{\alpha \to \beta} = \mathbf{B}_{\alpha \to \beta}$. Then

$$\Gamma_{\zeta+1} = \Gamma_\zeta \cup \{\neg(\mathbf{A}_{\alpha \to \beta}\!\downarrow \wedge \mathbf{B}_{\alpha \to \beta}\!\downarrow \wedge \mathbf{A}_{\alpha \to \beta}\,\mathbf{c}_\alpha \simeq \mathbf{B}_{\alpha \to \beta}\,\mathbf{c}_\alpha)\}$$

where $\mathbf{c}_\alpha$ is the first constant in $\mathcal{C}_\alpha$ that does not occur Γ_ζ or $\mathbf{S}_o^\zeta$.

Case 3: ξ is a limit ordinal. Then

$$\Gamma_\xi = \bigcup_{\zeta < \xi} \Gamma_\zeta.$$

It is easy to verify by transfinite induction that conditions (A) and (B) above are satisfied for all ordinals $\xi \leq \kappa$.

We will now prove by transfinite induction that Γ_ξ is consistent in $\mathfrak{A}$ for all ordinals $\xi \leq \kappa$. We have the same three subcases as above:

Case 1: $\xi = 0$. $T = (L, \Gamma)$ is consistent in $\mathfrak{A}$ by assumption. Hence $\Gamma_0 = \Gamma$ must be consistent in $\mathfrak{A}$.

Case 2: ξ is a successor ordinal $\zeta + 1$. We have the same three subcases as above:

Subcase 2a: $\Gamma_\zeta \cup \{\mathbf{S}_o^\zeta\}$ is consistent in $\mathfrak{A}$. Hence $\Gamma_{\zeta+1} = \Gamma_\zeta \cup \{\mathbf{S}_o^\zeta\}$ is trivially consistent in $\mathfrak{A}$.

Subcase 2b: $\Gamma_\zeta \cup \{\mathbf{S}_o^\zeta\}$ is inconsistent in $\mathfrak{A}$ and $\mathbf{S}_o^\zeta$ does not have the form $\mathbf{A}_{\alpha \to \beta} = \mathbf{B}_{\alpha \to \beta}$. Hence $\Gamma_{\zeta+1} = \Gamma_\zeta$ is consistent in $\mathfrak{A}$ by the induction hypothesis.

Subcase 2c: $\Gamma_\zeta \cup \{\mathbf{S}_o^\zeta\}$ is inconsistent in $\mathfrak{A}$ and $\mathbf{S}_o^\zeta$ has the form $\mathbf{A}_{\alpha \to \beta} = \mathbf{B}_{\alpha \to \beta}$. Suppose

$$\Gamma_{\zeta+1} = \Gamma_\zeta \cup \{\neg(\mathbf{A}_{\alpha \to \beta}{\downarrow} \wedge \mathbf{B}_{\alpha \to \beta}{\downarrow} \wedge \mathbf{A}_{\alpha \to \beta}\, \mathbf{c}_\alpha \simeq \mathbf{B}_{\alpha \to \beta}\, \mathbf{c}_\alpha)\}$$

is inconsistent in $\mathfrak{A}$. Then

$$\Gamma_\zeta \vdash_{\mathfrak{A}} \mathbf{A}_{\alpha \to \beta}{\downarrow} \wedge \mathbf{B}_{\alpha \to \beta}{\downarrow} \wedge \mathbf{A}_{\alpha \to \beta}\, \mathbf{c}_\alpha \simeq \mathbf{B}_{\alpha \to \beta}\, \mathbf{c}_\alpha$$

by the Deduction Theorem (Theorem A.50) and the definition of $\neg$. Let P be a proof of

$$\mathbf{A}_{\alpha \to \beta}{\downarrow} \wedge \mathbf{B}_{\alpha \to \beta}{\downarrow} \wedge \mathbf{A}_{\alpha \to \beta}\, \mathbf{c}_\alpha \simeq \mathbf{B}_{\alpha \to \beta}\, \mathbf{c}_\alpha$$

from a finite subset Δ of Γ_ζ, $(\mathbf{x} : \alpha)$ be a variable that does not occur in P or Δ, and P' be the result of replacing each occurrence of $\mathbf{c}_\alpha$ with $(\mathbf{x} : \alpha)$. P' is a proof of

$$\mathbf{A}_{\alpha \to \beta}{\downarrow} \wedge \mathbf{B}_{\alpha \to \beta}{\downarrow} \wedge \mathbf{A}_{\alpha \to \beta}\, (\mathbf{x} : \alpha) \simeq \mathbf{B}_{\alpha \to \beta}\, (\mathbf{x} : \alpha)$$

from Δ since $\mathbf{c}_\alpha$ does not occur in Γ_ζ, $\mathbf{A}_{\alpha\to\beta}$, or $\mathbf{B}_{\alpha\to\beta}$.

$$\Delta \vdash_{\mathfrak{A}} \mathbf{A}_{\alpha\to\beta}{\downarrow} \wedge \mathbf{B}_{\alpha\to\beta}{\downarrow} \wedge \mathbf{A}_{\alpha\to\beta}(\mathbf{x}:\alpha) \simeq \mathbf{B}_{\alpha\to\beta}(\mathbf{x}:\alpha) \tag{1}$$

$$\Delta \vdash_{\mathfrak{A}} \mathbf{A}_{\alpha\to\beta}{\downarrow} \tag{2}$$

$$\Delta \vdash_{\mathfrak{A}} \mathbf{B}_{\alpha\to\beta}{\downarrow} \tag{3}$$

$$\Delta \vdash_{\mathfrak{A}} \mathbf{A}_{\alpha\to\beta}(\mathbf{x}:\alpha) \simeq \mathbf{B}_{\alpha\to\beta}(\mathbf{x}:\alpha) \tag{4}$$

$$\Delta \vdash_{\mathfrak{A}} \forall \mathbf{x}:\alpha\,.\,\mathbf{A}_{\alpha\to\beta}\,\mathbf{x} \simeq \mathbf{B}_{\alpha\to\beta}\,\mathbf{x} \tag{5}$$

$$\Delta \vdash_{\mathfrak{A}} \mathbf{A}_{\alpha\to\beta} = \mathbf{B}_{\alpha\to\beta} \Leftrightarrow \forall \mathbf{x}:\alpha\,.\,\mathbf{A}_{\alpha\to\beta}\,\mathbf{x} \simeq \mathbf{B}_{\alpha\to\beta}\,\mathbf{x} \tag{6}$$

$$\Delta \vdash_{\mathfrak{A}} \mathbf{A}_{\alpha\to\beta} = \mathbf{B}_{\alpha\to\beta} \tag{7}$$

(1) is given; (2), (3), and (4) follow from (1) by the Tautology Rule (Corollary A.46); (5) follows from (4) by Universal Generalization (Theorem A.30) since $(\mathbf{x}:\alpha)$ does not occur in Δ; (6) follows from (2), (3), and Axiom A3 by the Substitution Rule (Theorem A.31); and (7) follows from (6) and (5) by Rule R2'.

Hence (a) $\Gamma_\zeta \vdash_{\mathfrak{A}} \mathbf{A}_{\alpha\to\beta} = \mathbf{B}_{\alpha\to\beta}$. However, $\Gamma_\zeta \cup \{\mathbf{A}_{\alpha\to\beta} = \mathbf{B}_{\alpha\to\beta}\}$ is inconsistent in $\mathfrak{A}$ in Subcase 2c, and so (b) $\Gamma_\zeta \cup \{\mathbf{A}_{\alpha\to\beta} = \mathbf{B}_{\alpha\to\beta}\} \vdash_{\mathfrak{A}} F_o$. (a) and (b) imply Γ_ζ is inconsistent, which contradicts the induction hypothesis. Therefore, $\Gamma_{\zeta+1}$ must be consistent in $\mathfrak{A}$.

Case 3: ξ is a limit ordinal. Γ_ζ is consistent in $\mathfrak{A}$ for all $\zeta < \xi$ by the induction hypothesis. Then each finite subset of Γ_ξ is a subset of some Γ_ζ with $\zeta < \xi$, and so Γ_ξ must be consistent in $\mathfrak{A}$.

Define $\Gamma' = \Gamma_\kappa$ and $T' = (L', \Gamma')$. Then $T \leq T'$ and T' is consistent in $\mathfrak{A}$. Therefore, conditions 1 and 2 are satisfied.

Now all we have left to show is that T' is syntactically and extensionally complete in $\mathfrak{A}$. Let $\mathbf{S}_o$ be any sentence of L'. Then $\mathbf{S}_o = \mathbf{S}_o^\zeta$ for some $\zeta < \kappa$.

If $\Gamma_\zeta \cup \{\mathbf{S}_o\}$ is consistent in $\mathfrak{A}$, then $\mathbf{S}_o \in \Gamma_{\zeta+1} \subseteq \Gamma'$ by Subcase 2a, and so $\Gamma' \vdash_{\mathfrak{A}} \mathbf{S}_o$. Otherwise $\Gamma_\zeta \cup \{\mathbf{S}_o\} \vdash_{\mathfrak{A}} F_o$, so $\Gamma_\zeta \vdash_{\mathfrak{A}} \neg\mathbf{S}_o$ by the Deduction Theorem (Theorem A.50) and the definition of $\neg$, and so $\Gamma' \vdash_{\mathfrak{A}} \neg\mathbf{S}_o$ Hence T' is syntactically complete in $\mathfrak{A}$. Therefore, condition 4 is satisfied.

Assume that $\mathbf{S}_o$ has the form $\mathbf{A}_{\alpha\to\beta} = \mathbf{B}_{\alpha\to\beta}$. If $\Gamma_\zeta \cup \{\mathbf{S}_o\}$ is consistent in $\mathfrak{A}$, then

$$T' \vdash_{\mathfrak{A}} (\mathbf{A}_{\alpha\to\beta}{\downarrow} \wedge \mathbf{B}_{\alpha\to\beta}{\downarrow}) \Rightarrow (\mathbf{A}_{\alpha\to\beta}\,\mathbf{C}_\alpha \simeq \mathbf{B}_{\alpha\to\beta}\,\mathbf{C}_\alpha \Rightarrow \mathbf{S}_o)$$

for all expressions $\mathbf{C}_\alpha \in \mathcal{E}(L')$ by the Tautology Rule (Corollary A.46). Notice that there is some expression $\mathbf{C}_\alpha$ that is closed with $T' \vdash_{\mathfrak{A}} \mathbf{C}_\alpha{\downarrow}$. If $\Gamma_\zeta \cup \{\mathbf{S}_o\}$ is inconsistent in $\mathfrak{A}$, then

$$T' \vdash_{\mathfrak{A}} \neg(\mathbf{A}_{\alpha\to\beta}{\downarrow} \wedge \mathbf{B}_{\alpha\to\beta}{\downarrow} \wedge \mathbf{A}_{\alpha\to\beta}\,\mathbf{c}_\alpha \simeq \mathbf{B}_{\alpha\to\beta}\,\mathbf{c}_\alpha)$$

for some $\mathbf{c}_\alpha \in \mathcal{C}'$ by Subcase 2c, and so

$$T' \vdash_{\mathfrak{A}} (\mathbf{A}_{\alpha\to\beta}{\downarrow} \wedge \mathbf{B}_{\alpha\to\beta}{\downarrow}) \Rightarrow (\mathbf{A}_{\alpha\to\beta}\, \mathbf{c}_\alpha \simeq \mathbf{B}_{\alpha\to\beta}\, \mathbf{c}_\alpha \Rightarrow \mathbf{S}_o)$$

by the Tautology Rule (Corollary A.46). Notice that $\mathbf{c}_\alpha$ is closed and $T' \vdash_{\mathfrak{A}} \mathbf{c}_\alpha{\downarrow}$ by Axiom A5.2. Hence T' is extensionally complete in $\mathfrak{A}$. Therefore, condition 5 is satisfied.

This completes the proof of the Extension Lemma. $\square$

Theorem C.3 (Henkin's Theorem) *Every theory of Alonzo that is consistent in* $\mathfrak{A}$ *has a frugal general model.*

Proof Let $T = (L, \Gamma)$ be a theory that is consistent in $\mathfrak{A}$, and let $T' = (L', \Gamma')$ be an extension of T as described in the Extension Lemma. For $\gamma \in \mathcal{T}(L')$, define $\overline{\mathcal{E}_\gamma} = \{\mathbf{A}_\gamma \mid \mathbf{A}_\gamma \in \mathcal{E}(L')$ that is closed$\}$.

We will simultaneously define, by recursion on the syntactic structure of the types in $\mathcal{T}(L')$, a frame $\mathcal{D} = \{D_\gamma \mid \gamma \in \mathcal{T}(L')\}$ and a partial function V on the closed expressions in $\mathcal{E}(L')$ so that the following conditions hold for all $\gamma \in \mathcal{T}(L')$:

(1^γ) $D_\gamma = \{V(\mathbf{A}_\gamma) \mid \mathbf{A}_\gamma \in \overline{\mathcal{E}_\gamma}$ and $V(\mathbf{A}_\gamma)$ is defined$\}$.

(2^γ) $V(\mathbf{A}_\gamma)$ is defined iff $\Gamma' \vdash_{\mathfrak{A}} \mathbf{A}_\gamma{\downarrow}$ for all $\mathbf{A}_\gamma \in \overline{\mathcal{E}_\gamma}$.

(3^γ) $V(\mathbf{A}_\gamma) = V(\mathbf{B}_\gamma)$ iff $\Gamma' \vdash_{\mathfrak{A}} \mathbf{A}_\gamma = \mathbf{B}_\gamma$ for all $\mathbf{A}_\gamma, \mathbf{B}_\gamma \in \overline{\mathcal{E}_\gamma}$.

Case 1: $\gamma = o$. Define $D_o = \{\text{F}, \text{T}\}$. For each $\mathbf{A}_o \in \overline{\mathcal{E}_o}$, define $V(\mathbf{A}_o) = \text{T}$ if $\Gamma' \vdash_{\mathfrak{A}} \mathbf{A}_o$ and $V(\mathbf{A}_o) = \text{F}$ if $\Gamma' \vdash_{\mathfrak{A}} \neg\mathbf{A}_o$. By the consistency and syntactic completeness of T' in $\mathfrak{A}$, exactly one of $\Gamma' \vdash_{\mathfrak{A}} \mathbf{A}_o$ and $\Gamma' \vdash_{\mathfrak{A}} \neg\mathbf{A}_o$ holds. Then the definition of V on $\overline{\mathcal{E}_o}$ is well-defined, (a) V is total on $\overline{\mathcal{E}_o}$, and (1°) is satisfied. (b) $\Gamma' \vdash_{\mathfrak{A}} \mathbf{A}_o{\downarrow}$ for all $\mathbf{A}_o \in \overline{\mathcal{E}_o}$ by Proposition A.8. (a) and (b) imply (2°) is satisfied. (3°) is satisfied by the definition of V on $\overline{\mathcal{E}_o}$ and the Tautology Rule (Corollary A.46). Therefore, (1°), (2°), and (3°) are satisfied.

Case 2: $\gamma = \mathbf{a}$. For each $\mathbf{A}_\mathbf{a} \in \overline{\mathcal{E}_\mathbf{a}}$, define

$$V(\mathbf{A}_\mathbf{a}) = \{\mathbf{B}_\mathbf{a} \mid \mathbf{B}_\mathbf{a} \in \overline{\mathcal{E}_\mathbf{a}} \text{ and } \Gamma' \vdash_{\mathfrak{A}} \mathbf{A}_\mathbf{a} = \mathbf{B}_\mathbf{a}\}$$

if $\Gamma' \vdash_{\mathfrak{A}} \mathbf{A}_\mathbf{a}{\downarrow}$, and otherwise define $V(\mathbf{A}_\mathbf{a})$ to be undefined. Define

$$D_\mathbf{a} = \{V(\mathbf{A}_\mathbf{a}) \mid \mathbf{A}_\mathbf{a} \in \overline{\mathcal{E}_\mathbf{a}} \text{ and } V(\mathbf{A}_\mathbf{a}) \text{ is defined}\}.$$

The definitions of $D_{\mathbf{a}}$ and V on $\overline{\mathcal{E}_{\mathbf{a}}}$ obviously satisfy $(1^{\mathbf{a}})$ and $(2^{\mathbf{a}})$. They also satisfy $(3^{\mathbf{a}})$ since $\Gamma' \vdash_{\mathfrak{A}} \mathbf{A_a} = \mathbf{B_a}$ is an equivalence relation over $\overline{\mathcal{E}_{\mathbf{a}}}$. Notice that $D_{\mathbf{a}}$ is nonempty by $(2^{\mathbf{a}})$ and the extensional completeness of T' in $\mathfrak{A}$.

Case 3: $\gamma = \alpha \to \beta$. For each $\mathbf{F}_{\alpha \to \beta} \in \overline{\mathcal{E}_{\alpha \to \beta}}$, define $V(\mathbf{F}_{\alpha \to \beta})$ to be undefined if $\Gamma' \vdash_{\mathfrak{A}} \mathbf{F}_{\alpha \to \beta}{\uparrow}$ and otherwise define $V(\mathbf{F}_{\alpha \to \beta})$ to be the (partial or total) function from D_{α} to D_{β} whose value at an argument $V(\mathbf{A}_{\alpha}) \in D_{\alpha}$ is $V(\mathbf{F}_{\alpha \to \beta} \mathbf{A}_{\alpha})$ if $V(\mathbf{A}_{\alpha})$ is defined and is undefined if $V(\mathbf{A}_{\alpha})$ is undefined. We must show that this definition does not depend on the choice of the particular closed $\mathbf{A}_{\alpha}$ to represent the argument. If $V(\mathbf{A}_{\alpha}) = V(\mathbf{B}_{\alpha})$, then $\Gamma' \vdash_{\mathfrak{A}} \mathbf{A}_{\alpha} = \mathbf{B}_{\alpha}$ by (3^{α}), and so $\Gamma' \vdash_{\mathfrak{A}} \mathbf{F}_{\alpha \to \beta} \mathbf{A}_{\alpha} \simeq \mathbf{F}_{\alpha \to \beta} \mathbf{B}_{\alpha}$ by part 5 of the Equality Rules (Lemma A.13), and so $V(\mathbf{F}_{\alpha \to \beta} \mathbf{A}_{\alpha}) \simeq V(\mathbf{F}_{\alpha \to \beta} \mathbf{B}_{\alpha})$ by (2^{β}), (3^{β}), and the definition of $\simeq$. Finally, define

$$D_{\alpha \to \beta} = \{V(\mathbf{A}_{\alpha \to \beta}) \mid \mathbf{A}_{\alpha \to \beta} \in \overline{\mathcal{E}_{\alpha \to \beta}} \text{ and } V(\mathbf{A}_{\alpha \to \beta}) \text{ is defined}\}.$$

$(1^{\alpha \to \beta})$ and $(2^{\alpha \to \beta})$ are satisfied by definitions of $D_{\alpha \to \beta}$ and V on $\overline{\mathcal{E}_{\alpha \to \beta}}$. We will now show that $(3^{\alpha \to \beta})$ is satisfied. Suppose $V(\mathbf{F}_{\alpha \to \beta}) = V(\mathbf{G}_{\alpha \to \beta})$. Then $\Gamma' \vdash_{\mathfrak{A}} \mathbf{F}_{\alpha \to \beta}{\downarrow}$ and $\Gamma' \vdash_{\mathfrak{A}} \mathbf{G}_{\alpha \to \beta}{\downarrow}$ by $(2^{\alpha \to \beta})$. Since T' is extensionally complete in $\mathfrak{A}$, there is some $\mathbf{C}_{\alpha} \in \overline{\mathcal{E}_{\alpha}}$ such that $\Gamma' \vdash_{\mathfrak{A}} \mathbf{C}_{\alpha}{\downarrow}$ and

$$\Gamma' \vdash_{\mathfrak{A}} (\mathbf{F}_{\alpha \to \beta}{\downarrow} \wedge \mathbf{G}_{\alpha \to \beta}{\downarrow}) \Rightarrow (\mathbf{F}_{\alpha \to \beta}\, \mathbf{C}_{\alpha} \simeq \mathbf{G}_{\alpha \to \beta}\, \mathbf{C}_{\alpha} \Rightarrow \mathbf{F}_{\alpha \to \beta} = \mathbf{G}_{\alpha \to \beta}).$$

Then

$$V(\mathbf{F}_{\alpha \to \beta}\, \mathbf{C}_{\alpha}) \simeq V(\mathbf{F}_{\alpha \to \beta})(V(\mathbf{C}_{\alpha})) \simeq V(\mathbf{G}_{\alpha \to \beta})(V(\mathbf{C}_{\alpha})) \simeq V(\mathbf{G}_{\alpha \to \beta}\, \mathbf{C}_{\alpha});$$

so $\Gamma' \vdash_{\mathfrak{A}} \mathbf{F}_{\alpha \to \beta}\, \mathbf{C}_{\alpha} \simeq \mathbf{G}_{\alpha \to \beta}\, \mathbf{C}_{\alpha}$ by (2^{β}), (3^{β}), and the definition of $\simeq$; and so $\Gamma' \vdash_{\mathfrak{A}} \mathbf{F}_{\alpha \to \beta} = \mathbf{G}_{\alpha \to \beta}$ by the Tautology Rule (Corollary A.46). Now suppose $\Gamma' \vdash_{\mathfrak{A}} \mathbf{F}_{\alpha \to \beta} = \mathbf{G}_{\alpha \to \beta}$. Then, for all $\mathbf{C}_{\alpha} \in \overline{\mathcal{E}_{\alpha}}$,

$$\Gamma' \vdash_{\mathfrak{A}} \mathbf{F}_{\alpha \to \beta}\, \mathbf{C}_{\alpha} \simeq \mathbf{G}_{\alpha \to \beta}\, \mathbf{C}_{\alpha}$$

by part 4 of the Equality Rules (Lemma A.13); so $V(\mathbf{F}_{\alpha \to \beta}\, \mathbf{C}_{\alpha}) \simeq V(\mathbf{G}_{\alpha \to \beta}\, \mathbf{C}_{\alpha})$ by (2^{β}), (3^{β}), and the definition of $\simeq$; and so

$$V(\mathbf{F}_{\alpha \to \beta})(V(\mathbf{C}_{\alpha})) \simeq V(\mathbf{F}_{\alpha \to \beta}\, \mathbf{C}_{\alpha}) \simeq V(\mathbf{G}_{\alpha \to \beta}\, \mathbf{C}_{\alpha}) \simeq V(\mathbf{G}_{\alpha \to \beta})(V(\mathbf{C}_{\alpha})).$$

Hence $V(\mathbf{F}_{\alpha \to \beta}) = V(\mathbf{G}_{\alpha \to \beta})$ by the definition of V on $\overline{\mathcal{E}_{\alpha \to \beta}}$. Therefore, $(3^{\alpha \to \beta})$ is satisfied, and this is the end of Case 3.

Case 4: $\gamma = \alpha \times \beta$. Define $D_{\alpha \times \beta} = D_{\alpha} \times D_{\beta}$. For each $\mathbf{A}_{\alpha \times \beta} \in \overline{\mathcal{E}_{\alpha \times \beta}}$, define

$$V(\mathbf{A}_{\alpha \times \beta}) = (V(\mathsf{fst}_{(\alpha \times \beta) \to \alpha}\, \mathbf{A}_{\alpha \times \beta}), V(\mathsf{snd}_{(\alpha \times \beta) \to \beta}\, \mathbf{A}_{\alpha \times \beta}))$$

if (a) $\Gamma' \vdash_{\mathfrak{A}} \mathbf{A}_{\alpha\times\beta}{\downarrow}$, and otherwise define $V(\mathbf{A}_{\alpha\times\beta})$ to be undefined. (a) implies $\Gamma' \vdash_{\mathfrak{A}} (\mathsf{fst}_{(\alpha\times\beta)\to\alpha} \mathbf{A}_{\alpha\times\beta}){\downarrow}$ and $\Gamma' \vdash_{\mathfrak{A}} (\mathsf{snd}_{(\alpha\times\beta)\to\beta} \mathbf{A}_{\alpha\times\beta}){\downarrow}$ by Axioms A5.5, A7.1, A7.2, A7.3, and A7.4, so $V(\mathsf{fst}_{(\alpha\times\beta)\to\alpha} \mathbf{A}_{\alpha\times\beta})$ and $V(\mathsf{snd}_{(\alpha\times\beta)\to\beta} \mathbf{A}_{\alpha\times\beta})$ are defined by (2^{α}) and (2^{β}). Hence the definition of V on $\overline{\mathcal{E}}_{\alpha\times\beta}$ is well-defined.

$(2^{\alpha\times\beta})$ is obviously satisfied by the definitions of $D_{\alpha\times\beta}$ and V on $\overline{\mathcal{E}}_{\alpha\times\beta}$.

Let $D = \{V(\mathbf{A}_{\gamma}) \mid \mathbf{A}_{\gamma} \in \overline{\mathcal{E}}_{\alpha\times\beta} \text{ and } V(\mathbf{A}_{\gamma}) \text{ is defined}\}$. We must show that $D_{\alpha\times\beta} = D$ to show that $(1^{\alpha\times\beta})$ is satisfied. Let $p \in D_{\alpha\times\beta}$. Then $p = (a, b)$ for some $a \in D_{\alpha}$ and $b \in D_{\beta}$, and so $a = V(\mathbf{A}_{\alpha})$ and $b = V(\mathbf{B}_{\beta})$ for some $\mathbf{A}_{\alpha} \in \overline{\mathcal{E}}_{\alpha}$ and $\mathbf{B}_{\beta} \in \overline{\mathcal{E}}_{\beta}$ by (1^{α}) and (1^{β}), respectively. Then

$$
\begin{aligned}
&(V(\mathbf{A}_{\alpha}), V(\mathbf{B}_{\alpha})) \\
&= (V(\mathsf{fst}_{(\alpha\times\beta)\to\alpha} (\mathbf{A}_{\alpha}, \mathbf{B}_{\beta})), V(\mathsf{snd}_{(\alpha\times\beta)\to\beta} (\mathbf{A}_{\alpha}, \mathbf{B}_{\beta}))) \quad (1) \\
&= V((\mathbf{A}_{\alpha}, \mathbf{B}_{\beta})) \quad\quad\quad\quad\quad\quad\quad\quad\quad\quad\quad\quad\quad\quad\quad\quad\quad (2)
\end{aligned}
$$

(1) is by (3^{α}), (3^{β}), and Lemma A.53; and (2) is by the definition of V on $\overline{\mathcal{E}}_{\alpha\times\beta}$ since $V((\mathbf{A}_{\alpha}, \mathbf{B}_{\beta}))$ is defined by (2^{α}), (2^{β}), $(2^{\alpha\times\beta})$, and Axiom A7.1. Hence $p \in D$, and so $D_{\alpha\times\beta} \subseteq D$. Now let $p \in D$. Then $p = V(\mathbf{A}_{\alpha\times\beta})$ for some $\mathbf{A}_{\alpha\times\beta} \in \overline{\mathcal{E}}_{\alpha\times\beta}$, and so

$$
p = (V(\mathsf{fst}_{(\alpha\times\beta)\to\alpha} \mathbf{A}_{\alpha\times\beta}), V(\mathsf{snd}_{(\alpha\times\beta)\to\beta} \mathbf{A}_{\alpha\times\beta})).
$$

Hence $p \in D_{\alpha\times\beta}$ by (1^{α}) and (1^{β}), and so $D \subseteq D_{\alpha\times\beta}$. Therefore, $(1^{\alpha\times\beta})$ is satisfied.

Now we will show that $(3^{\alpha\times\beta})$ is satisfied.

$$
\begin{aligned}
&V(\mathbf{A}_{\alpha\times\beta}) = V(\mathbf{B}_{\alpha\times\beta}) \\
\text{iff}\quad &(V(\mathsf{fst}_{(\alpha\times\beta)\to\alpha} \mathbf{A}_{\alpha\times\beta}), V(\mathsf{snd}_{(\alpha\times\beta)\to\beta} \mathbf{A}_{\alpha\times\beta})) = \\
&(V(\mathsf{fst}_{(\alpha\times\beta)\to\alpha} \mathbf{B}_{\alpha\times\beta}), V(\mathsf{snd}_{(\alpha\times\beta)\to\beta} \mathbf{B}_{\alpha\times\beta})) \quad\quad (1) \\
\text{iff}\quad &\Gamma' \vdash_{\mathfrak{A}} \mathsf{fst}_{(\alpha\times\beta)\to\alpha} \mathbf{A}_{\alpha\times\beta} = \mathsf{fst}_{(\alpha\times\beta)\to\alpha} \mathbf{B}_{\alpha\times\beta} \text{ and} \\
&\Gamma' \vdash_{\mathfrak{A}} \mathsf{snd}_{(\alpha\times\beta)\to\beta} \mathbf{A}_{\alpha\times\beta} = \mathsf{snd}_{(\alpha\times\beta)\to\beta} \mathbf{B}_{\alpha\times\beta} \quad\quad\quad (2) \\
\text{iff}\quad &\Gamma' \vdash_{\mathfrak{A}} (\mathsf{fst}_{(\alpha\times\beta)\to\alpha} \mathbf{A}_{\alpha\times\beta}, \mathsf{snd}_{(\alpha\times\beta)\to\beta} \mathbf{A}_{\alpha\times\beta}) = \\
&(\mathsf{fst}_{(\alpha\times\beta)\to\alpha} \mathbf{B}_{\alpha\times\beta}, \mathsf{snd}_{(\alpha\times\beta)\to\beta} \mathbf{B}_{\alpha\times\beta}) \quad\quad\quad\quad (3) \\
\text{iff}\quad &\Gamma' \vdash_{\mathfrak{A}} \mathbf{A}_{\alpha\times\beta} = \mathbf{B}_{\alpha\times\beta} \quad\quad\quad\quad\quad\quad\quad\quad\quad\quad\quad\quad (4)
\end{aligned}
$$

(1) is by the definition of V on $\overline{\mathcal{E}}_{\alpha\times\beta}$; (2) is by (3^{α}) and (3^{β}); (3) by Axioms A5.4, A5.5, A7.1, A7.2, A7.3, and A7.5; (4) is by $(2^{\alpha\times\beta})$ and Axiom A7.4. Therefore, $(3^{\alpha\times\beta})$ is satisfied, and this is the end of Case 4.

Therefore, we have shown that $\mathcal{D}$ is a frame and conditions (1^γ), (2^γ), and (3^γ) are satisfied for all $\gamma \in \mathcal{T}(L')$. V clearly maps each constant of L' of type γ to a member of D_γ. Hence $M' = (\mathcal{D}, I)$ is an interpretation of L' where $I = V \restriction_{\mathcal{C}'}$ and $\mathcal{C}'$ is the set of constants of L'.

We will next show that M' is a general model of L'. Choose an enumeration of $\mathcal{E}(L')$ and, for each $\varphi \in \mathsf{assign}(M')$ and variable $(\mathbf{x} : \alpha) \in \mathcal{E}(L')$, let $\theta_\varphi((\mathbf{x} : \alpha))$ be the first closed expression $\mathbf{E}_\alpha$ in this enumeration such that $\varphi((\mathbf{x} : \alpha)) = V(\mathbf{E}_\alpha)$. For each $\varphi \in \mathsf{assign}(M')$ and $\mathbf{C}_\gamma \in \mathcal{E}(L')$, let $(\mathbf{C}_\gamma)^\varphi = \mathbf{C}_\gamma^\varphi$ be the expression obtained by simultaneously replacing $(\mathbf{x}_i : \alpha_i)$ with $\theta_\varphi((\mathbf{x}_i : \alpha_i))$ for all i with $1 \le i \le n$ where $(\mathbf{x}_1 : \alpha_1), \ldots, (\mathbf{x}_n : \alpha_n)$ are the free variables of $\mathbf{C}_\gamma$. Define $V_\varphi^{M'}(\mathbf{C}_\gamma) = V(\mathbf{C}_\gamma^\varphi)$ if $V(\mathbf{C}_\gamma^\varphi)$ is defined, and otherwise define $V_\varphi^{M'}(\mathbf{C}_\gamma)$ to be undefined. Clearly, $\mathbf{C}_\gamma^\varphi \in \overline{\mathcal{E}_\gamma}$, and so $V_\varphi^{M'}(\mathbf{C}_\gamma) \in D_\gamma$ if $V_\varphi^{M'}(\mathbf{C}_\gamma)$ is defined. We are now ready to show that each of the seven conditions of the definition of a general model is satisfied.

1. Let $\mathbf{C}_\gamma$ be a variable $(\mathbf{x} : \alpha)$. Then

$$V_\varphi^{M'}(\mathbf{C}_\gamma) = V_\varphi^{M'}((\mathbf{x} : \alpha)) = V((\mathbf{x} : \alpha)^\varphi) = V(\theta_\varphi((\mathbf{x} : \alpha))) = \varphi((\mathbf{x} : \alpha)).$$

2. Let $\mathbf{C}_\gamma$ be a constant $\mathbf{c}_\alpha$. Then $\mathbf{c}_\alpha \in \overline{\mathcal{E}_\alpha}$, and so

$$V_\varphi^{M'}(\mathbf{C}_\gamma) = V_\varphi^{M'}(\mathbf{c}_\alpha) = V((\mathbf{c}_\alpha)^\varphi) = V(\mathbf{c}_\alpha) = I(\mathbf{c}_\alpha).$$

3. Let $\mathbf{C}_\gamma$ be an equality $\mathbf{A}_\alpha = \mathbf{B}_\alpha$. Then

$$V_\varphi^{M'}(\mathbf{C}_\gamma) = V_\varphi^{M'}(\mathbf{A}_\alpha = \mathbf{B}_\alpha) = V(\mathbf{A}_\alpha^\varphi = \mathbf{B}_\alpha^\varphi).$$

There are three cases to consider:

    a. $V_\varphi^{M'}(\mathbf{A}_\alpha)$ and $V_\varphi^{M'}(\mathbf{B}_\alpha)$ are defined with $V_\varphi^{M'}(\mathbf{A}_\alpha) = V_\varphi^{M'}(\mathbf{B}_\alpha)$. Then $V(\mathbf{A}_\alpha^\varphi) = V(\mathbf{B}_\alpha^\varphi)$, so $\Gamma' \vdash_\mathfrak{A} \mathbf{A}_\alpha^\varphi = \mathbf{B}_\alpha^\varphi$ by (3^α), and so $V(\mathbf{A}_\alpha^\varphi = \mathbf{B}_\alpha^\varphi) = \mathrm{T}$ by the definition of V on $\overline{\mathcal{E}_o}$. Hence $V_\varphi^{M'}(\mathbf{C}_\gamma) = \mathrm{T}$.

    b. $V_\varphi^{M'}(\mathbf{A}_\alpha)$ and $V_\varphi^{M'}(\mathbf{B}_\alpha)$ are defined with $V_\varphi^{M'}(\mathbf{A}_\alpha) \ne V_\varphi^{M'}(\mathbf{B}_\alpha)$. Then $V(\mathbf{A}_\alpha^\varphi) \ne V(\mathbf{B}_\alpha^\varphi)$, so $\Gamma' \vdash_\mathfrak{A} \neg(\mathbf{A}_\alpha^\varphi = \mathbf{B}_\alpha^\varphi)$ by (3^α), and so $V(\mathbf{A}_\alpha^\varphi = \mathbf{B}_\alpha^\varphi) = \mathrm{F}$ by the definition of V on $\overline{\mathcal{E}_o}$. Hence $V_\varphi^{M'}(\mathbf{C}_\gamma) = \mathrm{F}$.

    c. $V_\varphi^{M'}(\mathbf{A}_\alpha)$ or $V_\varphi^{M'}(\mathbf{B}_\alpha)$ is undefined. Then $V(\mathbf{A}_\alpha^\varphi)$ or $V(\mathbf{B}_\alpha^\varphi)$ is undefined, so $\Gamma' \vdash_\mathfrak{A} \mathbf{A}_\alpha \uparrow$ or $\Gamma' \vdash_\mathfrak{A} \mathbf{B}_\alpha \uparrow$ by (2^α), so $\Gamma' \vdash_\mathfrak{A} \neg(\mathbf{A}_\alpha^\varphi = \mathbf{B}_\alpha^\varphi)$ by Axioms A5.4 or A5.5, and so $V(\mathbf{A}_\alpha^\varphi = \mathbf{B}_\alpha^\varphi) = \mathrm{F}$ by the definition of V on $\overline{\mathcal{E}_o}$. Hence $V_\varphi^{M'}(\mathbf{C}_\gamma) = \mathrm{F}$.

4. Let $\mathbf{C}_\gamma$ be a function application $\mathbf{F}_{\alpha \to \beta} \mathbf{A}_\alpha$. There are two cases to consider:

   a. $V_\varphi^{M'}(\mathbf{F}_{\alpha \to \beta})$ and $V_\varphi^{M'}(\mathbf{A}_\alpha)$ are defined. Then $V(\mathbf{F}_{\alpha \to \beta}^\varphi)$ and $V(\mathbf{A}_\alpha^\varphi)$ are defined, and so

   $$V_\varphi^{M'}(\mathbf{C}_\gamma) \simeq V_\varphi^{M'}(\mathbf{F}_{\alpha \to \beta} \mathbf{A}_\alpha) \simeq V(\mathbf{F}_{\alpha \to \beta}^\varphi \mathbf{A}_\alpha^\varphi) \simeq$$
   $$V(\mathbf{F}_{\alpha \to \beta}^\varphi)(V(\mathbf{A}_\alpha^\varphi)) \simeq V_\varphi^{M'}(\mathbf{F}_{\alpha \to \beta})(V_\varphi^{M'}(\mathbf{A}_\alpha)).$$

   Hence, if $V_\varphi^{M'}(\mathbf{F}_{\alpha \to \beta})$ is defined at $V_\varphi^{M'}(\mathbf{A}_\alpha)$, then $V_\varphi^{M'}(\mathbf{C}_\gamma) = V_\varphi^{M'}(\mathbf{F}_{\alpha \to \beta})(V_\varphi^{M'}(\mathbf{A}_\alpha))$, and otherwise $V_\varphi^{M'}(\mathbf{C}_\gamma)$ is undefined.

   b. $V_\varphi^{M'}(\mathbf{F}_{\alpha \to \beta})$ or $V_\varphi^{M'}(\mathbf{A}_\alpha)$ is undefined. Then $V(\mathbf{F}_{\alpha \to \beta}^\varphi)$ or $V(\mathbf{A}_\alpha^\varphi)$ is undefined, and so (a) $\Gamma' \vdash_\mathfrak{A} \mathbf{F}_{\alpha \to \beta}^\varphi \uparrow$ by $(2^{\alpha \to \beta})$ or (b) $\Gamma' \vdash_\mathfrak{A} \mathbf{A}_\alpha^\varphi \uparrow$ by (2^α). If $\beta = o$, then $\Gamma' \vdash_\mathfrak{A} \neg(\mathbf{F}_{\alpha \to \beta}^\varphi \mathbf{A}_\alpha^\varphi)$ follows from (a) or (b) by Axioms A5.7 and A5.8, and so

   $$V_\varphi^{M'}(\mathbf{C}_\gamma) = V_\varphi^{M'}(\mathbf{F}_{\alpha \to \beta} \mathbf{A}_\alpha) = V(\mathbf{F}_{\alpha \to \beta}^\varphi \mathbf{A}_\alpha^\varphi) = \mathrm{F}$$

   by the definition of V on $\overline{\mathcal{E}_o}$. If $\beta \neq o$, then $\Gamma' \vdash_\mathfrak{A} (\mathbf{F}_{\alpha \to \beta}^\varphi \mathbf{A}_\alpha^\varphi) \uparrow$ follows from (a) or (b) by Axioms A5.9 and A5.10, and so

   $$V_\varphi^{M'}(\mathbf{C}_\gamma) \simeq V_\varphi^{M'}(\mathbf{F}_{\alpha \to \beta} \mathbf{A}_\alpha) \simeq V(\mathbf{F}_{\alpha \to \beta}^\varphi \mathbf{A}_\alpha^\varphi)$$

   is undefined by (2^β).

5. Let $\mathbf{C}_\gamma$ be a function abstraction $(\lambda \mathbf{x} : \alpha \, . \, \mathbf{B}_\beta)$. Let $V(\mathbf{A}_\alpha) \in D_\alpha$ and $\psi = \varphi[(\mathbf{x} : \alpha) \mapsto V(\mathbf{A}_\alpha)]$. Notice that (a) $\mathbf{A}_\alpha$ is closed and (b) $V(\mathbf{A}_\alpha)$ is defined. Then

   $$V_\varphi^{M'}(\mathbf{C}_\gamma)(V(\mathbf{A}_\alpha))$$

   $$\simeq V_\varphi^{M'}(\lambda \mathbf{x} : \alpha \, . \, \mathbf{B}_\beta)(V(\mathbf{A}_\alpha)) \tag{1}$$

   $$\simeq V(\lambda \mathbf{x} : \alpha \, . \, \mathbf{B}_\beta^\varphi)(V(\mathbf{A}_\alpha)) \tag{2}$$

   $$\simeq V((\lambda \mathbf{x} : \alpha \, . \, \mathbf{B}_\beta^\varphi) \mathbf{A}_\alpha) \tag{3}$$

   $$\simeq V(\mathbf{B}_\beta^\psi) \tag{4}$$

   $$\simeq V_\psi^{M'}(\mathbf{B}_\beta) \tag{5}$$

(1) is given; (2) and (5) are by the definition of $V_\varphi^{M'}$; (3) follows from (a), (b), Axiom A5.11, and the definition of V on $\overline{\mathcal{E}_{\alpha \to \beta}}$; and (4) follows from (a), (b), Axiom A4, (2^α), (2^β), (3^β), and the definition of $\simeq$. Hence the condition is satisfied.

6. Let $\mathbf{C}_\gamma$ be a definite description $(\mathbf{I}\mathbf{x} : \alpha \,.\, \mathbf{A}_o)$. Without loss of generality, we may assume that $(y : \alpha)$ is distinct from $(\mathbf{x} : \alpha)$ and does not occur in $\mathbf{A}_o$. There are two cases to consider:

a. $V_\varphi^{M'}(\lambda\mathbf{x} : \alpha \,.\, \mathbf{A}_o) = V_{\varphi[(y:\alpha)\mapsto d]}^{M'}(\lambda\mathbf{x} : \alpha \,.\, \mathbf{x} = (y : \alpha))$ for some $d \in D_\alpha$. Let $\mathbf{B}_\alpha$ be the first member of $\overline{\mathcal{E}_\alpha}$ in the enumeration of $\mathcal{E}(L')$ such that $d = V(\mathbf{B}_\alpha)$. Notice that (a) $\mathbf{B}_\alpha$ is closed and (b) $V(\mathbf{B}_\alpha)$ is defined.

$$V_\varphi^{M'}(\lambda\mathbf{x} : \alpha \,.\, \mathbf{A}_o) = V_{\varphi[(y:\alpha)\mapsto V(\mathbf{B}_\alpha)]}^{M'}(\lambda\mathbf{x} : \alpha \,.\, \mathbf{x} = (y : \alpha)) \tag{1}$$

$$V(\lambda\mathbf{x} : \alpha \,.\, \mathbf{A}_o^\varphi) = V(\lambda\mathbf{x} : \alpha \,.\, \mathbf{x} = \mathbf{B}_\alpha) \tag{2}$$

$$\Gamma' \vdash_\mathfrak{A} (\lambda\mathbf{x} : \alpha \,.\, \mathbf{A}_o^\varphi) = (\lambda\mathbf{x} : \alpha \,.\, \mathbf{x} = \mathbf{B}_\alpha) \tag{3}$$

$$\Gamma' \vdash_\mathfrak{A} \exists y : \alpha \,.\, (\lambda\mathbf{x} : \alpha \,.\, \mathbf{A}_o^\varphi) = (\lambda\mathbf{x} : \alpha \,.\, \mathbf{x} = y) \tag{4}$$

$$\Gamma' \vdash_\mathfrak{A} \exists ! \mathbf{x} : \alpha \,.\, \mathbf{A}_o^\varphi \tag{5}$$

$$\Gamma' \vdash_\mathfrak{A} (\lambda\mathbf{x} : \alpha \,.\, \mathbf{A}_o^\varphi)(\mathbf{I}\mathbf{x} : \alpha \,.\, \mathbf{A}_o^\varphi) \tag{6}$$

$$\Gamma' \vdash_\mathfrak{A} (\mathbf{I}\mathbf{x} : \alpha \,.\, \mathbf{A}_o^\varphi){\downarrow} \tag{7}$$

$$\Gamma' \vdash_\mathfrak{A} (\lambda\mathbf{x} : \alpha \,.\, \mathbf{x} = \mathbf{B}_\alpha)(\mathbf{I}\mathbf{x} : \alpha \,.\, \mathbf{A}_o^\varphi) \tag{8}$$

$$\Gamma' \vdash_\mathfrak{A} (\mathbf{I}\mathbf{x} : \alpha \,.\, \mathbf{A}_o^\varphi) = \mathbf{B}_\alpha \tag{9}$$

$$V(\mathbf{I}\mathbf{x} : \alpha \,.\, \mathbf{A}_o^\varphi) = V(\mathbf{B}_\alpha) \tag{10}$$

$$V_\varphi^{M'}(\mathbf{I}\mathbf{x} : \alpha \,.\, \mathbf{A}_o) = V(\mathbf{B}_\alpha) \tag{11}$$

(1) is given; (2) follows from (1) by the definition of $V_\varphi^{M'}$; (3) follows from (2) by $(3^{\alpha \to o})$; (4) follows from (a), (b), and (3) by Existential Generalization (Lemma A.51) and (2^α); (5) follows from (4) by the definition of $\exists !$; (6) follows from (5) and Axiom A6.1 by Rule R1$'$ and the definition of $\in$; (7) follows from (6) and Axiom A5.8; (8) follows from (3) and (6) by Rule R2$'$; (9) follows from (7) and (8) by Beta-Reduction Rule 1 (Lemma A.6); (10) follows from (9) by (3^α); and (11) follows from (10) by the definition of $V_\varphi^{M'}$. Hence the condition is satisfied for this case.

b. $V_\varphi^{M'}(\lambda\mathbf{x} : \alpha \,.\, \mathbf{A}_o) \neq V_{\varphi[(y:\alpha)\mapsto d]}^{M'}(\lambda\mathbf{x} : \alpha \,.\, \mathbf{x} = (\mathbf{y} : \alpha))$ for all $d \in D_\alpha$. Let $V(\mathbf{B}_\alpha) \in D_\alpha$. Notice that (a) $\mathbf{B}_\alpha$ is closed and (b)

$V(\mathbf{B}_\alpha)$ is defined.

$$V_\varphi^{M'}(\lambda \mathbf{x} : \alpha . \mathbf{A}_o) \neq V_{\varphi[(y:\alpha)\mapsto V(\mathbf{B}_\alpha)]}^{M'}(\lambda \mathbf{x} : \alpha . \mathbf{x} = (y : \alpha)) \tag{1}$$

$$V(\lambda \mathbf{x} : \alpha . \mathbf{A}_o^\varphi) \neq V(\lambda \mathbf{x} : \alpha . \mathbf{x} = \mathbf{B}_\alpha) \tag{2}$$

$$\Gamma' \vdash_\mathfrak{A} (\lambda \mathbf{x} : \alpha . \mathbf{A}_o^\varphi) \neq (\lambda \mathbf{x} : \alpha . \mathbf{x} = \mathbf{B}_\alpha) \tag{3}$$

$$\Gamma' \vdash_\mathfrak{A} (\lambda y : \alpha . (\lambda \mathbf{x} : \alpha . \mathbf{A}_o^\varphi) \neq (\lambda \mathbf{x} : \alpha . \mathbf{x} = y)) \, \mathbf{B}_\alpha \tag{4}$$

$$V(\lambda y : \alpha . (\lambda \mathbf{x} : \alpha . \mathbf{A}_o^\varphi) \neq (\lambda \mathbf{x} : \alpha . \mathbf{x} = y) \, \mathbf{B}_\alpha) = \mathrm{T} \tag{5}$$

$$V(\lambda y : \alpha . (\lambda \mathbf{x} : \alpha . \mathbf{A}_o^\varphi) \neq (\lambda \mathbf{x} : \alpha . \mathbf{x} = y))(V(\mathbf{B}_\alpha)) = \mathrm{T} \tag{6}$$

$$V(\lambda x : \alpha . T_o) =$$
$$V(\lambda y : \alpha . (\lambda \mathbf{x} : \alpha . \mathbf{A}_o^\varphi) \neq (\lambda \mathbf{x} : \alpha . \mathbf{x} = y)) \tag{7}$$

$$\Gamma' \vdash_\mathfrak{A} (\lambda x : \alpha . T_o) =$$
$$(\lambda y : \alpha . (\lambda \mathbf{x} : \alpha . \mathbf{A}_o^\varphi) \neq (\lambda \mathbf{x} : \alpha . \mathbf{x} = y)) \tag{8}$$

$$\Gamma' \vdash_\mathfrak{A} \forall y : \alpha . (\lambda \mathbf{x} : \alpha . \mathbf{A}_o^\varphi) \neq (\lambda \mathbf{x} : \alpha . \mathbf{x} = y) \tag{9}$$

$$\Gamma' \vdash_\mathfrak{A} \neg(\exists y : \alpha . (\lambda \mathbf{x} : \alpha . \mathbf{A}_o^\varphi) = (\lambda \mathbf{x} : \alpha . \mathbf{x} = y)) \tag{10}$$

$$\Gamma' \vdash_\mathfrak{A} \neg\exists! \mathbf{x} : \alpha . \mathbf{A}_o^\varphi \tag{11}$$

$$\Gamma' \vdash_\mathfrak{A} (I \mathbf{x} : \alpha . \mathbf{A}_o^\varphi)\!\uparrow \tag{12}$$

$$V(I \mathbf{x} : \alpha . \mathbf{A}_o^\varphi) \text{ is undefined} \tag{13}$$

$$V_\varphi^{M'}(I \mathbf{x} : \alpha . \mathbf{A}_o) \text{ is undefined} \tag{14}$$

(1) is given; (2) follows from (1) by the definition of $V_\varphi^{M'}$; (3) follows from (2) by $(3^{\alpha\to o})$; (4) follows from (a), (b), (3), and axiom A4 by Rule R2′; (5) follows from (4) by the definition of V on $\overline{\mathcal{E}_o}$; (6) follows from (5) by the definition of V on $\overline{\mathcal{E}_{\alpha\to o}}$; (7) follows from (6) since $V(\mathbf{B}_\alpha)$ has been arbitrarily chosen; (8) follows from (7) by $(3^{\alpha\to o})$; (9) follows from (8) by the definition of $\forall$; (10) follows from (9) by the Tautology Rule (Corollary A.46) and the definition of $\exists$; (11) follows from (10) by the definition of $\exists!$; (12) follows from (11) and Axiom A6.2 by Rule R1′; (13) follows from (12) by (2^α); and (14) follows from (13) by the definition of $V_\varphi^{M'}$. Hence the condition is satisfied for this case.

7. Let $\mathbf{C}_\gamma$ be an ordered pair $(\mathbf{A}_\alpha, \mathbf{B}_\beta)$. Then

$$V_\varphi^{M'}(\mathbf{C}_\gamma) \simeq V_\varphi^{M'}((\mathbf{A}_\alpha, \mathbf{B}_\beta)) \simeq V((\mathbf{A}_\alpha^\varphi, \mathbf{B}_\beta^\varphi)).$$

There are two cases to consider:

a. $V_\varphi^{M'}(\mathbf{A}_\alpha)$ and $V_\varphi^{M'}(\mathbf{B}_\beta)$ are defined. Then $V(\mathbf{A}_\alpha^\varphi)$ and $V(\mathbf{B}_\beta^\varphi)$ are defined, so $\Gamma' \vdash_{\mathfrak{A}} \mathbf{A}_\alpha^\varphi{\downarrow}$ and $\Gamma' \vdash_{\mathfrak{A}} \mathbf{B}_\beta^\varphi{\downarrow}$ by (2^α) and (2^β), and so $\Gamma' \vdash_{\mathfrak{A}} (\mathbf{A}_\alpha^\varphi, \mathbf{B}_\beta^\varphi){\downarrow}$ is defined by Axiom A7.1.

$$V((\mathbf{A}_\alpha^\varphi, \mathbf{B}_\beta^\varphi))$$
$$= (V(\mathsf{fst}_{(\alpha\times\beta)\to\alpha}(\mathbf{A}_\alpha^\varphi, \mathbf{B}_\beta^\varphi)), V(\mathsf{snd}_{(\alpha\times\beta)\to\beta}(\mathbf{A}_\alpha^\varphi, \mathbf{B}_\beta^\varphi))) \quad (1)$$
$$= (V(\mathbf{A}_\alpha^\varphi), V(\mathbf{B}_\beta^\varphi)) \quad (2)$$
$$= (V_\varphi^{M'}(\mathbf{A}_\alpha), V_\varphi^{M'}(\mathbf{B}_\beta)) \quad (3)$$

(1) is by the definition of V on $\overline{\mathcal{E}_{\alpha\times\beta}}$; (2) is by (3^α), (3^β), and Lemma A.53; and (3) is by the definition of $V_\varphi^{M'}$. Hence $V_\varphi^{M'}(\mathbf{C}_\gamma) = (V_\varphi^{M'}(\mathbf{A}_\alpha), V_\varphi^{M'}(\mathbf{B}_\beta))$.

b. $V_\varphi^{M'}(\mathbf{A}_\alpha)$ or $V_\varphi^{M'}(\mathbf{B}_\beta)$ is undefined. Then $V(\mathbf{A}_\alpha^\varphi)$ or $V(\mathbf{B}_\beta^\varphi)$ is undefined, so $\Gamma' \vdash_{\mathfrak{A}} \mathbf{A}_\alpha^\varphi{\uparrow}$ by (2^α) or $\Gamma' \vdash_{\mathfrak{A}} \mathbf{B}_\beta^\varphi{\uparrow}$ by (2^β), so $\Gamma' \vdash_{\mathfrak{A}} (\mathbf{A}_\alpha^\varphi, \mathbf{B}_\beta^\varphi){\uparrow}$ is undefined by Axioms A7.2 or A7.3 and the Tautology Rule (Corollary A.46), and so $V((\mathbf{A}_\alpha^\varphi, \mathbf{B}_\beta^\varphi))$ is undefined by $(2^{\alpha\times\beta})$. Hence $V_\varphi^{M'}(\mathbf{C}_\gamma)$ is undefined.

Therefore, M' is a general model of L'.

Let $\mathbf{A}_o \in \Gamma'$. Then $\mathbf{A}_o \in \overline{\mathcal{E}_o}$ and $\Gamma' \vdash_{\mathfrak{A}} \mathbf{A}_o$, and so $V(\mathbf{A}_o) = \mathrm{T}$. Hence $V_\varphi^{M'}(\mathbf{A}_o) = V(\mathbf{A}_o^\varphi) = V(\mathbf{A}_o) = \mathrm{T}$ for all $\varphi \in \mathsf{assign}(M')$, and so $M' \vDash \mathbf{A}_o$. Therefore, M' is a model of T', and so the reduct M of M' to L is a model of T.

For all $\gamma \in \mathcal{T}(L')$, $|D_\gamma| \le |\overline{\mathcal{E}_\gamma}| \le |\mathcal{E}(L')| = \|L'\|$ since V maps $\overline{\mathcal{E}_\gamma}$ onto D_γ, and so $\|M'\| \le \|L'\|$. However, $\|L'\| = \|L\|$ by assumption, and so $\|M\| = \|M'\| \le \|L\|$. Therefore, M is frugal.

At last, this completes the proof of Henkin's Theorem for $\mathfrak{A}$! $\square$

Bibliography

[1] W. Ackermann. Begründung des "tertium non datur" mittels der Hilbertschen Theorie der Widerspruchsfreiheit. *Mathematische Annalen*, 93:1–36, 1924.

[2] P. B. Andrews. A reduction of the axioms for the theory of propositional types. *Fundamenta Mathematicae*, 52:345–350, 1963.

[3] P. B. Andrews. General models and extensionality. *Journal of Symbolic Logic*, 37:395–397, 1972.

[4] P. B. Andrews. General models, descriptions, and choice in type theory. *Journal of Symbolic Logic*, 37:385–394, 1972.

[5] P. B. Andrews. *An Introduction to Mathematical Logic and Type Theory: To Truth through Proof.* Springer, second edition, 2002.

[6] P. B. Andrews, M. Bishop, S. Issar, D. Nesmith, F. Pfennig, and H. Xi. TPS: A theorem proving system for classical type theory. *Journal of Automated Reasoning*, 16:321–353, 1996.

[7] E. Astesiano, M. Bidoit, H. Kirchner, B. Krieg-Brückner, P. D. Mosses, D. Sannella, and A. Tarlecki. CASL: The Common Algebraic Specification Language. *Theoretical Computer Science*, 286:153–196, 2002.

[8] F. Baader and T. Nipkow. *Term Rewriting and All That.* Cambridge University Press, 1999.

[9] C. Ballarin. Locales: A module system for mathematical theories. *Journal of Automated Reasoning*, 52:123–153, 2014.

[10] J. Barwise and J. Seligman. *Information Flow: The Logic of Distributed Systems*, volume 44 of *Cambridge Tracts in Computer Science*. Cambridge University Press, 1997.

© The Editor(s) (if applicable) and The Author(s), under exclusive license to Springer Nature Switzerland AG 2023
W. M. Farmer, *Simple Type Theory*, Computer Science Foundations and Applied Logic, https://doi.org/10.1007/978-3-031-21112-6

[11] M. J. Beeson. Formalizing constructive mathematics: Why and how? In F. Richman, editor, *Constructive Mathematics*, volume 873 of *Lecture Notes in Mathematics*, pages 146–190. Springer, 1981.

[12] M. J. Beeson. *Foundations of Constructive Mathematics*. Springer, 1985.

[13] C. Benzmüller and P. Andrews. Church's type theory. In E. N. Zalta, editor, *The Stanford Encyclopedia of Philosophy*. Metaphysics Research Lab, Stanford University, Summer 2019 edition, 2019. Available at `http://plato.stanford.edu/archives/sum2019/entries/type-theory-church/`.

[14] C. Benzmüller, C. E. Brown, and M. Kohlhase. Higher-order semantics and extensionality. *Journal of Symbolic Logic*, 69:1027–1088, 2004.

[15] A. Bove, P. Dybjer, and U. Norell. A brief overview of Agda — A functional language with dependent types. In S. Berghofer, T. Nipkow, C. Urban, and M. Wenzel, editors, *Theorem Proving in Higher Order Logics*, volume 5674 of *Lecture Notes in Computer Science*, pages 73–78. Springer, 2009.

[16] C. T. Burge. *Truth and Some Referential Devices*. PhD thesis, Princeton University, 1971.

[17] T. Burge. Truth and singular terms. *Noûs*, 8:309–325, 1974.

[18] R. M. Burstall and J. A. Goguen. Putting theories together to make specifications. In R. Reddy, editor, *Proceedings of the Fifth International Joint Conference on Artificial Intelligence*, pages 1045–1058. William Kaufmann, 1977.

[19] R. M. Burstall and J. A. Goguen. The semantics of CLEAR, A specification language. In D. Bjørner, editor, *Abstract Software Specifications*, volume 86 of *Lecture Notes in Computer Science*, pages 292–332. Springer, 1979.

[20] G. Cantor. *Gesammelte Abhandlungen mathematischen und philosophischen Inhalts*. Springer, 1932.

[21] J. Carette, W. M. Farmer, and M. Kohlhase. Realms: A structure for consolidating knowledge about mathematical theories. In S. M. Watt, J. H. Davenport, A. P. Sexton, P. Sojka, and J. Urban, editors,

Intelligent Computer Mathematics, volume 8543 of *Lecture Notes in Computer Science*, pages 252–266. Springer, 2014.

[22] J. Carette, W. M. Farmer, M. Kohlhase, and F. Rabe. Big math and the one-brain barrier: A position paper and architecture proposal. *The Mathematical Intelligencer*, 43:78–87, 2021.

[23] J. Carette, W. M. Farmer, and Y. Sharoda. Leveraging the information contained in theory presentations. In C. Benzmüller and B. R. Miller, editors, *Intelligent Computer Mathematics*, volume 12236 of *Lecture Notes in Computer Science*. Springer, 2020.

[24] J. Carette and R. O'Connor. Theory presentation combinators. In J. Jeuring, J. A. Campbell, J. Carette, P. Sojka G. Dos Reis, M. Wenzel, and V. Sorge, editors, *Intelligent Computer Mathematics*, volume 7362 of *Lecture Notes in Computer Science*. Springer, 2012.

[25] J. Carette, R. O'Connor, and Y. Sharoda. Building on the diamonds between theories: Theory presentation combinators. *Computing Research Repository (CoRR)*, abs/1812.08079, 2018.

[26] R. Carnap. *Die Logische Syntax der Sprache*. Springer, 1934.

[27] R. Carnap. *The Logical Syntax of Language*. Routledge, 1937.

[28] C. C. Chang and H. J. Keisler. *Model Theory*. Dover, third edition, 2012.

[29] A. Church. A formulation of the simple theory of types. *Journal of Symbolic Logic*, 5:56–68, 1940.

[30] L. Chwistek. Antynomje logiki formalnej (Antinomies of formal logic). *Przeglad Filozoficzny*, 24:164–171, 1921.

[31] J. Collins. *A History of the Theory of Types: Developments After the Second Edition of 'Principia Mathematica'*. LAP LAMBERT Academic Publishing, 2012.

[32] R. L. Constable, S. F. Allen, H. M. Bromley, W. R. Cleaveland, J. F. Cremer, R. W. Harper, D. J. Howe, T. B. Knoblock, N. P. Mendler, P. Panangaden, J. T. Sasaki, and S. F. Smith. *Implementing Mathematics with the Nuprl Proof Development System*. Prentice-Hall, 1986.

[33] T. Coquand and G. Huet. The calculus of constructions. *Information and Computation*, 76:95–120, 1988.

[34] L. de Moura, S. Kong, J. Avigad, F. van Doorn, and J. von Raumer. The Lean theorem prover (system description). In A. P. Felty and A. Middeldorp, editors, *Automated Deduction — CADE-25*, volume 9195, pages 378–388, 2015.

[35] R. Dedekind. *Was sind und was sollen die Zahlen?* Braunschweig, 1888.

[36] Epigram: Marking Dependent Types matter. `http://www.e-pig.or g/`. Access on 17 August 2022.

[37] F*. `https://www.fstar-lang.org/`. Access on 17 August 2022.

[38] W. M. Farmer. A partial functions version of Church's simple theory of types. *Journal of Symbolic Logic*, 55:1269–91, 1990.

[39] W. M. Farmer. A simple type theory with partial functions and subtypes. *Annals of Pure and Applied Logic*, 64:211–240, 1993.

[40] W. M. Farmer. Theory interpretation in simple type theory. In J. Heering, K. Meinke, B. Möller, and T. Nipkow, editors, *Higher-Order Algebra, Logic, and Term Rewriting*, volume 816 of *Lecture Notes in Computer Science*, pages 96–123. Springer, 1994.

[41] W. M. Farmer. Theory interpretation in simple type theory. In J. Heering, K. Meinke, B. Möller, and T. Nipkow, editors, *Higher-Order Algebra, Logic, and Term Rewriting*, volume 816 of *Lecture Notes in Computer Science*, pages 96–123. Springer, 1994.

[42] W. M. Farmer. A scheme for defining partial higher-order functions by recursion. In *3rd Irish Workshop on Formal Methods*, Electronic Workshops in Computing (eWiC). BCS, 1999. `https://www.scienc eopen.com/hosted-document?doi=10.14236/ewic/IWFM1999.5`.

[43] W. M. Farmer. A proposal for the development of an interactive mathematics laboratory for mathematics education. In E. Melis, editor, *CADE-17 Workshop on Deduction Systems for Mathematics Education*, pages 20–25, 2000.

[44] W. M. Farmer. STMM: A Set Theory for Mechanized Mathematics. *Journal of Automated Reasoning*, 26:269–289, 2001.

[45] W. M. Farmer. Formalizing undefinedness arising in calculus. In D. A. Basin and M. Rusinowitch, editors, *Automated Reasoning — IJCAR*

2004, volume 3097 of *Lecture Notes in Computer Science*, pages 475–489. Springer, 2004.

[46] W. M. Farmer. Andrews' type system with undefinedness. In C. Benzmüller, C. E. Brown, J. Siekmann, and R. Statman, editors, *Reasoning in Simple Type Theory: Festschrift in Honor of Peter B. Andrews on his 70th Birthday*, volume 17 of *Studies in Logic*, pages 223–242. College Publications, 2008.

[47] W. M. Farmer. The seven virtues of simple type theory. *Journal of Applied Logic*, 6:267–286, 2008.

[48] W. M. Farmer. The formalization of syntax-based mathematical algorithms using quotation and evaluation. In J. Carette, D. Aspinall, C. Lange, P. Sojka, and W. Windsteiger, editors, *Intelligent Computer Mathematics*, volume 7961 of *Lecture Notes in Computer Science*, pages 35–50. Springer, 2013.

[49] W. M. Farmer. Theory morphisms in Church's type theory with quotation and evaluation. In H. Geuvers, M. England, O. Hasan, F. Rabe, and O. Teschke, editors, *Intelligent Computer Mathematics*, volume 10383 of *Lecture Notes in Computer Science*, pages 147–162. Springer, 2017.

[50] W. M. Farmer. Incorporating quotation and evaluation into Church's type theory. *Information and Computation*, 260:9–50, 2018.

[51] W. M. Farmer and J. D. Guttman. A set theory with support for partial functions. *Studia Logica*, 66:59–78, 2000.

[52] W. M. Farmer, J. D. Guttman, and F. J. Thayer. Little theories. In D. Kapur, editor, *Automated Deduction — CADE-11*, volume 607 of *Lecture Notes in Computer Science*, pages 567–581. Springer, 1992.

[53] W. M. Farmer, J. D. Guttman, and F. J. Thayer. IMPS: An Interactive Mathematical Proof System. *Journal of Automated Reasoning*, 11:213–248, 1993.

[54] W. M. Farmer, J. D. Guttman, and F. J. Thayer. Contexts in mathematical reasoning and computation. *Journal of Symbolic Computation*, 19:201–216, 1995.

[55] R. O. Gandy. The simple theory of types. In R. O. Gandy and J. M. E. Hyland, editors, *Logic Colloquium 76*, volume 87 of *Studies in Logic and the Foundations of Mathematics*, pages 173–181. North-Holland, 1977.

[56] K. Gödel. Über formal unentscheidbare Sätze der Principia Mathematica und verwandter Systeme I. *Monatshefte für Mathematik und Physik*, 38:173–198, 1931.

[57] J. Goguen, T. Winkler, J. Meseguer, K. Futatsugi, and J.-P. Jouannaud. Introducing OBJ. In J. Goguen and G. Malcolm, editors, *Software Engineering with OBJ: Algebraic Specification in Action*, volume 2 of *Advances in Formal Methods*, pages 3–167. Springer, 2000.

[58] M. J. C. Gordon and T. F. Melham. *Introduction to HOL: A Theorem Proving Environment for Higher Order Logic*. Cambridge University Press, 1993.

[59] D. R. Grayson. An introduction to univalent foundations for mathematicians. *Bulletin of the American Mathematical Society*, 55:427–450, 2018.

[60] J. Harrison. HOL Light: An overview. In S. Berghofer, T. Nipkow, C. Urban, and M. Wenzel, editors, *Theorem Proving in Higher Order Logics*, volume 5674 of *Lecture Notes in Computer Science*, pages 60–66. Springer, 2009.

[61] L. Henkin. Completeness in the theory of types. *Journal of Symbolic Logic*, 15:81–91, 1950.

[62] L. Henkin. A theory of propositional types. *Fundamenta Mathematicae*, 52:323–344, 1963.

[63] L. Henkin. Identity as a logical primitive. *Philosophia*, 5:31–45, 1975.

[64] J. R. Hindley. *Basic Simple Type Theory*. Cambridge University Prss, 1997.

[65] W. Hodges. *A Shorter Model Theory*. Cambridge University Press, 1997.

[66] W. Hodges. Model theory. In E. N. Zalta, editor, *The Stanford Encyclopedia of Philosophy*. Metaphysics Research Lab, Stanford University, Spring 2022 edition, 2022. Available at `http://plato.stanford.edu/archives/spr2022/entries/model-theory/`.

[67] W. Hodges and T. Scanlon. First-order model theory. In E. N. Zalta, editor, *The Stanford Encyclopedia of Philosophy*. Metaphysics Research Lab, Stanford University, Winter 2018 edition, 2018. Available at `http://plato.stanford.edu/archives/win2018/entries/modeltheory-fo/`.

[68] M. Hofmann. *Extensional Constructs in Intensional Type Theory*, chapter 2. Syntax and Semantics of Dependent Types, pages 13–54. Springer, 1997.

[69] C. Hort and H. Osswald. On nonstandard models in higher order logic. *Journal of Symbolic Logic*, 49:204–219, 1984.

[70] W. A. Howard. The formulae-as-types notion of construction. In J. P. Seldin and J. R. Hindley, editors, *To H. B. Curry: Essays on Combinatory Logic, Lambda Calculus and Formalism*, pages 479–490. Academic Press, 1980.

[71] E. V. Huntington. Complete sets of postulates for the theory of real quantities. *Transactions of the American Mathematical Society*, 4:358–370, 1903.

[72] Idris: A Language for Type-Driven Development. `https://www.idris-lang.org/`. Access on 17 August 2022.

[73] B. Jacobs and T. Melham. Translating dependent type theory into higher order logic. In M. Bezem and J. F. Groote, editors, *Typed Lambda Calculi and Applications*, volume 664 of *Lecture Notes in Computer Science*, pages 209–229. Springer, 1993.

[74] H. J. Keisler. Fundamentals of model theory. In J. Barwise, editor, *Handbook of Mathematical Logic*, volume 90 of *Studies in Logic and the Foundations of Mathematics*, pages 47–103. North-Holland, 1977.

[75] J. Kirby. *An Invitation to Model Theory*. Cambridge University Press, 2019.

[76] M. Kohlhase. The Flexiformalist Manifesto. In A. Voronkov, V. Negru, T. Ida, T. Jebelean, D. Petcu, S. M. Watt, and D. Zaharie, editors, *14th International Workshop on Symbolic and Numeric Algorithms for Scientific Computing (SYNASC 2012)*, pages 30–35. IEEE Computer Society, 2012.

[77] M. Kohlhase. Mathematical knowledge management: Transcending the one-brain-barrier with theory graphs. *European Mathematical Society (EMS) Newsletter*, 92:22–27, June 2014.

[78] M. Kohlhase, F. Rabe, and V. Zholudev. Towards MKM in the large: Modular representation and scalable software architecture. In S. Autexier, J. Calmet, D. Delahaye, P. D. F. Ion, L. Rideau, R. Rioboo, and A. P. Sexton, editors, *Intelligent Computer Mathematics*, volume 6167 of *Lecture Notes in Computer Science*, pages 370–384. Springer, 2010.

[79] G. Kreisel. Informal rigour and completeness proofs. In I. Lakatos, editor, *Problems in the Philosophy of Mathematics*, pages 138–157. North-Holland, 1967.

[80] E. Landau. *Foundations of Analysis*. Chelsea, 1951.

[81] Maple: The Essential Tool for Mathematics. `https://www.maplesof t.com/products/Maple/`. Access on 11 September 2022.

[82] D. Marker. Introduction to model theory. In D. Haskell, A. Pillary, and C. Steinhorn, editors, *Model Theory, Algebra, and Geometry*, volume 39 of *Mathematical Sciences Research Institute Publications*, pages 15–35. Cambridge University Press, 2000.

[83] D. Marker. *Model Theory : An Introduction*, volume 217 of *Graduate Texts in Mathematics*. Springer, 2002.

[84] P. Martin-Löf. *Intuitionistic Type Theory*. Bibliopolis, 1984.

[85] Metamath Proof Explorer Home Page. `http://us.metamath.org/mp euni/mmset.html`. Access on 17 August 2022.

[86] L. G. Monk. PDLM: A Proof Development Language for Mathematics. Technical Report M86-37, The MITRE Corporation, Bedford, Massachusetts, 1986.

[87] R. Nakajima and T. Yuasa, editors. *The IOTA Programming System*, volume 160 of *Lecture Notes in Computer Science*. Springer, 1982.

[88] A. Naumowicz and A. Korniłowicz. A brief overview of Mizar. In S. Berghofer, T. Nipkow, C. Urban, and M. Wenzel, editors, *Theorem Proving in Higher Order Logics*, volume 5674 of *Lecture Notes in Computer Science*, pages 67–72. Springer, 2009.

[89] R. Nederpelt and H. Geuvers. *Type Theory and Formal Proof: An Introduction.* Cambridge University Press, 2014.

[90] R. P. Nederpelt, J. H. Geuvers, and R. C. De Vrijer, editors. *Selected Papers on Automath*, volume 133 of *Studies in Logic and the Foundations of Mathematics.* North Holland, 1994.

[91] R. Nickson, O. Traynor, and M. Utting. Cogito Ergo Sum — Providing structured theorem prover support for specification formalisms. In K. Ramamohanarao, editor, *Proceedings of the Nineteenth Australasian Computer Science Conference*, volume 18 of *Australian Computer Science Communications*, pages 149–158, 1997.

[92] J. Nolt. Free logic. In E. N. Zalta, editor, *The Stanford Encyclopedia of Philosophy.* Metaphysics Research Lab, Stanford University, Fall 2021 edition, 2021. Available at `http://plato.stanford.edu/arc hives/fall2021/entries/logic-free/`.

[93] U. Norell. *Towards a Practical Programming Language based on Dependent Type Theory.* PhD thesis, Chalmers University of Technology, 2007.

[94] S. Orey. Model theory for the higher order predicate calculus. *Transactions of the American Mathematical Society*, 92:72–84, 1959.

[95] S. Owre, S. Rajan, J. M. Rushby, N. Shankar, and M. Srivas. PVS: Combining specification, proof checking, and model checking. In R. Alur and T. A. Henzinger, editors, *Computer Aided Verification*, volume 1102 of *Lecture Notes in Computer Science*, pages 411–414. Springer, 1996.

[96] S. Owre and N. Shankar. Theory interpretations in PVS. Technical Report NASA/CR-2001-211024), NASA, 2001.

[97] D. L. Parnas. Predicate logic for software engineering. *IEEE Transactions on Software Engineering*, 19:856–862, 1993.

[98] L. C. Paulson. *Isabelle: A Generic Theorem Prover*, volume 828 of *Lecture Notes in Computer Science.* Springer, 1994.

[99] G. Peano. *Arithmetices Principia: Nova Methodo Exposita.* Bocca, Turin, Italy, 1889.

[100] B. Poizat. *A Course in Model Theory:: An Introduction to Contemporary Mathematical Logic*. Universitext. Springer, 2000.

[101] Proof Power. `http://www.lemma-one.com/ProofPower/index/index.html`. Access on 17 August 2022.

[102] W. V. Quine. New foundations for mathematical logic. *American Mathematical Monthly*, 44:70–80, 1937.

[103] F. Rabe. Generic literals. In M. Kerber, J. Carette, C. Kaliszyk, F. Rabe, and V. Sorge, editors, *Intelligent Computer Mathematics*, volume 9150 of *Lecture Notes in Computer Science*, pages 102–117. Springer, 2015.

[104] F. Rabe and M. Kohlhase. A scalable module system. *Information and Computation*, 230:1–54, 2013.

[105] F. Rabe and Y. Sharoda. Diagram combinators in MMT. In C. Kaliszyk, E. C. Brady, A. Kohlhase, and C. Sacerdoti Coen, editors, *Intelligent Computer Mathematics*, volume 11617 of *Lecture Notes in Computer Science*, pages 211–226. Springer, 2019.

[106] F. P. Ramsey. The foundations of mathematics. *Proceedings of the London Mathematical Society, Series 2*, 25:338–384, 1926.

[107] R. M. Robinson. An essentially undecidable axiom system. In *Proceedings of the International Congress of Mathematics, Cambridge, Massachusetts, 1950*, volume I, pages 729–730. American Mathematical Society, 1952.

[108] J. Rushby, S. Owre, and N. Shankar. Subtypes for specifications: Predicate subtyping in PVS. *IEEE Transactions on Software Engineering*, 24:709–720, 1998.

[109] J. Rushby, F. von Henke, and S. Owre. An introduction to formal specification and verification using EHDM. Technical Report SRI-CSL-91-02, SRI International, 1991.

[110] B. Russell. Mathematical logic as based on the theory of types. *American Journal of Mathematics*, 30:222–262, 1908.

[111] SageMath. `https://www.sagemath.org/`. Access on 11 September 2022.

[112] R. Schock. *Logics without Existence Assumptions*. Almqvist & Wiksells, 1968.

[113] Y. Sharoda. *Leveraging the Information Contained in Theory Presentations*. PhD thesis, McMaster University, 2021.

[114] D. R. Smith. KIDS: A knowledge-based software development system. *IEEE Transactions on Software Engineering*, 16:483–514, 1991.

[115] D. R. Smith. Constructing specification morphisms. *Journal of Symbolic Computation*, 15:571–606, 1993.

[116] D. R. Smith. Mechanizing the development of software. In M. Broy and R. Steinbrueggen, editors, *Calculational System Design*, pages 251–292. IOS Press, 1999.

[117] M. Spivak. *Calculus*. Publish or Perish, fourth edition, 2008.

[118] Y. V. Srinivas and R. Jüllig. Specware: Formal support for composing software. In B. Möller, editor, *Mathematics of Program Construction*, volume 947 of *Lecture Notes in Computer Science*, pages 399–422. Springer, 1995.

[119] A. Tarski. Pojęcie prawdy w językach nauk dedukcyjnych (The concept of truth in the languages of the deductive sciences). *Prace Towarzystwa Naukowego Warszawskiego*, 1933.

[120] A. Tarski. Der Wahrheitsbegriff in den formalisierten Sprachen. *Studia Philosophica*, 1:261–405, 1935.

[121] A. Tarski. The concept of truth in formalized languages. In J. Corcoran, editor, *Logic, Semantics, Metamathematics*, pages 152–278. Hackett, second edition, 1983.

[122] N. Tennant. *Natural Logic*. Edinburgh University Press, 1978.

[123] K. Tent and M. Ziegler. *A Course in Model Theory*, volume 40 of *Lecture Notes in Logic*. Cambridge University Press, 2012.

[124] The Coq Proof Assistant. `https://coq.inria.fr/`. Access on 17 August 2022.

[125] The On-Line Encyclopedia of Integer Sequences. `https://oeis.org/`. Access on 11 September 2022.

[126] The Univalent Foundations Program. Homotopy type theory: Univalent foundations of mathematics. `https://homotopytypetheory.org/book`, 2013.

[127] A. N. Whitehead and B. Russell. *Principia Mathematica,* 3 volumes. Cambridge University Press, 1910–13.

[128] Wolfram Mathematica. `https://www.wolfram.com/mathematica/`. Access on 11 September 2022.

List of Figures

© The Editor(s) (if applicable) and The Author(s), under exclusive license
to Springer Nature Switzerland AG 2023
W. M. Farmer, *Simple Type Theory*, Computer Science Foundations
and Applied Logic, https://doi.org/10.1007/978-3-031-21112-6

List of Tables

© The Editor(s) (if applicable) and The Author(s), under exclusive license   275
to Springer Nature Switzerland AG 2023
W. M. Farmer, *Simple Type Theory*, Computer Science Foundations
and Applied Logic, https://doi.org/10.1007/978-3-031-21112-6

List of Theorems, Examples, Remarks, and Modules

W. M. Farmer, *Simple Type Theory*, Computer Science Foundations
and Applied Logic, https://doi.org/10.1007/978-3-031-21112-6

Index

This index begins with the symbols and notation employed in this book divided into the following groups:

1. Logical symbols in order of appearance.

2. Symbols and notation for sets in order of appearance.

3. Symbols and notation for tuples, lists, and relations in order of appearance.

4. Symbols and notation for functions in order of appearance.

5. Symbols and notation for Boolean values, predicates, binders, equalities, and structures in order of appearance.

6. Symbols and notation for syntax in order of appearance.

7. Symbols and notation for semantics in order of appearance.

8. Symbols for proof systems in order of appearance.

9. Symbols and notation for theories and morphisms in order of appearance.

10. Symbols in blackboard font in alphabetical order.

11. Selected names in sans sarif font in alphabetical order.

12. Symbols in calligraphic font in alphabetical order.

13. Symbols in fraktur font in alphabetical order.

14. Alonzo type and expression constructors in alphabetical order.

15. Alonzo compact notation introduced via notational definitions in order of appearance and divided into groups.

W. M. Farmer, *Simple Type Theory*, Computer Science Foundations and Applied Logic, https://doi.org/10.1007/978-3-031-21112-6

Printed in the United States
by Baker & Taylor Publisher Services